AF597680

SOLID - LIQUID SEPARATION

SOLID-LIQUID SEPARATION

Editor:
J. GREGORY
Department of Civil Engineering
University College London

Published for the
SOCIETY OF CHEMICAL INDUSTRY
London, by

ELLIS HORWOOD LIMITED
Publishers · Chichester

First published in 1984 by

ELLIS HORWOOD LIMITED

Market Cross House, Cooper Street, Chichester, West Sussex, PO19 1EB, England

The publisher's colophon is reproduced from James Gillison's drawing of the ancient Market Cross, Chichester.

Distributors:

Australia, New Zealand, South-east Asia:
Jacaranda-Wiley Ltd., Jacaranda Press,
JOHN WILEY & SONS INC.,
G.P.O. Box 859, Brisbane, Queensland 40001, Australia

Canada:
JOHN WILEY & SONS CANADA LIMITED
22 Worcester Road, Rexdale, Ontario, Canada.

Europe, Africa:
JOHN WILEY & SONS LIMITED
Baffins Lane, Chichester, West Sussex, England.

North and South America and the rest of the world:
Halsted Press: a division of
JOHN WILEY & SONS
605 Third Avenue, New York, N.Y. 10016, U.S.A.

British Library Cataloguing in Publication Data
Gregory, John
Solid-liquid Separation.
1. Separation (Technology)
I. Title
660.2'842 TP156.S45

Library of Congress Card No. 83-22731

ISBN 0-85312-684-4 (Ellis Horwood Limited)
ISBN 0-470-20021-9 (Halsted Press)

Typeset by Ellis Horwood Limited
Printed in Great Britain by Unwin Brothers of Woking

Table of Contents

Preface

Solid-liquid separation processes are used in a very wide range of industries and are under active investigation by research workers all over the world. A symposium on 'Advances in Solid-Liquid Separation', organized by the Society of Chemical Industry and held at University College London, 19–21 September 1983, attracted papers on most aspects of the subject by many international experts in the field. This book is a collection of the symposium papers.

The diversity of topics posed some problems over organization of the material, but it was decided that the sequence of chapters should follow the order of presentation at the symposium, where the papers were grouped, as far as possible, into sessions with a common theme. Chapters on electrical separation methods, flotation, thickening and dewatering are followed by contributions on flocculation, including several dealing with test methods. The final group of papers (Chapters 17 to 22) have to do with filtration of one kind or another, ranging from virus adsorption on sand to non-woven cloths as filter media. The order is not entirely logical and several rather arbitrary choices had to be made, for instance in the case of Chapter 11, which is a comprehensive account of protein recovery methods and involves a whole range of processes. Nevertheless, by using the Index, it should be possible for the reader to locate material of interest fairly quickly.

Contributions were invited on both fundamental and applied aspects of solid-liquid separation and the present volume certainly covers a very broad spectrum of activity, from pilot-plant and full-scale work, through laboratory investigations to theoretical studies. This has led to a variety of approaches and styles but, since many of the chapters will be read as self-contained units, there should not be any serious difficulties of interpretation.

The fact that this book has been published within a few months of the meeting is due, in large part, to the cooperation of authors, the majority of whom submitted manuscripts well in advance. The few late papers have been processed very efficiently by the publishers and now appear in their rightful places, quite indistinguishable from the more punctual contributions. Contributors to future symposia should not take this as an encouragement to miss deadlines!

My thanks are due to fellow members of the organizing committee for their help and encouragement, which led to a very successful Symposium and, we hope, to a useful addition to the literature of Solid-Liquid Separation.

John Gregory
University College London
January 1984

CHAPTER 1

Electrical Separation Processes in the Treatment of Radioactive Wastes

W. R. BOWEN and A. D. TURNER, Applied Electrochemistry Group, Materials Development Division, AERE Harwell, Oxon, UK

1. INTRODUCTION

During the past two years, a section of the Applied Electrochemistry Group at Harwell has been investigating the application of electrical processes to the treatment of liquid radioactive wastes of medium activity. The potential attractiveness of electrical methods arises from the possibility of remotely controlling a variety of processes at ambient temperature and pressure by the extra variable of electrical potential difference. The present paper will consider one aspect of the programme, the use of electrically controlled membrane processes in solid–liquid separation.

There is no rigid definition of the activity of the wastes which comprise the category Medium Active Liquid Wastes (MALW), but they are such as require biological protection during handling, though not controlled cooling [1]. Such wastes arise primarily in spent fuel reprocessing, nuclear reactor operation and to a lesser extent from nuclear research centres. The most general method for the removal of dissolved radioactive ions or colloids from such waste is by adsorption on, or occlusion in, a suitable floc, for example, iron hydroxide [2]. After settling, such floc slurries generally average ~5% solids content. Ideally, this should be increased, maybe to 30–40% solids, so as to minimize the volume occupied by the waste after immobilization in a form suitable for disposal. This dewatering must be achieved with very high solids separation. Sludges which may arise in the treatment of spent fuel claddings may also require dewatering before disposal. The present paper shows that electro-osmotic dewatering provides a suitable treatment technique for both of these waste types. Wastes treated by the addition of a small amount of colloidal material as an adsorbant, and those already containing active colloidal material, may be dewatered directly by electro-osmosis. This latter group of wastes may also be separated by ultrafiltration, and it is shown that the application of an electric field is effective in the control of fouling at ultrafiltration membranes.

The paper begins with a brief description of the nature of the fundamental processes exploited. This is followed by a description of the experimental techniques used in the investigation of electro-osmotic dewatering and an outline of the results obtained, firstly with colloidal materials and secondly with flocculated materials. The use of a continuous flow electro-osmotic dewatering cell is described. Sections are then devoted to the use of an electric field in the control of fouling at ultrafiltration membranes, and to the development of a simple device for the determination of the electrokinetic or zeta potential of concentrated colloidal dispersions and flocs. The final section discusses the significance of the results.

2. FUNDAMENTAL PROCESSES

Most substances acquire a surface electrical charge when brought into contact with a polar (e.g. aqueous) medium [3]. This may arise by ion dissociation, ion adsorption or ion dissolution. The charge tends to produce an ordering of the surrounding solution – in particular, ions of opposite charge are attracted towards the surface. When combined with the randomizing effect of thermal motion, this leads to the formation of an 'electrical double layer' comprising the charged surface and the neutralizing excess of counter ions (Fig. 1).

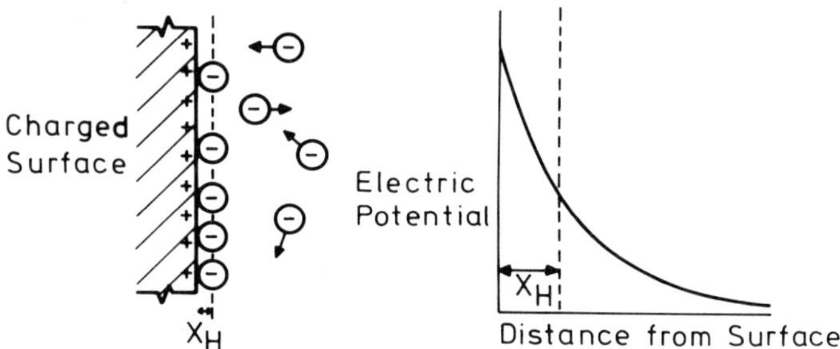

Fig. 1 – The 'electrical double layer' and potential distribution at a charged surface.

If an electric field is applied parallel to such a charged surface, forces are exerted on both the solution part of the double layer and the surface. These forces are opposite in direction, due to the separation of charge between the two phases. The mobile part of the double layer will move under the influence of the field, carrying solvent with it. If the charged particle is mobile it too will migrate, but in the opposite direction. These events are conventionally divided into two limiting cases:

(i) Electro-osmosis – defined as the transport of liquid relative to an immobile charged surface by an electric field, for example, the movement of water through a capillary under the influence of a potential gradient (Fig. 2a).

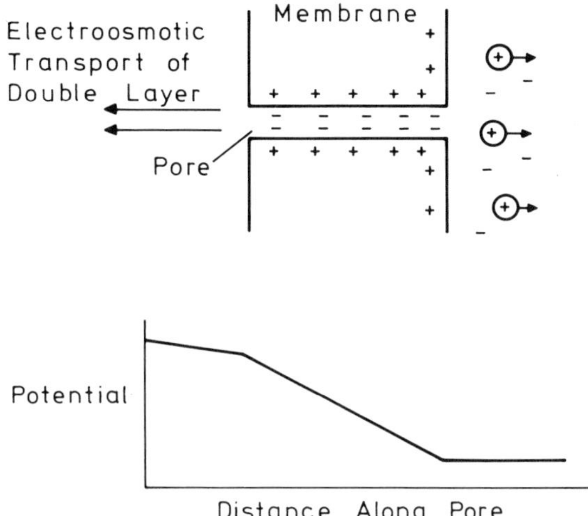

Fig. 2a – Electro-osmotic transport through a capillary under the action of an electric field across a positively charged membrane.

(ii) Electrophoresis – the transport of a charged surface relative to a stationary liquid by an electric field, for example, the movement of ions or particles between electrodes under the influence of a potential gradient (Fig. 2b).

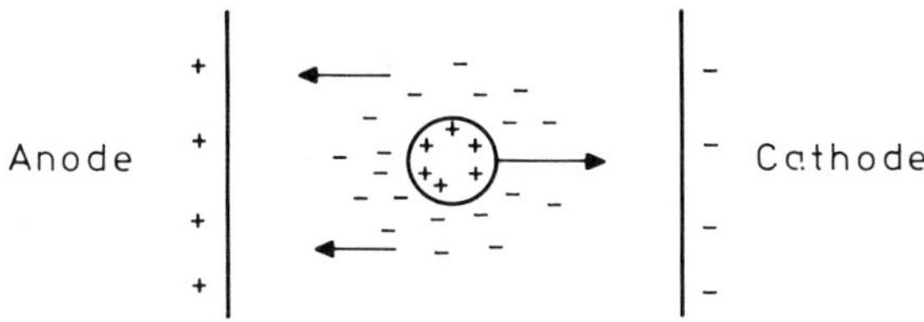

Fig. 2b – Electrophoretic transport of a positively charged particle in an electric field.

Both effects are operative in the work described in the present paper. For aqueous solutions the effects may be mathematically described by an equation developed by Smoluchowski [4]. For the particular case of electro-osmosis this equation may be written

$$V_{EO} = I\epsilon\zeta/\eta k_0 \qquad (1)$$

where V_{EO} is the volume flow rate per unit area, I is the current density, ϵ the permittivity of the electrolyte, η the viscosity of the solvent, and k_0 the specific conductivity of the electrolyte. ζ is termed the electrokinetic or zeta potential and is the electrical potential at the surface of shear between the immobile and mobile parts of the double layer.

Thus, if an electric field is applied across a microporous membrane, solvent will be transported through the membrane. That is, electro-osmotic dewatering takes place. The effect may be subdivided into:

(i) Filter-medium electro-osmosis, where the charged surface is that of the membrane itself.
(ii) Filter-cake electro-osmosis, where the charged particles to be separated deposit on or in a membrane, giving the surface charge, or in the case of thicker deposits, actually form a microporous membrane themselves.

This second case is most common in the practical separation of dispersed solids from liquids.

3. ELECTRO-OSMOTIC DEWATERING

3.1 Equipment

A simple experimental apparatus [5, 6] which allows the measurement of electro-osmotic flow rates is shown in Fig. 3. A central tube, the dipped cell, is faced at one end by a mesh working electrode. Membranes were positioned in front of this electrode by means of a Quick-fit screwcap. A cylindrical counter electrode completed the electrochemical cell. A Watson-Marlow Delta B series flow inducer removed extracts from the dipped cell to a collector, at the same time maintaining the two compartments of the electrochemical cell in hydrostatic equilibrium. A constant voltage was applied between the electrodes by means of a stabilized power supply (125 V, 5 A).

For convenience, platinum electrodes were generally used in the test cell. However, stainless steel cathodes and platinized titanium anodes were also shown to be effective. Microporous membranes with mean pore sizes in the range 8.0 μm to 0.1 μm were used in the treatment of colloidal materials. These were composed of mixed esters of cellulose acetate and cellulose nitrate. Cambric cotton was used as a membrane in the treatment of flocculated material.

3.2 Treatment of Colloidal Material

Three examples showing the application of electro-osmotic dewatering in the treatment of colloidal waste materials are presented below.

3.2.1 Iron hydroxide colloids

Iron hydroxide flocs and colloids are effective in the treatment of radioactive waste. Subsequent dewatering can only achieve very high decontamination

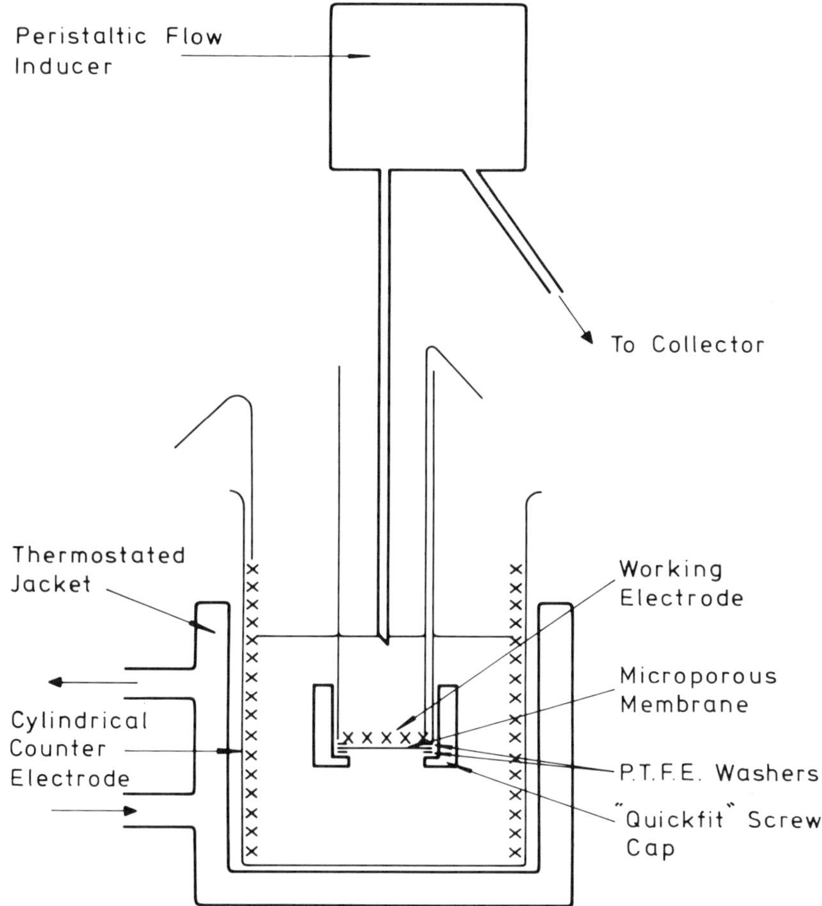

Fig. 3 – Dipped cell apparatus.

factors if the smallest colloidal particles are excluded from the extract. Of importance in the initial study of iron hydroxide colloids was an assessment of the interdependence of the achieved solids separation factor on applied voltage and membrane pore size.

For these experiments the colloids were prepared by boiling a 10^{-2} M solution of $FeCl_3$ for 1 hour. Microporous membranes were precoated by soaking in the colloid containing solution for 20 minutes. Electron probe X-ray microanalysis of such treated membranes (Fig. 4), showed a relatively high concentration of iron at the surface which faced the solution, and a reasonably uniform distribution of iron throughout the depth of the membrane. It is this latter occluded material which provides the charged surfaces for electro-osmotic flow. Below a certain critical pore size there was no occlusion of colloidal material in the depth of the membrane, and no electro-osmotic flow on the application of a voltage (lower trace in Fig. 4).

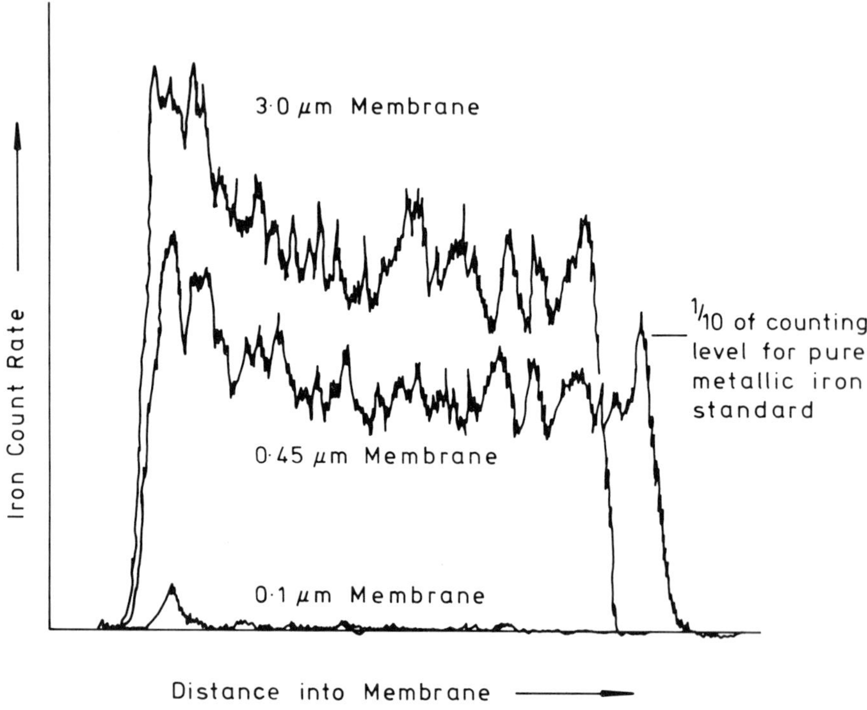

Fig. 4 – Electron probe X-ray microanalysis of microporous membranes of different pore size after soaking in an iron hydroxide colloid.

The dependence of the rate of electro-osmotic extraction on the applied current density is shown in Fig. 5. Electro-extraction occurred when the working electrode was cathodic. It may be seen that the rate of electro-extraction was substantial. The power consumption was modest, being in the range 0.06 kWh L^{-1} (at 10 V) to 0.18 kWh L^{-1} (at 30 V). This may be compared with the requirement of ~2 kWh L^{-1} for dewatering by evaporation. Solids separation factors achieved for the data points shown in Fig. 5 are given in Table 1. The data show that the application of a potential significantly improves the solids separation achieved with a given membrane. Thus, an important advantage of electro-osmotic extraction is that the remotely controlled variation of applied voltage allows the matching of membrane performance to variable feed properties and solids retention requirements. In practice, to achieve a given degree of solids separation it is most energy efficient to choose a membrane just above the critical pore size.

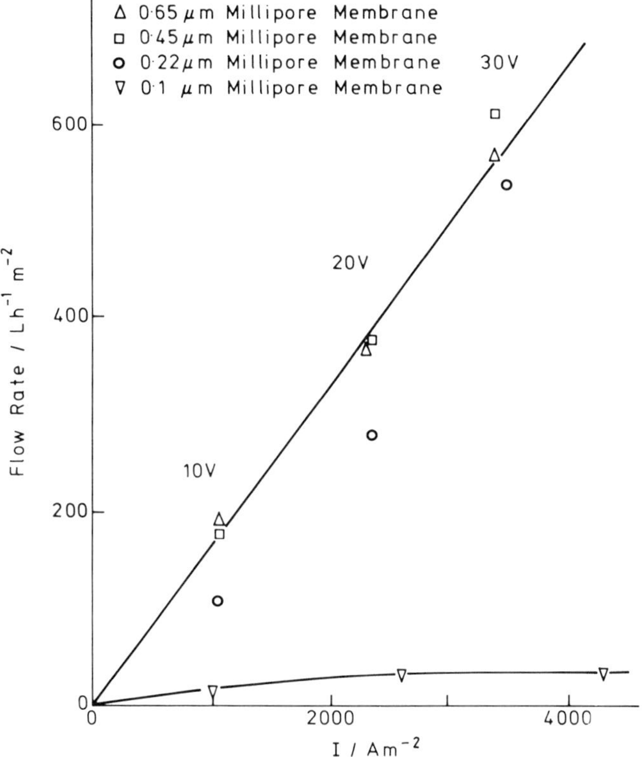

Fig. 5 – Electro-extraction from iron hydroxide colloid as a function of current density.

Table 1

Dependence of solids separation factors on applied voltage and membrane pore size for electroextraction from iron hydroxide colloids

Membrane pore size (μm)	Applied potential (V)			
	0	10	20	30
0.22	3.0	290	>2500*	>2500*
0.45	1.7	10	42	870
0.65	1.2	4	5	12

* Limit of spectrophotometric detection.

3.2.2 Titanium hydroxide colloids

It has been shown that the addition of $TiCl_4$ in HCl to an alkaline effluent gives a submicron $Ti(OH)_4$ precipitate which efficiently adsorbs Sr, U and Th from the effluent. High decontamination factors may be achieved when this precipitation is followed by ultrafiltration. However, a feature of the operation of UF modules is the formation of deposits on the membrane surface which may only be removed by chemical backwashing. In electro-osmotic extraction, membrane blockage is reduced by the electrophoretic transport of colloidal particles away from the membrane surface.

In the present experiments the solution contained 200 mg L^{-1} NaOH and 0.05 mg L^{-1} $^{85}Sr^{2+}$. A $Ti(OH)_4$ colloid was precipitated at a concentration of 4 mg L^{-1}. Membranes were precoated by soaking in the colloid-containing solution for 20 minutes prior to electro-extraction, which occurred when the working electrode was cathodic. Results for microporous membranes of three different pore sizes are shown in Fig. 6. Power consumption was in the range 0.02 kWh L^{-1} (at 25 V) to 0.06 kWh L^{-1} (at 100 V).

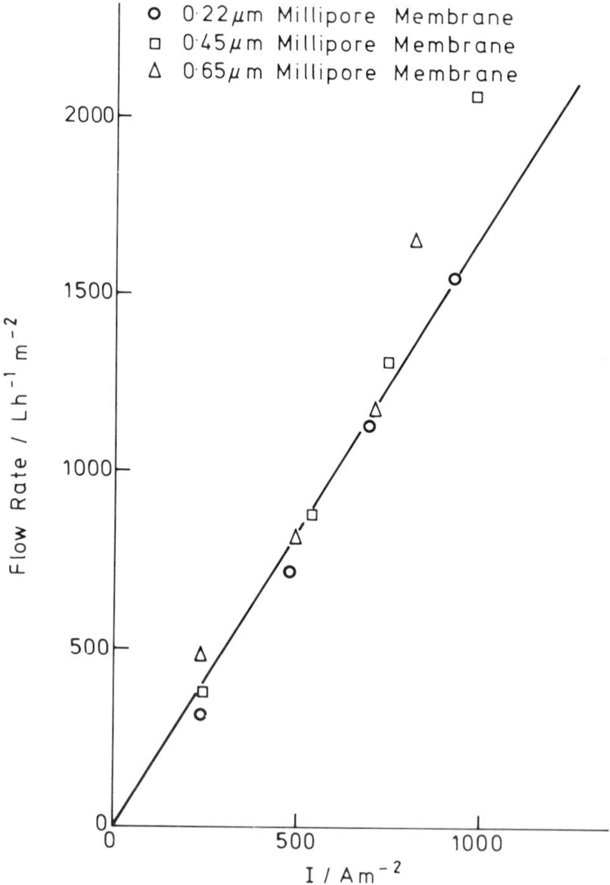

Fig. 6 – Electro-extraction from titanium hydroxide colloid as a function of current density.

The dependence of the achieved decontamination factor (DF) on the applied voltage and membrane pore size is shown in Table 2. The DF is defined as [^{85}Sr] in the feed/[^{85}Sr] in the extract.

Table 2

Dependence of decontamination factor on applied voltage and membrane pore size for electroextraction from $Ti(OH)_4$ colloids

Membrane pore size (μm)	Applied potential (V)				
	0	25	50	75	100
0.22	20	310	~900	120	260
0.45	13	80	220	120	80
0.65	8	40	40	40	70

Application of a voltage greatly increases the DF at a given membrane. The DF apparently reaches a maximum value at intermediate voltages. From work on iron hydroxide colloids it is known that the fraction of colloidal material passing through a membrane decreases continuously with application of increased voltage. The maximum observed here may therefore reflect a displacement of the adsorption equilibrium of $^{85}Sr^{2+}$ on $Ti(OH)_4$ at the highest applied voltages. This is an interesting observation but not of critical importance as in electro-extraction it is best to use the lowest voltage required to achieve a required DF so as to minimize power consumption.

3.2.3 Plutonium colloids

The term plutonium 'colloid' or 'polymer' has been used to describe solutions containing highly hydrolysed plutonium(IV). A study of the electro-osmotic dewatering of such solutions is in progress using a specially modified version of the dipped cell designed for use in a glove box. It has been shown, by α counting of the extracts and feed, that improved DFs are achieved on the application of a voltage. It is proposed to continue this work by making decontamination measurements on Pu colloid containing solutions at lower electrolyte concentration, these being simulants for rinse waste waters.

3.3 Treatment of Flocculated Material and Sludges

Examples showing the application of electro-osmotic dewatering in the treatment of settled flocculated material and fuel cladding sludges are presented below.

3.3.1 Waste treatment flocs

A number of settled waste treatment flocs may be further dewatered electro-osmotically. These flocs include $Fe(OH)_3$, $Fe(OH)_3/Al(OH)_3$, $CuSO_4/K_4Fe(CN)_6$ and $CuSO_4/K_4Fe(CN)_6/FeSO_4/Ca(NO_3)_2/Na_3PO_4$.

As an example, Fig. 7 shows the dependence of the rate of electro-osmotic extraction on the current density for an alumino-ferric floc. As supplied, the floc contained ~6% solids and high concentrations of dissolved salts (nitrates and sulphates) giving it an electrical conductivity of ~200 mS cm^{-1}. Before electro-osmotic dewatering it was washed with water until the conductivity was ~6mS cm^{-1}. Electro-extraction took place at precoated microporous membranes when the working electrode was anodic. Clear extracts obtained had iron concentrations in the range 10^{-2}—10^{-3} M, corresponding to solids separation factors of 10^2—10^3 and gave no positive result with the alizarin test for aluminium. Power consumption was in the range 0.13 kWh L^{-1} (at 10 V) to 0.22 kWh L^{-1} (at 20 V). It was shown that the floc could be electro-osmotically dewatered to give a sludge of solids content ~20%, even with the simple non-optimum cell used.

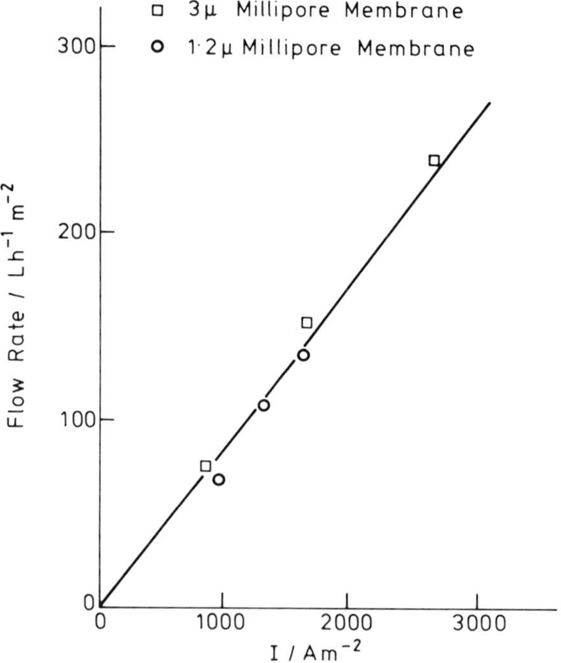

Fig. 7 – Electro-extraction from settled alumino-ferric floc as a function of current density.

The retention of activity during electro-osmotic dewatering of slurries has been tested in measurements on Harwell low level effluent slurry. This results from a simple alkaline ferric hydroxide precipitation. The settled slurry was electro-osmotically dewatered at a pre-coated cambric cotton membrane. It was found that there was essentially complete retention of actinides and ^{60}Co and ~55% retention of ^{137}Cs. These figures represent an improvement in performance compared to conventional vacuum filtration.

3.3.2 Fuel cladding sludges

Simulants for two sludges which may arise in the treatment of spent fuel claddings have been successfully treated electro-osmotically, a Magnox cladding sludge simulant and a zirconia floc. Data for the extraction from Magnox cladding sludge simulant are shown in Fig. 8. Electro-osmotic dewatering at precoated cambric cotton membranes gave extracts of high clarity, no colloidal sludge particles being detected spectrophotometrically. Power consumption was in the range 0.07 kWh L^{-1} (at 40 V) to 0.13 kWh L^{-1} (at 80 V). With the simple dipped cell it was found that the solids content of the sludge could be increased from 16% to at least 32%.

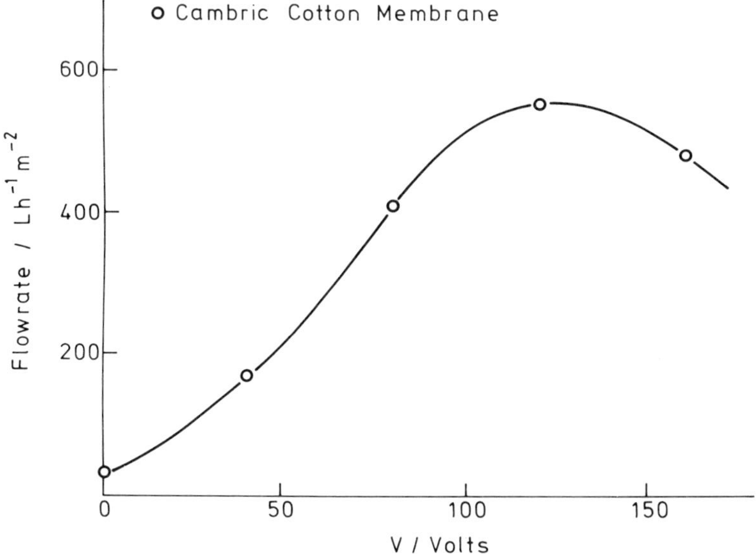

Fig. 8 – Electro-extraction from Magnox cladding sludge simulant as a function of voltage.

3.4 Development of a Continuous Flow Electro-extraction Cell

In practice, flocs and colloids could be dewatered either in a batch process, in equipment similar to a dipped cell, or in a continuous process [7]. To test the principle of continuous electro-osmotic extraction, a laboratory scale prototype cross-flow electro-extraction cell has been designed, constructed and tested under recycling flow conditions.

The cell is shown in Fig. 9. The material to be treated was pumped through a compartment of diameter 2.5 cm and depth adjustable between 0.4 cm and 1 cm. A membrane, backed by the working electrode, formed one face of the compartment. The counter electrode was positioned against the opposite face.

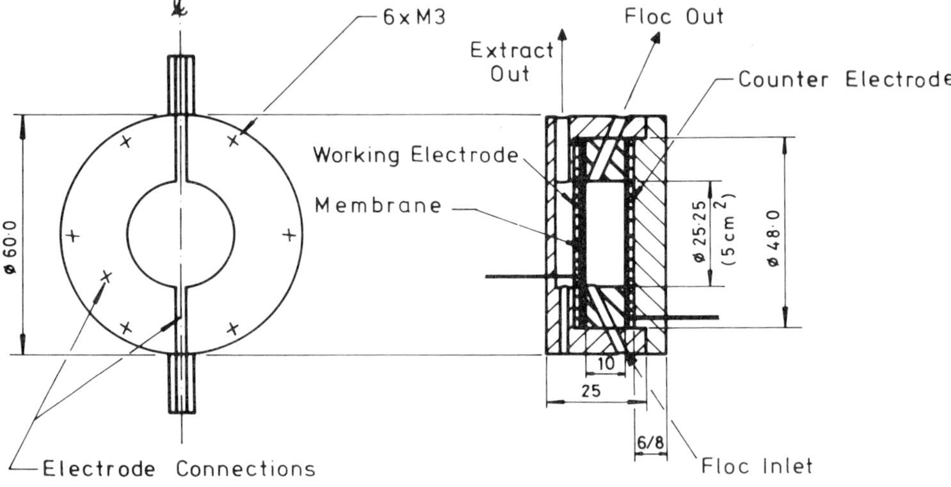

Fig. 9 – Continuous flow electro-extraction cell.

Extracts could pass through the membrane into another compartment and hence to a collection vessel. The cell was constructed in perspex and for these tests the electrodes were of platinum mesh. Cambric cotton membranes were used in most experiments. An overall view of the apparatus is given in Fig. 10. The material to be treated was contained in the 1-l beaker and the extract was collected in a measuring cylinder. The sludge was pumped by a Watson-Marlow MHRE flow inducer.

An alumino-ferric floc, zirconia floc and a Magnox cladding sludge simulant have been treated by the continuous flow cell. Data for the Magnox cladding sludge simulant are presented here, the main aim of the experiments being to compare the overall performance of the flow cell with that of the dipped cell. Precoated cotton cambric membranes were used and the experiments were run at a constant current density of 20 mA cm^{-2} (corresponding to 200 A m^{-2}). During the course of a complete dewatering run the voltage across the cell dropped from 100 V to 30 V due to an increase in feed conductivity. A final solids content of 35% was achieved, the limiting factor being the ability of the flow inducer to pump the sludge through the cell. Power consumption was similar to that for the dipped cell. The principle of continuous flow electro-extraction has thus been successfully demonstrated.

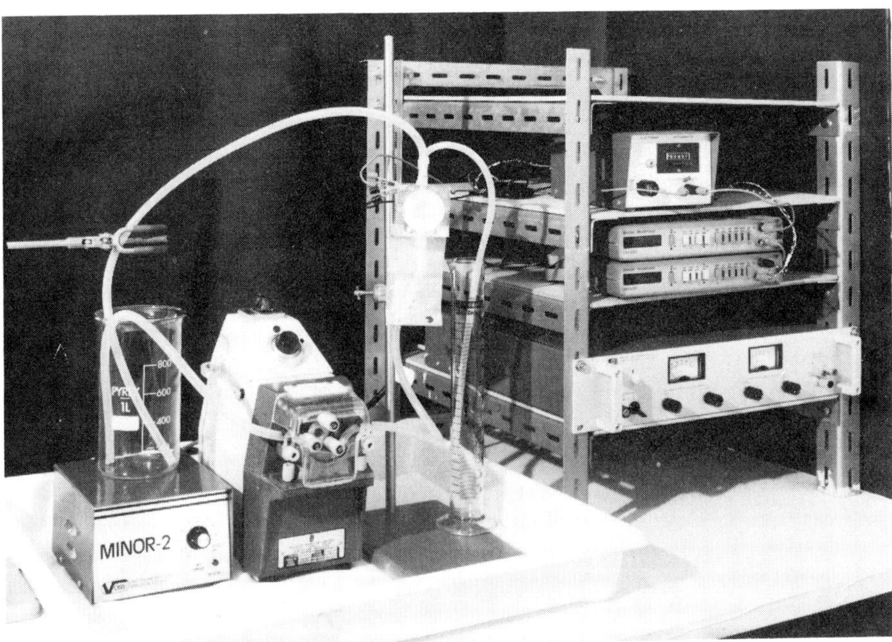

Fig. 10 – Continuous flow electro-extraction apparatus.

4. ELECTRICAL PREVENTION OF FOULING AT UF MEMBRANES

It is possible to treat medium and low active liquid wastes by the use of ultrafiltration to remove activity in the form of colloidal or larger particles which are either already present or are deliberately formed in the stream [8]. An undesirable feature of the operation of UF modules is the formation of deposits on the membrane surface which decrease the efficiency of operation and may conventionally only be removed by chemical backwashing. It was therefore decided to study the effect of an electric field on the rate of deposition at a UF membrane in operation.

A resin-coated stanless steel flat sheet membrane test cell was used (Fig. 11). Amicon Diaflo PMIO flat sheet UF membranes were used. One electrode, a platinum wire, was mounted along the bottom of the labyrinthine flow channel. A second electrode was mounted directly behind the membrane. A γ-probe positioned behind a perspex window directly behind this electrode and membrane allowed continuous direct measurement of the build-up of activity at the membrane, and hence the degree of fouling.

The data presented here are for a ^{85}Sr-containing solution which had been treated by the deliberate formation of a $Ti(OH)_4$ colloid, onto which the $^{85}Sr^{2+}$ ions adsorbed. Figure 12 shows how the build-up of γ activity due to adsorption of colloid at the membrane surface during ultrafiltration depended on the

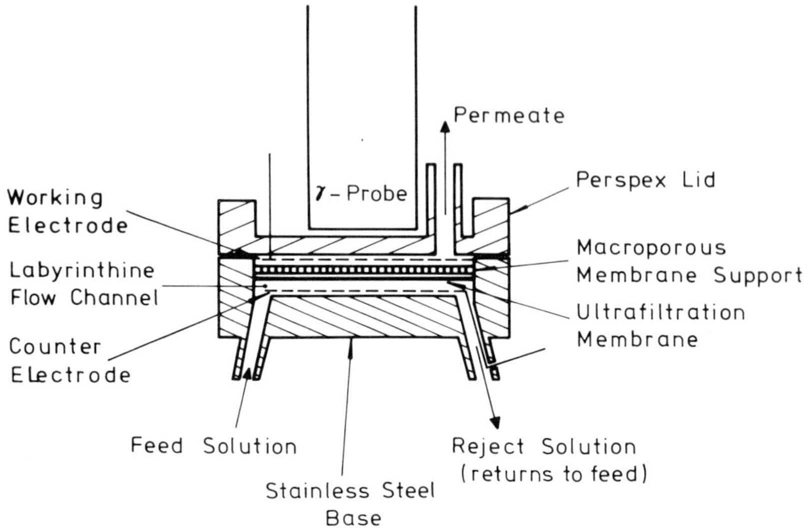

Fig. 11 – Modified flat sheet UF membrane test cell.

applied voltage. Table 3 shows the data of Fig. 12 expressed as percentage deposition rates. (A rate of deposition of 70% indicates that 70% of the colloidal material filtered from a given volume of extract adhered to the membrane.) The application of the electric field across the membrane dramatically reduced the rate of fouling. The electrical power consumption was 0.015 kWh L^{-1} at 50 V, compared to a typical power consumption of 0.03 kWh L^{-1} (at 50 V) for electro-osmosis alone and to 0.001–0.01 kWh L^{-1} for ultrafiltration alone. The polarity of the applied voltage was such that colloid particles were transported away from the membrane electrophoretically, preventing fouling, while at the same time electro-osmotic flow took place at the membrane, enhancing the extraction rate by about 25%. Table 3 also shows that the decontamination factor decreased on the application of the voltage. This is mainly due to a small percentage of free Sr^{2+}, existing as a result of the presence of a large excess of competing ions (Mg^{2+} and Ca^{2+}) being electrophoretically accelerated toward the

Table 3

The effect of an applied voltage on the rate of deposition at a UF membrane

Voltage applied	Rate of deposition	DF obtained
0	70%	200
50	8%	100
100	4%	12–35

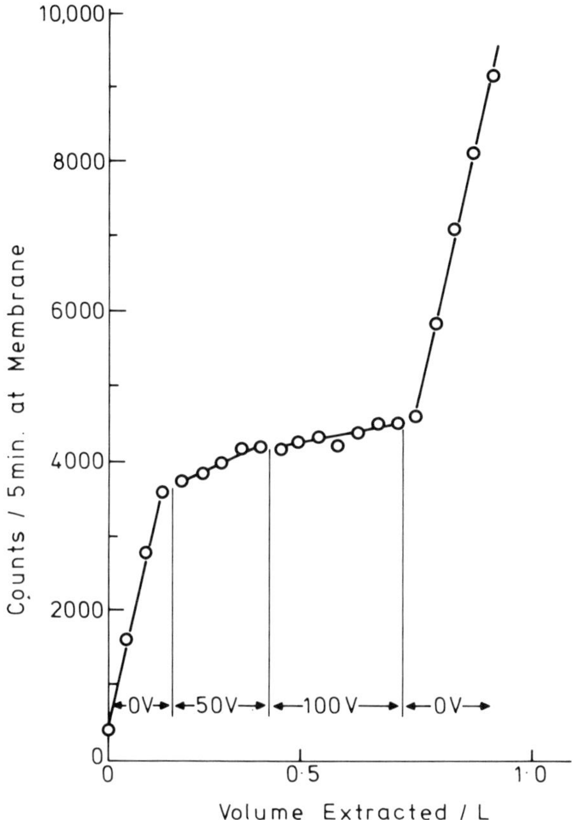

Fig. 12 – The build-up of γ activity at a UF membrane showing the effect of applied voltage.

membrane. Use of a higher concentration of colloid would have improved the DF obtained. Thus, the electrical prevention of fouling at UF membranes shows considerable promise if care is taken in the selection of the applied voltage.

The cleaning of a fouled membrane by applying an electric field across it has also been investigated. For a feed containing NaOH, $Ti(OH)_4$ and $^{85}Sr^{2+}$, it has been shown that 20% of the adsorbed colloid may be removed on the application of 125 V across the membrane, the adsorbed material being transported back into the feed electrophoretically. However, if the feed contained Ca^{2+} or Mg^{2+} no such cleaning occurred. In the former case the $Ti(OH)_4$ forming the fouling layer presumably retains a residual negative charge, whereas in the presence of Ca^{2+} or Mg^{2+} incorporation of these ions into the fouling layer neutralizes this charge or increases the binding at the membrane surface. Electrokinetic cleaning is clearly only viable in the case of charged fouling layers.

5. DETERMINATION OF ZETA POTENTIALS

The zeta potential of flocs or colloids may be calculated from the slopes of plots such as Figs. 5, 6 or 7 by the use of eqn. (1). The zeta potential is an important parameter in determining the DF obtainable by adsorption of active species onto these treatment media. The zeta potential is also important in determining the rate of settling and the final solids content of a settled floc or sedimented suspension. A knowledge of the zeta potential and its variation with feed parameters thus allows the optimization of waste treatment processes. A knowledge of zeta potentials also helps the prediction of rates of fouling at filtration membranes.

Microelectrophoresis [9] is the most common method for determining zeta potential. It has the disadvantage of being limited to very dilute sols and is inappropriate for more concentrated colloidal dispersions and for flocs. Conventional streaming potential measurements require the preparation of a porous plug of the material to be studied, the surface properties of which may be different to those of the dispersed material. The pre-coat used in the dipped cell measurements should have surface properties more closely related to those of the dispersed material than such a plug. The dipped cell requires a simpler apparatus and instrumentation than either micro-electrophoresis or streaming potential measurements. Further it allows *in situ* measurements to be made.

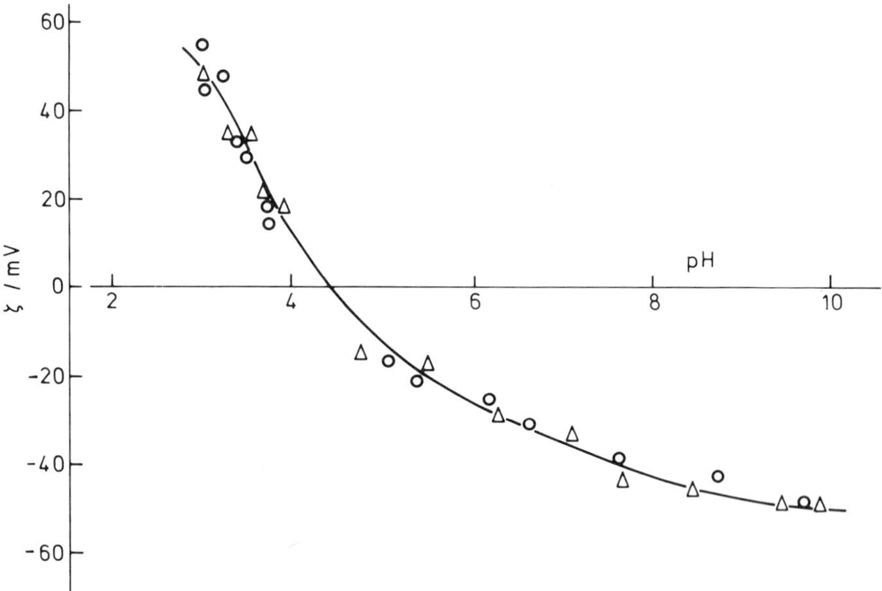

Fig. 13 – The variation of the zeta potential of dispersed α-Fe_2O_3 powder as a function of pH. Comparison of data obtained using the dipped cell (○) with data obtained by microelectrophoresis (△).

A comparison of the zeta potential of a dispersed α-Fe_2O_3 powder over the pH range 3–10 as determined with the dipped cell and microelectrophoresis is shown in Fig. 13. Very good agreement was obtained over the whole pH range studied, confirming the viability of the dipped cell occluded membrane technique. It is possible to make measurements on dispersion containing up to 50% solids with this technique.

6. DISCUSSION

The data presented in this paper show that electrical separation processes have a potential role to play in the treatment of radioactive wastes.

Realistic goals for an electro-osmotic dewatering treatment are shown in Table 4. Electro-osmotic dewatering is viable for feeds with conductivity in the range 0.1–10 mS cm^{-1}. Other workers [10] have compared the relative costs of the electro-osmotic dewatering, centrifugation, rotary vacuum filtration, filter pressing, winkle pressing and band pressing. In their assessment, the costs of electro-osmosis became competitive if the initial solids content of the feed was greater than 5%. Electro-osmosis has the advantages of combining very high solids retention with high final solids content and the possible matching of membrane performance to variable feed properties and solids retention requirements by the remotely controlled variation of applied voltage. High solids retention with high final solids content is important in radioactive waste management and in the treatment of other toxic wastes, for example, plating wastes.

Table 4

Realistic goals for an electro-osmotic dewatering treatment

Final solids content	35%
Solids retention factor	>2500
Power consumption	0.1 kWh L^{-1}

The dramatic reduction in the rate of fouling at UF membranes when an electric field is applied across the membrane indicates that there is much scope for future developments in this field. The equipment described in this paper is uniquely valuable in the study of the control of fouling by electric fields. Results so far indicate that it is better to prevent fouling than to try to displace an already existing fouling layer electrically.

Finally, measurement of the rate of electro-osmosis at a dipped cell provides a useful addition to the presently available methods for determination of the zeta potential of membranes and colloids. The technique permits direct measurement on flocs and concentrated colloidal dispersions of industrial importance.

7. ACKNOWLEDGEMENTS

This work has been commissioned by the Department of the Environment, as part of its radioactive waste management research programme. The results will used in the formulation of Government policy, but at this stage they do not necessarily represent Government policy. This work was also performed in the frame of the indirect action programme of the European Atomic Energy Community.

We wish to thank Dr R. G. Gutman for collaboration and use of laboratory facilities and Mr R. A. Clark for performing some of the experiments.

REFERENCES

[1] Chauver, P., in Carley-Macaulay, K. W. (ed.), *Radioactive waste: Advanced Management Methods for Medium Active Liquid Waste,* Harwood Academic Publishers, Brussels (1981).

[2] Amphlett, C. B., *Treatment and Disposal of Radioactive Wastes,* Pergamon Press, Oxford (1961).

[3] Bockris, J. O'M., and Reddy, A. K. N., *Modern Electrochemistry,* Vol. 2, p. 629, Macdonald, London (1970).

[4] Smoluchowski, M., in Graetz, B. (ed.), *Handbuch der Elektrizitat und des Magnetismus,* Vol. 2, p. 366, Leipzig (1914).

[5] Sunderland, J. G., *ECRC/N* 973 (1976).

[6] Turner, A. D., Bowen, W. R., Bridger, N. J., and Harrison, K. T., *AERE-G2599,* (1982).

[7] Henry, J. D., Lawler, L. F., and Kuo, C. H. A., *AIChE Journal,* **23**, 851 (1977).

[8] Smyth, M. J. and Cumberland, R. F., UK Patent 1 590 828 (1981).

[9] Shaw, D. J., *Electrophoresis,* p. 43, Academic Press, London (1969).

[10] Ellis, D., and Sunderland, J. G., *ECRC/N 1022* (1977).

CHAPTER 2

Laboratory Studies of Electrolytic Flotation as a Separation Technique

KOSTAS A. MATIS, Lab. Gen & Inorg. Chem. Techn., Aristotelian University (114), Thessaloniki, Greece and **JOHN R. BACKHURST**, Dept. Chem. Eng., The University, Newcastle upon Tyne, UK

ABSTRACT

The flotation process has been applied in effluent treatment to separate dispersions and emulsions, and to reduce suspended solids, oil and BOD content, usually as a secondary treatment step, so that effluent criteria may be met. The generation of gas bubbles by electrolysis and their application to flotation was the method used in the present work. The effects of feed concentration, residence time, and coagulant addition are described. In continuous flow, on a laboratory scale, concentrations approaching zero were obtained, while in batch systems, reductions of up to 99.9% have been recorded. The problem of electrode material and performance was examined. Results of experiments where the electrodes were separated by a membrane, work on bubble size measurements and consideration of the scale-up, kinetics and economics of the process are included.

1. INTRODUCTION

Dissolved-air flotation, which is the dominant flotation technique in waste treatment, has been used mainly for secondary treatment to reduce suspended solids and oil content of effluents as well as COD and BOD levels. One such application is the clarification of oily wastes [19]. Volesky and Agathos published a literature review on oil removal by air flotation methods in refinery wastes [22].

As primary separation in refinery waste treatment, the API gravity separator is often used, its primary function being to separate free oil from the refinery waste waters. The Design Manual [1], however, gives no prediction of the effluent quality, but only provides a separator size for specified gravity and flow rate. Such a unit does not break or separate emulsions, nor is it specified for these purposes. Generalized approaches previously acceptable are no longer adequate because of more stringent regulations which demand covered gravity-type oil/water separators, in addition to further waste water treatment downstream. It is in this field that the flotation process is expected to operate, so that the effluent criteria of around 10 p.p.m. oil content can be met.

Electrolytic flotation is the most recent flotation technique and has been extensively studied by the authors [13] amongst others. In this paper, an attempt is made to give a comprehensive summary of the literature found on the process as shown in Table 1.

2. EXPERIMENTAL WORK

The laboratory experiments were carried out both in batch and continuous flow operation, in an apparatus whose capacity was approximately 17 litres and which has been described in detail elsewhere [13]. Preliminary information on electrolytic flotation characteristics was obtained when the technique was applied as a solid/liquid separation method on dilute emulsion paint dispersions for the removal of suspended solids. An oil dispersion used in workshop machines diluted with water as a cooling oil, was also examined. This oil contained an emulsifier and when diluted, produced a white and very stable emulsion.

A spectrophotometric method was used for the analysis, by measuring the ultraviolet absorption of the samples and comparing it to that of a set of standards of known concentration. It was empirically determined that 225 nm was an optimum wavelength to use for analytical purposes, since around that value the maximum absorbance of the various chromophoric groups, contained in an oil, occur [17].

The pilot plant experiments were carried out batchwise in a 0.61 m square tank. A set of two stainless steel electrodes with an interelectrode gap of 4 mm was constructed and fitted near the bottom of the tank. The lower electrode was a plain sheet, 1.6 mm thick, and the upper an expanded mesh electrode, having an effective working area of approximately 40% of the total cross-sectional area. The applied current density was 95–100 A/m^2 at a voltage of 17.5 V. A sampling point 0.5 m from the bottom of the tank was available for the withdrawal of samples.

After establishing the initial parameters, the process was tried on industrial liquid wastes. A waste from a power station, consisting of oil and pulverized fuel ash in an aqueous solution, was treated with promising results. An initial oil concentration of 18.4 kg/m^3 in the raw sample was measured, which reduced to 9 p.p.m. after treatment.

An oil–water sludge sample from a refinery interceptor was also treated successfully by electrolytic flotation after it was found that no phase separation

Table 1

Summary of the literature on electrolytic flotation

Author	Ref.	Effluent	Electrodes	Other details
Ryvkin *et al.*	12	Minerals	(a) Horiz. mesh cath. (b) vert. rod cath.	In a froth flotation cell
Zozulya	12	Minerals	Vert. stain. steel	Use of diaphragm
Matov and Fursov	15	Molasses residues	Graph. anode. sta. st. wire screen cathode	Extraction of yeast. c. d. 200–250 A/m^2
Matov *et al.*	16	Meat proces. industry	–	Use of ferric chloride and lime
Matov	14a	Silk proces. plants	–	Use of flocculants, c. d. 200 A/m^2
Matov	14b	Sewage	Graph. plate an., wire grid cath.	Cells with consumable aluminium anodes
Matov and Gasyuk	11	Fruit–berry liquors	Slanting electr., and a diaphragm	Multisection apparatus, c. d. 200–230 A/m^2
Kharlan *et al.*	9	Electrochem. treatm. of metals	Horizontal	c. d. 200–400 A/m^2
Faynshteyn and Mamakov	6a	Sugar fact., tanneries	–	Influence of pH
Faynshteyn and Mamakov	6b	Dairy, cellulose combine, oil refinery	Horizontal	Power consumption 1.8–7.2 MJ/m^3, reduction of BOD
Føyn	7	Sewage	Diaphragm cell	Mixing with seawater
Axell	4	Sewage	Diaphragm cell	Mixing with seawater
Smith	21	Sewage	Paral. plate system of different material	Sacrificial electrodes
Armstrong	3	–	Vertical	Patent
Mckenna *et al.*	17	Bilge, ballast water	Pt–Ir wire, spot welded to Cb as anode	Mixing with seawater, adjustment of pH, use of polyelectrolyte
Clark	5	Activated sludge	Horizontal, stain. steel	Feasible for bulking sludges
PD Process Engineering	18	Food industry, oils and s.s.	Horizontal	Advisable on units of 5 m^2 area or less
Morgett Electrochemicals	2	Domestic sewage	Horizontal	Pilot plant, primary treatment
Simon-Hartley	20	Activated sludge, paper mill, resin recovery	PbO_2 covered expanded Ti mesh an., st. st. cat.	Inclined bottom and electrodes
ICI	8	Printing units, meat pack. ind., road tanker washings	Horizontal	Power consumption 1.44 MJ/m^3

was achieved by dispersed-air flotation as the bubble size was too large. Finally, in another group of experiments, flotation was tried as the separation method for organics entrained in the aqueous phase from copper solvent extraction. In this case dispersed-air flotation was particularly successful.

3. RESULTS AND DISCUSSION

3.1 Batchwise Experiments

The batch tests on the emulsified oil–water dispersions are shown in Fig. 1 where the current density was 100 A/m^2 at a voltage of around 13 V. The electrolytic cell was made from two horizontal and parallel stainless steel electrodes near the base of the tank, the upper one being perforated. The total liquid height was 1 m and the initial vessel concentration was in the range of 50–100 p.p.m., whilst final concentrations as low as 7 p.p.m. were achieved. The residence time was in the range of 7–9 ks.

The whole process occurred in two stages. In the first, little appeared to be happening from the separation point of view as the concentration remained constant, so that after the initial experiments no samples were taken until a froth layer was apparent. The second stage formed the flotation process and had a duration of around 2.4 ks, depending upon the initial feed concentration.

Observations of the graphical results shows that the tests are not quite identical as some of the lines cross the others. This questioned the reproducibility of the experiments and it was felt that the main source of error lay in the mixing of the dispersion at the start, before the current was switched on, although every attempt was made to ensure thorough mixing by recirculation of the solutions. It was noticed that longer mixing tended to shorten the first stage in particular. For example, the run with an intial concentration 71.5 p.p.m. was mixed for 1.8 ks, in comparison to 0.3 ks of most of the rest (see Fig. 1).

The next round of experiments examined the effect of a coagulant on electrolytic flotation. Aluminium sulphate was chosen for this purpose. The voltage was around 11 V, and the liquid height was 0.8 m. The results, shown as Fig. 2, proved that the initial stage of the process was shortened, and with 100 p.p.m. of coagulant, it disappeared completely. It seems that alum not only accelerated the process rate, but also somehow counteracted the effect of the emulsifier.

The use of coagulants is considered essential in dissolved-air flotation. Due to the inherent increase in running costs however, their use in electrolytic flotation should depend on an optimization of the process, as it often works successfully without their addition.

The gradient of the electric field between the electrodes aids flotation of suspended matter, with the result that clarification can be effected with effluents that previously would not have been considered suitable for treatment by flotation. The electrode grids can be arranged to provide a good coverage of the whole surface area of the flotation tank, so that uniform mixing between the effluent and the tiny gas bubbles is achieved with minimum turbulence.

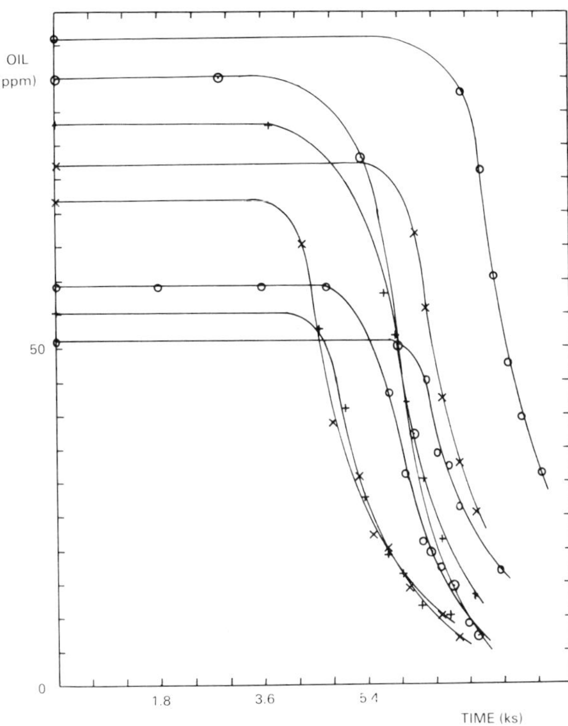

Fig. 1 – Batchwise experiments of electrolytic flotation. Preliminary tests with different initial tank concentration. Effect of time of treatment.

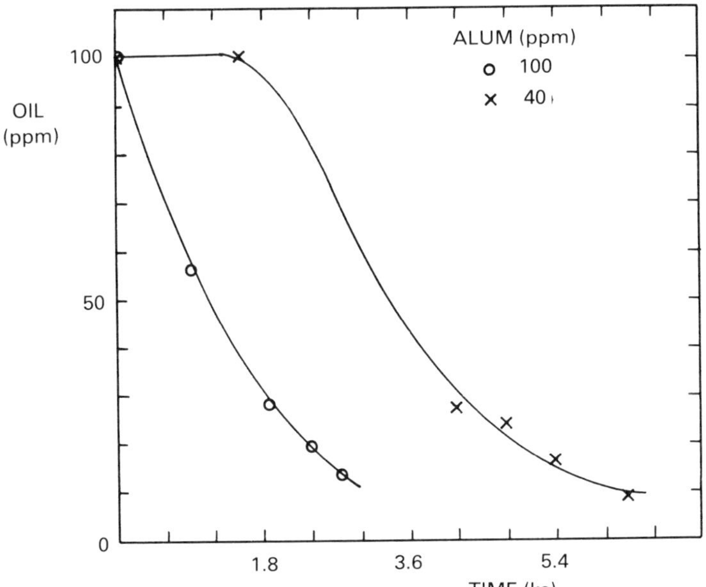

Fig. 2 – Use of flocculants and effect on retention time.

3.2 Pilot Plant Tests

The scale-up experiments used initial oil concentrations in the range of 60–65 p.p.m. The duration of the first stage of the process, not shown in Fig. 3, was approximately 16.2 to 18 ks. This initial retardation was greater with increased liquid heights. A final oil concentration very near zero was measured on one occasion. The results in the above mentioned figure are drawn on a semi-logarithmic graph paper and the straight line shows an agreement with the known first-order equation of chemical kinetics.

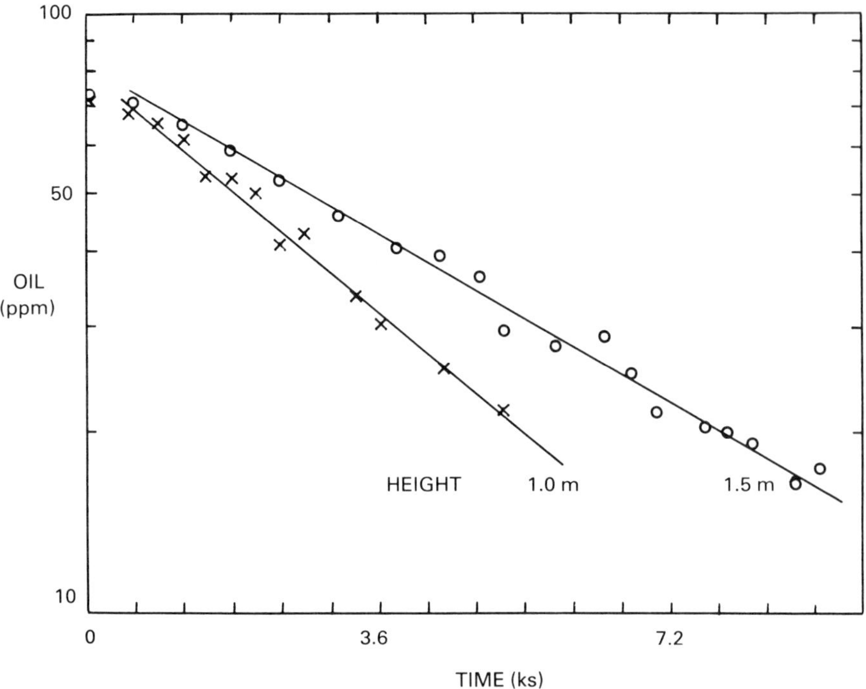

Fig. 3 – Pilot plant work – First order equation.

Another observation was a break-away of large agglomerated particles, due to partial deaeration, from the bottom of the floated layer, which, because of the size of the agglomerate, sank to the bottom of the tank. It is believed that in a continuous flow unit where a scraping mechanism would be used this would not present any problem as the paddle would scrape the floated sludge before a significant fraction of the entrapped gas has escaped.

The same phenomenon was promoted by turbulence created from the periodic formation of large hydrogen bubbles, due to coalescence, under the bottom electrode. These bubbles had, as the only means of escape, the gap between the edge of the electrode and the perspex wall of the tank. A hydrodynamic study of the system was undertaken in an attempt to clarify this aspect of the observed behaviour.

An interesting proposal as far as the placing of grid electrodes is concerned, was to locate them at an angle so that any settling particles would accumulate in a trough arrangement between the electrodes, from where they could be conveyed elsewhere [20]. Otherwise, it is possible to have a layer form on the electrodes, which will hinder the bubble evolution, or possibly allow the particles to be carried off in the outlet.

In the pilot flotation tests, it was observed that the rate was slower. It is believed that this was due to two main reasons: (a) the wall effects in the smaller scale, and (b) the internal reflux of suspended matter from the floated layer.

3.3 Continuous Flow Operation

The continuous flow tests were based on previously gained experience and the effect of flow rate was studied when the success of the process was shown to be affected by high throughputs. In these tests the flow rate was in the range of 1.1–2.1 cm^3/s, with inlet concentrations in the range 30–70 p.p.m. Outlet concentrations were usually less than 10 p.p.m. and on one occasion a concentration of 0.6 p.p.m. was measured. The optimum residence time was approximately 6.6 ks. At the maximum flow rate used of 2.5 cm^3/s, a concentration reduction reduction of around 80% was achieved using a current density of 100 A/m^2.

The operation was countercurrent as there were no suspended solid particles that could pass to the outlet while settling. The advantages of this mode of operation are a quiescent exit under the electrodes and a better possibility of contact between the particulate matter and the bubbles.

Based on the familiar equation:

$$\ln (C_0/C) = kt$$

where k is the flotation rate coefficient and C_0 is the initial concentration. The flotation rate coefficient was calculated as a function of time enabling the flotation rate, dC/dt, to be presented as shown in Fig. 4.

It is observed that the flotation rate reaches a maximum which increases with higher feed concentrations. The time shown represents only the flotation stage, and the start was found by extrapolation of the line of the first-order equation.

Information about the bubble size produced by electrolysis was thought to be helpful in flotation design, as there is a notable confusion in the literature on this aspect. Average sizes of as small as 5 μm and as large as 200 μm were reported

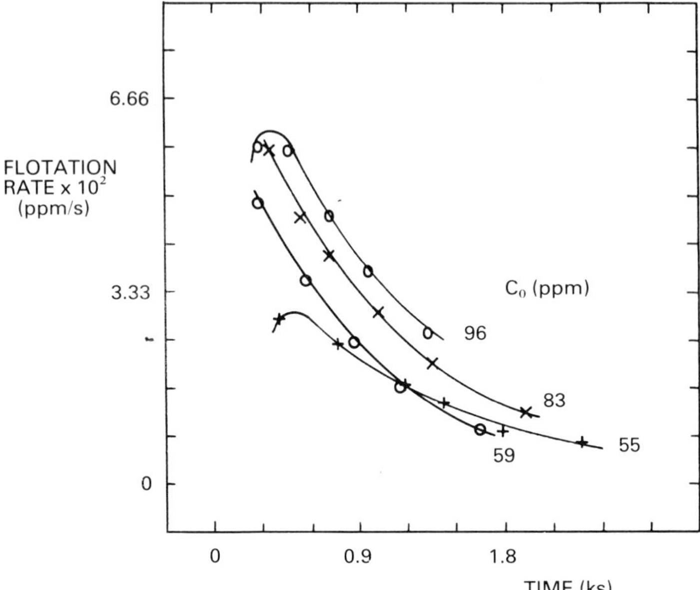

Fig. 4 – Variation of flotation rate in batch tests.

in different papers. Figure 5 shows the measured bubble diameter plotted against current density in the range investigated. Surprisingly, the bubble size was found to decrease with increased current density as the number of bubbles increases.

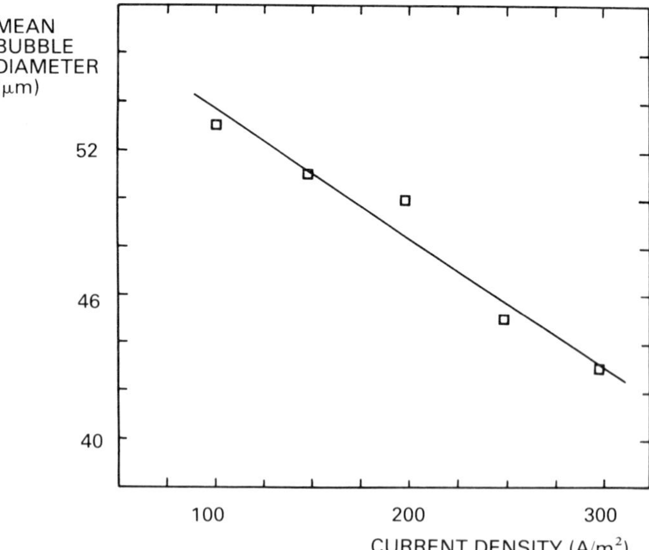

Fig. 5 – Effect of current density on bubble size.

3.4 The Problem of Electrodes

Before a viable electrolytic flotation process could be developed, it was necessary to find an electrode material of adequate mechanical and electrical properties with an extended lifetime [17]. The basic requirement for the electrodes is to evolve very fine bubbles by electrolysis of the solution. To accomplish this in practice a number of criteria must be satisfied: (1) no electrode corrosion, (2) avoidance of scaling. (3) capability of operating at high current densities, and (4) there should be no objectionable gaseous products. The corrosion of the anode could be a problem in electrolytic flotation, mainly with high anion concentration in the effluent, and particularly with chloride solutions.

Lead dioxide has the required high resistance to oxidation but has little mechanical strength. A carrier is therefore necessary and titanium has been found to be suitable and methods were developed for the deposit of lead dioxide on lightweight titanium sheets and grids [10]. Accelerated life tests by the Electricity Council proved that lead dioxide covered titanium was still operational after 2 years in 0.1% sodium chloride solution.

If gases that severely attack the anode, such as chlorine, are produced (or, if they are not desired in the clarified liquid) the anode can be separated by a membrane and the gas can be removed from the space between the membrane and the anode by another liquid introduced into this space.

Work done on this aspect is presented in Fig. 6. Another aim of these tests was to see if the chemical difference of the gases had any effect on the process. In the figure, a comparison is shown between the process with both electrodes contributing, and with that of the anode alone. The flotation coefficient was calculated and found to be of the same order.

3.5 Economic Criteria

A comparison of running costs between flotation systems [18] showed that electrolytic flotation required significantly less power on units of 5 m^2 area or lower, i.e. units able to accept flows in the range up to 13.9 m^3/ks. The advantage decreased with increase in the size of the cell, with approximately 33% of the power needed for dissolved air flotation being used by 1 m^2 cell, to 85% of the power being used on the 5 m^2 cell. Progressing up the range, there was a significant increase in power consumption, and for a 10 m^2 cell the power requirement for electrolytic flotation was 65% greater than by air flotation.

There is little difference in other costs between the two systems and it has been reported [8] that for a 5.55 m^3/ks stream of easily floatable material, an electrolytic flotation plant would cost between £12,000 and £18,000.

By-product recovery is a potentially profitable aspect of effluent treatment, and as the decade of the 'technology of non-disposal' approaches, so proteins, fats, oil, organics and metals could be recovered from wastes, and these are possible interesting fields for the application of future flotation processes. Finally, electrolytic flotation could be applied successfully in the recovery of fines at present lost from mineral processing, especially in the case of valuable ores.

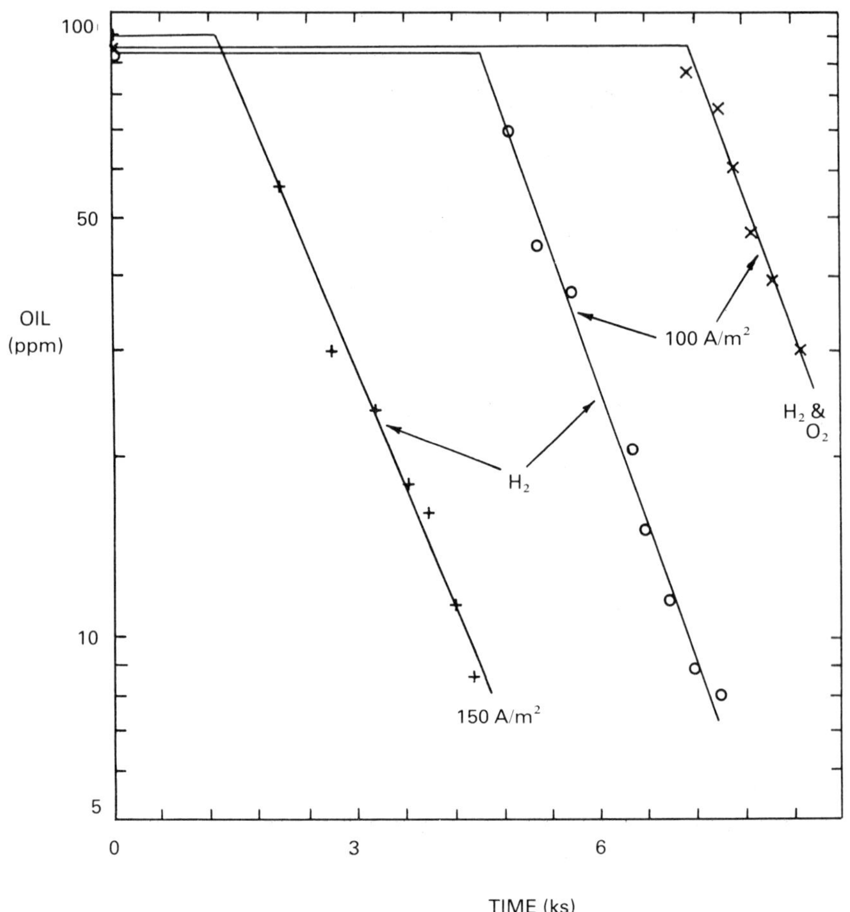

Fig. 6 – Experiments with separated electrodes.

4. CONCLUSIONS

The current density of 100 A/m^2 at 13 V was applied to the horizontal cell and final oil concentrations approached zero. The use of flocculants, which is considered essential in other flotation processes can be avoided, as the electric field itself aids flotation. The electrodes can be arranged to cover the whole surface area of the tank, so that uniform mixing between the effluents and the bubbles in counter-current operation is achieved.

The relative quantities of gases produced by electrolysis are a function of current density and salinity of the solution. A modest blower system would eliminate any hazard from hydrogen escaping from the tank in any potentially dangerous situations.

The scale-up experiments were successful although the flotation rate was lower. It reached a maximum, which increased with higher feed concentrations. The work with separated gases showed that hydrogen promoted the oil flotation process, while oxygen depressed this particular process. By using atomic hydrogen and oxygen separately, it was established that their chemical action on the suspended particle surfaces, changed the physicochemical properties.

It is appreciated that by separating the cathodic and anodic spaces of the electrolytic cell in the flotation chamber, the whole design becomes more complex. However, the advantage of operating with either electrodes can be applied, when moving from the laboratory scale to an industrial plant, by using the effect of the differences in the surface areas of the working electrodes.

Information on bubble size was obtained and an average size of around 50 μm was measured with bubble diameter decreasing with increasing current density.

An electrode material of adequate mechanical and electrical properties and extended lifetime is needed. Lead dioxide-covered titanium is suggested and has been found suitable to avoid anode corrosion and give long life.

Economic aspects have been considered and valuable by-product recovery could in future be a profitable aspect of flotation treatment of effluents.

REFERENCES

[1] Amer. Petr. Inst., *Manual on Disposal of Refinery Wastes – Volume on Liquid Wastes,* Chapt. 5: Oil–water separator process design (1969).

[2] Appleton, B., *New Civ. Engr.,* Jun. 20, 34 (1974).

[3] Armstrong, L. B., *U.S. Pat.* 3,664,951, May 23 (1972).

[4] Axell, J. P., *Inst. Publ. Health Engrs., N. Zealand,* **64**, 218 (1965).

[5] Clark, W. A., *M.Sc. Dissert.,* Civ. Eng. Dept., Univ. Newcastle upon Tyne (1973).

[6] Faynshteyn, L. B. and Mamakov, A. A. (a) *Appl. Electr. Phenomena,* **3**, 50 (1970); (b) *Appl. Electr. Phenomena,* **1**, 46 (1970).

[7] Føyn, E., *Proc. Ist Intl. Conf., Waste Disposal in the Marine Environment,* 279 (1959).

[8] ICI Pollution Control Systems, *Electroflotation,* Brochure.

[9] Kharlan, N. G., *et al., Appl. Electr. Phenomena,* **5**, 374 (1969).

[10] Kuhn, A. T., *Industrial Electrochemical Processes,* Elsevier, Amsterdam (1971).

[11] Mamakov, A. A., *Appl. Electr. Phenomena,* **4**, 226 (1968).

[12] Mamakov, A. A. and Avvakumov M. I., *Appl. Electr. Phenomena,* **5**, 357 (1968).
[13] (a) Matis, K. A. and Backhurst, J. R., *J. Chem. Tech. Biotechn.,* **31**, 431 (1981), (b) Matis, K. A., *Water Pollut. Control,* 136 (1980).
[14] Matov, B. M. (a) *Appl. Electr. Phenomena,* **3**, 220 (1966); (b) *Appl. Electr. Phenomena,* **5**, 317 (1966).
[15] Matov, B. M. and Fursov, S. P., *Appl, Electr. Phenomena,* **2**, 149 (1965).
[16] Matov, B. M., *et al, Appl. Electr. Phenomena,* **5–6**, 465 (1965).
[17] McKenna, Q. H., *et al., Dept. Transp., U.S. Coast Guard, Proj. No. 4101,* Jan. (1973).
[18] PD Process Eng. (a) *Brit. Pat.,* 1,194,850, Jun. 10 (1970); (b) *Water and Wastewater Clarification,* Technical Brochure.
[19] Quigley, R. E. and Hoffman E. L., *Ind. Water Eng.,* Jan., 22 (1967).
[20] Simon-Hartley, *Electrolytic sludge thickener,* Publication TP72.
[21] Smith, C. E., *Prog. Water Technol.,* **I**, 325 (1972).
[22] Volesky, B. and Agathos, S., *Wat. Pollut. Res. Can.,* **9**, 325 (1974).

CHAPTER 3

The Use of Colloidal Gas Aphrons (CGAs) for Removal of Slimes from Water by Floc Flotation

W. L. AUTEN and F. SEBBA, Department of Chemical Engineering, Virginia Polytechnic Institute and State University, Blacksburg, Virginia 24061, USA

ABSTRACT

By combining flocculation with the use of gas bubbles of about 25 μm diameter, in the form of colloidal gas aphrons, it is possible to entrain the bubbles in the floc, thus conferring buoyancy. In this way, it is possible to separate, by flotation, very finely divided particles (ultrafines) quickly and completely from the water in which they are suspended. This technique of bubble entrained floc flotation opens up the possibility of floating particles too small to be separated by conventional flotation.

1. INTRODUCTION

Without question, one of the most important developments in separating solid particles from suspension in an aqueous medium has been that of froth flotation in which the non-wettability, natural or induced, of mineral particles has been used to attach them to bubbles generated in the water, which buoy the solids to the surface, thus recovering them and also separating them from the unwanted gangue. The technology, however, has a limitation in that the effectiveness of the flotation is very much reduced when the particle size is less than approximately 10 μm in diameter, and for particles smaller than 1 μm it fails completely. These are the very particles that present difficulties in filtration, and suspensions of clays, often referred to as slimes, fall within this category. These can become a considerable nuisance, as in the so-called black waters of coal washings, the Florida phosphate slimes produced in the preliminary wash of the ore, and in the slimes produced in the hot water extraction of tar sands in Alberta. Furthermore, the elimination of reserves of high grade ores necessitates the need to process lower grade ores and this requires crushing more finely in order to liberate the desired minerals. There is therefore an urgent need for an improved technology which would enable the flotation of fine particles to be economically achieved.

There appear to be several reasons for the failure of conventional flotation to recover very fine particles, including the fact that tiny particles do not have enough momentum to penetrate the thin water layer that has to be pierced before the solid can enter the bubble. There are hydrodynamic considerations which keep very small particles away from the interface, turbulence due to the agitation in conventional methods for generating the bubbles, and the problems associated with the formation of a high contact angle on a particle whose radius of curvature is very small. The development of colloidal gas aphrons (CGA) overcomes some of these problems and shows promise of providing a new technology which should be of interest to mineral processors, civil and chemical engineers, and those involved in treatment of waste waters.

2. COLLOIDAL GAS APHRONS (CGA)

CGA, first described as 'microfoam' in 1971 [1], is a dispersion in water of bubbles of about 25 μm diameter. Later, the term 'aphron' was coined for a spherical phase which is encapsulated in a shell of a thin soapy film. As these are gas aphrons, which because of their size have colloidal properties, the present name was considered more appropriate. Dispersions of this sort are really a form of 'kugelschaum', but with smaller bubbles. These are true bubbles in that the gas is encapsulated in just the same way as is a free floating soap bubble in air, and the shell has the same elastic properties which are a consequence of the Gibbs adsorption isotherm. These bubbles must be clearly distinguished from what could better be described as gas-filled cavities which are introduced by sparging, pressure reduction (gas precipitation) or by electrolysis. Such cavities are generally believed to have a single interface separating the gas and the water, with a monolayer of the surfactant adsorbed at the interface should there be a surfactant present. The gas aphron, like the free floating soap bubble, has two surfaces of the encapsulating shell with surfactant adsorbed at each, and with aqueous surfactant solution separating the two surfaces. This bubble shell has an important function in that it prevents coalescence of the bubbles which thus retain their size, except for changes due to the diffusion of gas from the smaller to the larger bubbles because of the higher pressure inside the smaller bubbles. The rate of diffusion increases as the bubbles get smaller. For that reason, bubbles smaller than 25 μm soon contract and disappear. This can be observed under a microscope [2].

CGAs can readily be produced by passing a stream of surfactant solution at high velocity past a very narrow gas entry point situated at a venturi throat. The velocity has to exceed a critical velocity described in [1]. This has produced reproducible CGAs for many years but does not seem to be suitable for scale-up.

A newer method just developed, which is faster, uses a fast spinning disc immersed in the surfactant solution. This produces a surface wave which hits a vertical baffle causing it to re-enter the water, thus entraining the gas as colloidal-sized bubbles. However, such bubbles are not as uniform in size as those produced by the venturi generator. Bubbles of 25 μm can persist as long as the water in which they are dispersed is kept in sufficient motion to prevent the creaming that would otherwise occur due to gravity forces. If there is no movement, creaming occurs over about 10 minutes and the surface becomes so crowded that the bubbles change from spheres to polyhedra. The thin film between the bubbles, now with parallel surfaces, thins further so that it eventually bursts, producing coalescence of the bubbles and then a conventional foam.

A typical CGA can hold up to 65% by volume of gas dispersed in this way and because the bubbles are tiny spheres, the CGA flows as easily as the water in which the bubbles are dispersed. This means that a CGA can be pumped into a floatation cell with very little turbulence. This is one of the properties which gives flotation by CGA an advantage for fine particle flotation. Virtually any surfactant whose HLB number is high enough to make it highly soluble in water can be used to make a CGA. The concentration should be just enough to allow for adsorption at both interfaces and is usually about 10^{-3} M. Such a CGA with bubbles of about 30 μm diameter will provide a very large surface area of about 100 m^2 per litre of CGA.

A preliminary estimate for the power needed to produce CGA by the baffle method is slightly more than 1 kWh per 1000 l. This would contain approximately 5×10^{13} bubbles.

3. THE APPLICATIONS OF CGA IN FLOTATION

Testing of CGA at the laboratory level for flotation has shown that it can be very effective in floating ultrafines, but that the mechanism of flotation is very different from that of conventional flotation. Because the elastic shell surrounding the bubble is very difficult to penetrate, the particle does not enter the gas phase at all. For that reason, hydrophobicity is no longer a requirement for flotation. It is for this reason that Barnett *et al.* [3] were able to float colloidal particles as hydrophilic as the colloids which exist in fish wastes from aquaculture systems. Sebba and Barnett [4] list some of the different modes of operation of CGA in flotation. These range from ion flotation and precipitate flotation to floc flotation. When it comes to flotation of ultrafine particles, it appears that there are three methods of attachment of the particle to the bubble. In one case, there is adherence of the charged particle to the outside of the bubble, which is oppositely charged, by coulombic attraction. In such cases, a charged surfactant is required to make the CGA. This mechanism is similar to that of precipitate flotation and has been used by Gregory, Barnett and De Luise [5] to float precipitated manganic hydroxide using an anionic surfactant. In other cases, particles seem to be attached to the outer shell, perhaps held by the elasticity of the shell. This is shown by use of CGA to float graphite and coal of micrometre size. In these cases, aluminium ions are needed to serve as a conditioner and an anionic surfactant is needed to attract the aluminium cations which seem to act as a bridge between the bubble and the particle. Microscopic observation of the particles revealed obvious flocculation of the aluminium and a single point of attachment of the particles to the bubbles. The most extraordinary feature of this is that particles much too large to be levitated by the bubbles are, in fact, buoyed to the surface [2]. Plate 1 is a photomicrograph of a particle of

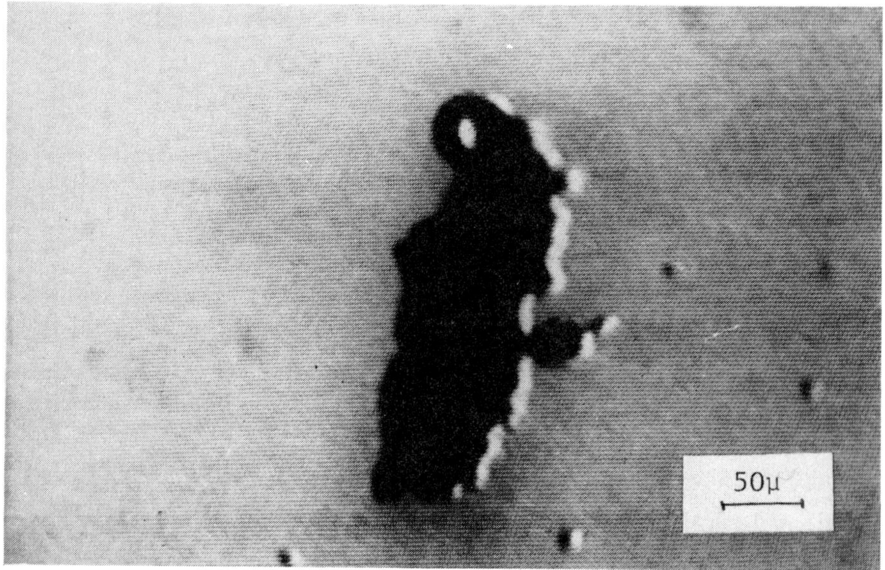

Plate 1 – Photomicrograph of a coal particle being buoyed by a CGA.

coal buoyed by a much smaller CGA, while in Plate 2 a graphite particle is shown also being lifted. The only explanation that can be suggested is that there must be coverage of the particle by myriads of invisible bubbles. This follows because, as mentioned, gas aphrons smaller than 25 μm rapidly diminish in size until they are no longer visible under high-powered microscopy. Nevertheless, it is unlikely that they no longer exist because as bubbles diminish in size, the thickness of the shell must increase, making diffusion of the gas slower. Such bubbles, although invisible, would tend to stick to irregularities on the solid, thereby increasing the buoyancy.

However, one mode of flotation and one especially applicable to flotation of slimes is when flocculation of the particles occurs in such a way as to entrap the bubbles and in this way induce in the floc a buoyancy which will convey it to the surface. The gas aphrons appear to act as nuclei for the flocculation, and for this reason the technique is being named bubble-entrained floc flotation.

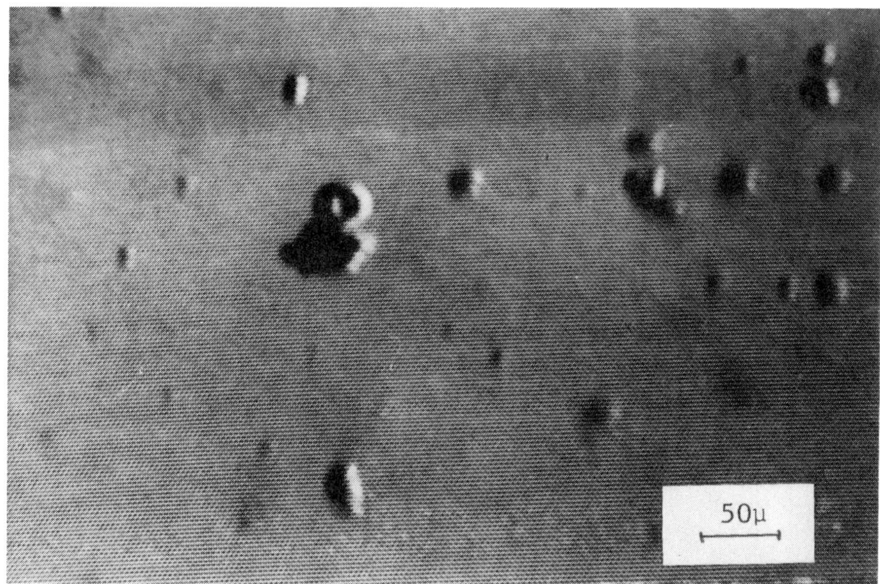

Plate 2 – Photomicrograph of a graphite particle being buoyed by a CGA.

This is to distinguish it from the technique in which a floc is formed and then floated in a conventional way by adhering to subsequently provided bubbles. Plate 3 is a photomicrograph of such bubbles entrained in flocculated clay.

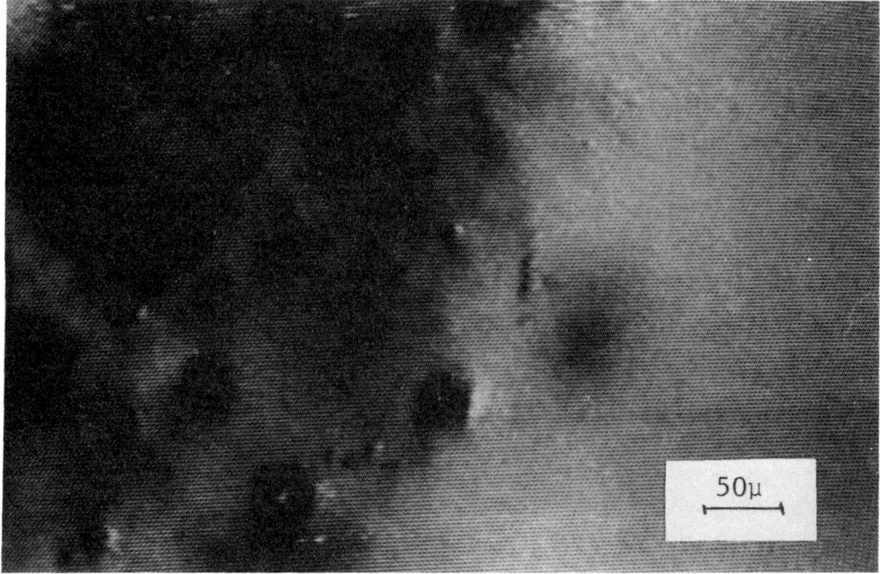

Plate 3 – Photomicrograph showing CGAs entrained in a phosphate slimes floc.

In some cases, the surfactant can itself serve as a flocculating agent; in others, it is desirable to add another flocculant in quantities much below that needed for conventional flocculation, and then to add the CGA which causes complete flocculation. Selective flotation by CGAs has been found possible for removing coal wetted with kerosine from ash [6] and a few experiments on separations of minerals by selective bubble-entrained floc flotation have shown encouraging results.

3.1 Bubble-entrained Floc Flotation of Phosphate Slimes

The CGAs used in the flotation tests were prepared by the venturi method described earlier. Cationic, anionic, and non-ionic surfactants were investigated as frothers, at concentrations ranging from 0.03 to 1.0 g l^{-1} solution. The flotation cell was simply a glass conical cell 120 mm inside diameter (ID) at the top, 5 mm ID at the bottom, and 175 mm in height. A glass stopcock was affixed to the bottom of the cell and CGA were pumped into the cell at a controlled rate of 90–100 ml min^{-1} by a 'Masterflex' peristaltic pump.

The phosphate clay wastes, or slimes, used in the flotation tests were run-of-mine effluent samples at 4.7–5.1 weight % solids. In some tests, the slimes were made more dilute by addition of distilled water to determine the effect of solids concentration on flotation performance. Flotation performance was evaluated on the basis of completeness of flotation and by the extent of consolidation of solids in the flotation froth as expressed on a weight percentage basis by the formula,

$$\text{weight \% solids} = \frac{\text{net weight solids in froth}}{\text{weight of froth}} \times 100 \ .$$

The quality of the CGA used in each flotation was assured both by microscopic observation and by measuring the volume percentage of gas incorporated. With the venturi method of CGA generation, it was found that the size distribution of the CGA bubbles was essentially invariant over the range of experimental conditions. However, the number of bubbles formed, and hence the total volume of gas incorporated, was found to be a function of gas supply pressure, surfactant concentration, and generation time. For a given gas pressure and surfactant concentration, a plateau representing the maximum volume of gas which could be incorporated was reached within 5–10 minutes. For all of the flotation tests conducted, high-purity nitrogen was used as the gas source.

Slimes samples of 50 to 100 ml were usually used for the flotation tests. In some runs, the volume of CGA added to the flotation cell was dictated by the amount needed to effect complete flotation of the slimes. In other tests, the volume of CGA added was held constant while other flotation parameters were varied. Under these latter conditions, known CGA volumes equivalent to one, two, or three times the sample volume were used to float the slimes.

Prior to flotation, slimes samples were conditioned by addition of flocculants and/or pH modifiers. Because floc formation is time dependent, slimes samples were given exactly 10 minutes to flocculate before flotation was effected. During this time, the slimes were agitated at a fixed rate using a magnetic stirrer and pH adjustments were made using either 1.0 M HCl or 1.0 M NaOH. Because of the tendency for protons and hydroxyl ions to adsorb onto clay surfaces and in so doing change solution pH, the pH of the slimes samples was maintained within 0.10 pH units of the desired value throughout the flocculation period. After the allotted flocculation time, each sample was transferred to the flotation cell and CGA was added at a fixed rate into the bottom of the cell. After flotation, entrapped water in the flotation froth was allowed to drain back into the water phase and then a minimum of 90% of the froth was removed and weighed. After drying the froth in a forced-convection oven for 4 h at 110–120 °C, the sample was reweighed and the solids concentration in the flotation froth was calculated.

3.2 Flotation with Anionic CGAs

Use of the anionic surfactant sodium dodecyl benzene sulphonate (SDBS) at a concentration of 0.75 g l^{-1} for generation of CGA achieved negligible flotation without prior treatment of the slimes. As the primary clay constituents in Florida phosphate slimes are attapulgite and montmorillonite, both of which have a negative net surface charge, this is not surprising. However, flotation was effected with this frother if aluminium ions were added prior to flotation at concentrations of 10^{-3} to 10^{-2}M using 0.1 M $Al_2(SO_4)_3 \cdot 18\ H_2O$ solution with a CGA volume equivalent to that of the sample. Presumably the adsorbed aluminium cations impart a net positive charge to the clay particles enabling them to attach to the CGA interface. This flotation was considerably improved if the slimes were first diluted with water as much as threefold. An explanation for this may be provided by the observation that addition of aluminium sulphate to undiluted slimes, even at concentrations as low as 10^{-3} M, produced a marked increase in viscosity. At normal plant effluent concentrations, the particles are sufficiently close to each other that formation of discrete flocs is unlikely. Instead a type of network between flocs arises, promoted by the aluminium ions which act as bridges. In this way, the flocs become effectively too heavy for the small bubbles to buoy to the surface. By diluting the slimes, discrete flocs can form which are floatable. There is thus a profound difference between this type of flocculation and flocculation prior to sedimentation where the objective is to produce flocs of maximum density so that they settle rapidly. Table 1 shows the effect on flotation of diluting the slimes with three parts distilled water, resulting in a sixfold consolidation of solids in the froth. However, based on the original concentration, this is only a twofold improvement which is hardly useful. Above a pH of 4.8, flotation efficiency dropped sharply.

Table 1

Flotation results for phosphate slimes in 1:3 dilution with water, floated using an anionic CGA (SDBS at 0.75 g l^{-1}, sample concentration at 1.2% solids)

Test	pH	Al flocculant concentration ($M \times 10^3$)	% Solids in froth	Flotation efficiency
1	9.0	1.0	4.3	Negligible
2	8.0	1.0	5.0	Negligible
3	7.2	1.0	5.7	Negligible
4	6.0	1.0	5.3	Incomplete
5	5.0	1.0	6.0	Incomplete
6	4.8	1.0	5.5	Complete
7	4.6	1.0	5.5	Complete
8	4.3	1.0	8.0	Complete
9	4.2	1.0	8.0	Complete
10	4.1	1.0	7.2	Complete

3.3 Dewatering in Froth

One of the problems is to achieve maximum dewatering in the froth. Table 2 shows how pretreatment of the slimes with either an anionic frother or the polyethylene oxide surfactant Tergitol 15-S-3, (a non-ionic oil soluble surfactant) improves the consolidation in the froth. For example, at pH 4.1 to 4.2, an aluminium ion concentration of 2 to 3×10^{-3} M produced a froth with 10.3 weight % solids.

Table 2

Flotation results for phosphate slimes (undiluted) floated with an anionic CGA (SDBS at 0.75 g l^{-1}, sample concentration at 4.7% solids)

Test	pH	Al flocculant concentration ($M \times 10^3$)	% Solids in froth	% Water removed by air drying*
1	4.49	2.6†	10.4	99.9
2	4.10	2.6‡	10.3	99.8
3	4.10	2.6	9.4	99.7
4	4.25	4.8	9.3	99.7

* Air dried for 24 hours at 22–24 °C, ambient humidity
† Added 5.0 ml of 0.75 g l^{-1} SDBS prior to flotation
‡ Added 3.0 ml of Tergitol 15-S-3 prior to flotation

3.4 Flotation with Cationic CGAs

The surfactant used was hexadecyl trimethyl ammonium chloride (Arquad 16–50). Because of the negative charge on the slimes, a cationic CGA acted both as collector and bubble producer and no other reagent was necessary to effect complete flotation. Flocculation did not appear to be as important in this case, which resembles more a precipitation flotation, so dilution of the slimes before flotation made no difference. Table 3 shows the results. In all cases, except run 10, all the solids were completely floated leaving a completely clear liquid layer under the froth. Run 10 reflects the lowest surfactant concentration, representing the lowest concentration to achieve a stable CGA with this particular surfactant. Another advantage of cationic CGA is that the flotations were independent of pH for all pH values below about 10.

Table 3

Flotation results for phosphate slimes floated with a cationic CGA (Arquad 16–50 used as frother, 100 ml of slimes floated per sample at 4.6 weight % solids)

Test	pH	Surfactant concentration ($g\,l^{-1}$ soln)	Collector (Arquad 16–50) concentration ($g\,l^{-1}$ soln)	CGA quality (ml N_2 per 100 ml CGA)	Volume CGAs required to float (ml)	% Solids in froth
1	7.44	1.0	0.0	61	233	8.81
2	7.43	1.0	0.1	61	167	7.91
3	7.43	1.0	0.2	60	152	7.84
4	7.44	0.5	0.0	56	179	8.60
5	7.43	0.5	0.1	58	134	8.55
6	7.43	0.5	0.2	57	160	8.03
7	7.44	0.1	0.0	39	283	8.01
8	7.43	0.1	0.1	40	275	9.23
9	7.42	0.1	0.2	39	206	7.71
10	7.43	0.05	0.1	36	–	8.49

The results in Table 3 seem to show that pre-treatment of the slimes with collector to a concentration of about 0.1 $g\,l^{-1}$ enhanced solids consolidation over that achieved with no pre-treatment at all. However, doubling the collector concentration to about 0.2 $g\,l^{-1}$ actually resulted in poorer solids consolidation than achieved with no additions prior to flotation. These observations are apparently reflections of the nature and degree of floc formation resulting from the different collector concentrations.

3.5 Use of a Non-ionic Frother

In flotations using anionic or cationic frothers, the resulting froths proved to be stable over long periods and maintained basically the same lamellar structures even after complete water removal by oven-drying. Because the capillary forces in the lamellar regions of the froth were believed to retard dewatering significantly, it was felt that a less stable froth might possibly dewater more quickly and completely. To this end, the use of a non-ionic frother was investigated. The surfactant chosen for consideration was Tergitol 15-S-9, an ethylene oxide-based secondary alcohol with an average molecular weight of 596 and an HLB of 13.3, at a concentration of 0.05 volume %.

The use of Tergitol alone was not sufficient to effect flotation and so a number of charged and uncharged flocculants were investigated in conjunction with the non-ionic frother. Among these was an oil-soluble Tergitol of relatively low molecular weight and HLB, an amine acetate salt, a polyethoxylated quaternary ammonium salt, ultra-high molecular weight polyethylene oxides, starches, and treated guar gums. Two collectors, a cationic guar gum and an amine acetate salt (cocoamine acetate), not only served to produce complete flotations but also seemed to produce a significantly drier froth than had previously been achieved. In fact, with a cationic guar gum addition of only 0.2 mg per 75 ml of slimes sample, the froth dewatered to over 17% solids in 40 minutes. Table 4 lists some of the data. The results were very encouraging, especially in view of the completeness of flotation which was achieved, the degree of solids consolidation obtained, and the remarkably low concentrations of reagents (frother and collector) needed to achieve the results.

Table 4

Flotation results for phosphate slimes floated with a non-ionic CGA (Tergitol 15-S-9 used as frother, 75 ml of slimes floated per sample at 4.9 weight % solids)

Test	pH	Flocculant	Flocculant concentration ($g\,l^{-1}$)	Volume CGAs added (ml)	% Solids in froth	Flotation efficiency
1	10.1	Armac C	0.04	225	15.1	Incomplete
2	6.97	Armac C	0.01	225	13.9	Complete
3	4.01	Armac C	0.04	225	13.6	Complete
4	10.05	Treated guar gum	0.007	225	14.4	Negligible
5	7.32	Treated guar gum	0.003	275	17.5	Complete
6	7.01	Treated guar gum	0.007	300	17.0	Complete
7	4.10	Treated guar gum	0.013	225	15.1	Complete
8	3.97	Treated guar gum	0.007	225	14.3	Complete
9	3.95	Treated guar gum	0.003	225	14.2	Complete

3.6 Effect of Drainage Time on Solids in Froth

Improved dewatering in a non-ionic CGA system as compared with a charged system apparently hinges on the observed instability of the froth after flotation. As the CGA slowly decomposes into a foam, the lamellae of the foam also break down. As a result, the capillary rise forces which normally prevent drainage of water entrapped in the lamellar regions of the froth disappear and gravitational draining is no longer impaired. It was also observed, especially with the guar gum, that the collector could also act to retard the tendency for solids in the froth to re-enter the water phase as the froth breaks down.

The rate of solids consolidation in the froth is dictated by the conficting forces of (1) gravity to promote draining and (2) capillary and chemical forces which act to retard draining. Unfortunately, small close-packed spheres such as CGAs lend themselves to capillary action more so than larger spheres because of the relatively larger surface area. By designing a flotation system so that the CGA will break down quickly upon froth formation, the action of capillary forces could be minimized. Still remaining, however, is any water physically or chemically adsorbed onto the slimes particles. This 'water of hydration' is not subject to normal gravitational draining and hence represents a practical upper limit to the extent of solids consolidation which could normally be achieved.

In Table 5 are results from a series of tests conducted to determine the pattern of draining for a typical flotation system. In this particular system, 1.0 ml of 1.0 g l^{-1} cationic guar gum stock solution was added to 75 ml of slimes (at 4.9 weight % solids), solution pH was adjusted to 4.25, and the slimes were then floated with CGA made with Tergitol 15-S-9, at a concentration of 0.05% by volume. The flotation was repeated varying the draining times allowed for the froth. In this series, the solids consolidation appears to approach a plateau of 15 weight %.

Table 5

Dewatering behaviour of a phosphate slimes flotation froth with a non-ionic CGA (Tergitol 15-S-9 used as frother, 75 ml of slimes floated per sample at 4.9 weight % solids

Test	pH	Collector concentration (guar gum, g l^{-1})	CGA quality (ml N_2 per 100 ml CGA)	Volume CGAs added (ml)	Draining time (min)	% Solids in froth
1	4.20	0.013	59	225	0	5.23
2	4.26	0.013	60	225	5	7.45
3	4.23	0.013	59	225	10	9.58
4	4.23	0.013	59	225	20	11.71
5	4.23	0.013	58	225	30	13.35
6	4.25	0.013	59	225	40	14.26
7	4.24	0.013	58	225	60	14.39

4. CONCLUSIONS

(i) It has been shown that ultrafine particles can be completely removed from suspension in water by flotation using CGA.

(ii) In many cases, there has to be flocculation entraining the bubbles which convey buoyancy.

(iii) The mechanism is completely different from that operating in conventional froth flotation and it would be a mistake to apply froth flotation principles to bubble-entrained floc-flotation.

(iv) Though only tentative experiments have been attempted to achieve selective separation of particles, indications are that this is possible. Such selectivity should be possible wherever selective flocculation is possible.

(v) In many cases, the amount of flocculating agent required is less than that needed for sedimentation flocculation as smaller flocs are desirable and in some cases flocculation can be enhanced by prior sensitization with very small quantities of reagent.

5. ACKNOWLEDGEMENTS

Thanks are expressed to the Florida Institute of Phosphate Research for a grant which made this investigation possible, and to Mr. G. L. Devan for his technical assistance.

REFERENCES

[1] Sebba, F., *J. Coll. and Interface Sc.*, **35**, 643 (1971).

[2] Sebba, F., *Report OWRT/RU–82/10* to U.S. Dept. of Interior, Office of Water Research and Technology (1982).

[3] Barnett, S. M. and Liu, S. F., *Proc. Conf. on Seafood Waste Management in the 1980's* (1980).

[4] Sebba, F., and Barnett, S. M., *Proc. 2nd World Congress of Chemical Engineering,* IV, 27 (1981).

[5] Gregory, O. J., Barnett, S. M. and De Luise, F., *Sep. Sci. Technol.*, **15**, 1499 (1980).

[6] Halsey, G. S., Yoon, R. H., and Sebba, F., *Proc. Tech. Program, Int. Powder and Bulk Solids Handling and Processing,* **67** (1982).

CHAPTER 4

Design and Operation of Final Settlement Tanks for Activated Sludge Systems

D. W. M. JOHNSTONE, Thames Water – Western Division, Witney area, Ducklingham Lane, Witney, Oxon, OX8 7JH, UK

---

1. INTRODUCTION

Biological treatment of sewage and other wastewaters involves the use of processes which bring the waste into contact with sufficient numbers of micro-organisms in the presence of dissolved oxygen. Some of the most efficient of these processes involve the use of activated sludge, which is a flocculent mass of micro-organisms produced when the wastewater is continuously aerated. Activated sludge systems are used extensively to treat the sewage from communities ranging from a few hundred people to those having several million.

Successful operation of activated sludge systems depends on efficient separation of the flocculent biomass from the surrounding liquid. Indeed, failure of the solid–liquid separation stage, i.e. the final settlement tank, means failure of the whole system. Efficient performance demands that clarification is sufficient to meet the standards required of the final effluent, and that the separated solids be consolidated (or thickened) into a concentrated suspension to be recycled for maintenance of the process. Therefore, any theory used to describe the performance of a final settlement tank should consider both the clarification and thickening functions, and the tank should be designed and operated according to which of the two functions is limiting.

Traditionally, clarification has been the main concern in the design of final tanks. Nominal retention time and surface loading (or overflow rate) are the most commonly used design parameters. Yet, in practice, unsatisfactory performance is most often caused by overloading with solids, a result of an inadequate consideration of thickening. The introduction of solids loading as a design criterion did attempt to relate performance to thickening, but the methods involved limiting the applied solids to some arbitrary value, and did not consider the effects of sludge withdrawal by underflow. The limitations of traditional design methods and the folklore associated with the design and operation of final settlement tanks has been adequately documented by Dick [1].

Over the last decade, one significant development in design philosophy has been the introduction of classical solids flux analysis by Dick [1–5], White [6, 7] and others [8, 9]. While this approach has some inherent limitations (e.g. it does not consider inlet and outlet conditions) it does attempt to overcome the more important limitations of traditional methods. It considers both the clarification function and the thickening function (although for activated sludge systems, thickening generally governs the design); it predicts the maximum solids-handling capacity from measurement of settling velocity at different solids concentrations and it considers the effect of the underflow on sludge withdrawal. In practice, good agreement has been found between theoretical predictions based on laboratory measurements and full-scale performance, although in one study [9] large discrepancies were found.

A particular problem concerning the performance and operation of settlement tanks for activated sludge concerns the variability of the sludge settlement properties. Although much research has been carried out into the settlement of activated sludge no theory exist to account for all the variations in these properties. Therefore the practical operation of a final settlement tank requires some measurement of settlement properties. Traditionally, this was carried out by allowing the sludge to settle for 30 minutes in a standard one litre measuring cylinder. This produces a measure named sludge volume index (SVI). The disadvantage of SVI is that it is influenced both by the presence of the walls of the small measuring cylinder in which the test is carried out, and by the concentration of solids, especially at high values of the latter.

Although SVI can be useful and is a widely accepted parameter for control of final settlement tanks it does not provide a useful basis for design [1, 7]. It is hardly realistic to translate the conditions in the small, unstirred cylinder used in SVI measurement to the dynamic conditions found in full-scale tanks. In addition, it only represents one point on a settlement curve and could equally be achieved by a number of sludges with different settlement curves. Changes in settleability can be masked by coincidental changes in solids concentration, and, generally, results are not comparable from one plant to another.

In an attempt to overcome the deficiencies in SVI, the Water Research Centre has developed and evaluated an alternative parameter for measurement of sludge settling characteristics [6, 7]. This parameter, the Stirred Specific Volume Index (SSVI) is measured in a 10-cm diameter tube which incorporates a low speed stirrer [7, 10] thus reducing particle-bridging and wall-effects. Measurements are made simultaneously at three solids concentrations but by interpolation the results are always quoted at a standard concentration. Evaluation of SSVI [6, 7, 11, 12] at 3.5 gl^{-1} shows that it correlates particularly well with sludge settling characteristics, and that it can predict (± 20%) the maximum solids loading which can be applied to a final settlement tank, except in cases where sludge settlement is very bad.

The object of this paper is to discuss the use of the mass flux theory along with SSVI in the design and operation of final tanks for activated sludge systems.

2. MASS FLUX ANALYSIS

Mass flux analysis examines the flow of solids passing through each cross-section of the settlement tank and considers that the total flux of solids per unit surface area is comprised of two components; one due to gravitational settlement of the sludge, and one due to withdrawal of sludge by continuous underflow from the base of the tank. Thus the total mass flux (F) is given by:

$$F = VC + \frac{Q_u}{A} C \quad (1)$$

As long as the underflow rate is not too high, the curve of F against C will have a maximum followed by a minimum as shown in Fig. 1. The value of F at the

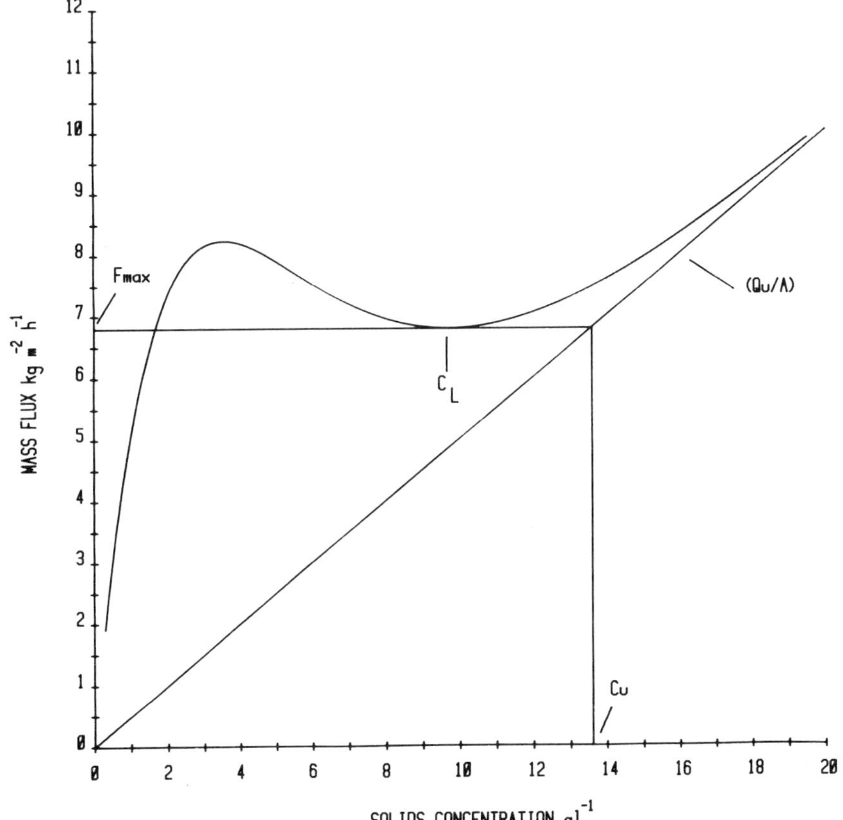

Fig. 1 – Total mass flux v. solids concentration.

minimum, F_{max}, determines the maximum predicted solids loading which can be applied to the tank before solids are lost in the overflow. The theory assumes that the rate of mass flow is constant throughout the depth of the tank, despite the increasing solids concentration. That is, in the long term there should be no accumulation of sludge in the tank. Loss of suspended solids in the effluent is assumed to be negligible.

The rate of gravity settlement is assumed to be a simple function of the concentration of suspended solids only, and is assumed not to depend on the depth of sludge and hence the weight of any overlying sludge blanket. It has been found convenient [6, 7] to use an exponential function to describe the hindered settling velocity against concentration,

$$V = V_0 \exp(-kC) \tag{2}$$

so that, overall, the mass flux of solids can be described by:

$$F = V_0\, C \exp(-kC) + \frac{Q_u}{A} C \tag{3}$$

where V_0 and k are arbitrary constants. V_0 can be considered to be the settling velocity of an individual floc particle at infinite dilution and k is dependent on the settlement properties of the sludge.

The rate of sludge return, or underflow rate (Q_u/A), is obtained from the slope of the straight line on Fig. 1. As the rate of return increases, the shape of the total mass flux curve changes until, at a certain critical underflow rate, there is a point of inflexion and no maximum or minimum. Under these conditions the tank ceases to behave like a thickener and merely acts as a clarifier.

3. MASS FLUX ANALYSIS AND SSVI

The operation of a settlement tank using a mass flux approach means keeping the applied solids loading, F_{app}, below the maximum flux rate, F_{max}, predicted from theory, i.e.

$$F_{app} = (Q_0/A + Q_u/A)\, C_f < F_{max} \tag{4}$$

In practical terms it is not possible to measure linear settling rates at different concentrations and derive V_0 and k on a daily basis. However, it has been shown [6, 7, 11, 12] that it is possible to use the information derived from SSVI measurements in an operational environment. White [6, 7] has shown that F_{max} can be obtained from the following empirical relationship

$$F_{max} = 310\,(\text{SSVI})^{-0.77}\,(Q_u/A)^{0.68}\ \text{kgm}^{-2}\text{h}^{-1} \tag{5}$$

Thus for a given value of (Q_u/A), the operator can use equations (4) and (5) to obtain a maximum MLSS concentration for a given rate of flow, or a maximum rate of flow for a given MLSS. The operator can then (if possible) alter the rate

Table 1

Maximum MLSS levels to prevent sludge washout

SSVI (at 3.5 g/l) Range	Return sludge flow		Maximum predicted solids-handling capacity kg/m² h	Maximum maintainable MLSS levels (g/l) for a maximum flow of:				
	X DWF	$\frac{Q_u}{A}$		2 DWF $\frac{Q_o}{A} = 0.5$	2.5 DWF $\frac{Q_o}{A} = 0.625$	3 DWF $\frac{Q_o}{A} = 0.75$	3.5 DWF $\frac{Q_o}{A} = 0.875$	4 DWF $\frac{Q_o}{A} = 1.0$
60–79	0.5	0.125	2.4	3.75	3.00	2.75	2.40	2.10
80–99	0.5	0.125	2.2	3.50	2.75	2.50	2.20	1.90
100–119	0.5	0.125	2.0	3.00	2.50	2.25	2.00	1.75
120–139	0.5	0.125	1.8	2.75	2.25	2.00	1.80	1.60
60–79	1.0	0.25	4.1	5.50	4.75	4.25	3.75	3.25
80–99	1.0	0.25	3.8	5.00	4.50	4.00	3.50	3.00
100–119	1.0	0.25	3.4	4.50	4.00	3.50	3.00	2.50
120–139	1.0	0.25	3.1	4.00	3.50	3.00	2.50	2.00
60–79	1.5	0.375	5.5	6.00	5.50	4.75	4.25	4.00
80–99	1.5	0.375	5.1	5.50	5.00	4.50	4.00	3.50
100–119	1.5	0.375	4.6	5.00	4.50	4.00	3.50	3.25
120–139	1.5	0.375	4.1	4.50	4.00	3.50	3.25	3.00
60–79	2.0	0.5	6.8	7.00	6.00	5.50	5.00	4.50
80–99	2.0	0.5	6.4	6.50	5.50	5.00	4.50	4.00
100–119	2.0	0.5	5.7	5.50	5.00	4.50	4.00	3.50
120–139	2.0	0.5	5.0	5.00	4.50	4.00	3.50	3.00
60–79	2.5	0.625	8.0	7.00	6.50	5.75	5.25	4.75
80–99	2.5	0.625	7.4	6.50	6.00	5.25	4.75	4.25
100–119	2.5	0.625	6.5	5.75	5.00	4.50	4.25	3.75
120–139	2.5	0.625						
60–79	3.0	0.75	9.0	7.00	6.50	6.00	5.50	5.00
80–99	3.0	0.75	8.3	6.50	6.00	5.50	5.00	4.50
100–119	3.0	0.75						
120–139	3.0	0.75						

of return to accommodate the plant conditions. However, since (Q_u/A) appears in both equations it is difficult to envisage the effect of such a change. In addition many operators do not like using equations. An alternative is to use the nomograph given by White [6].

A further alternative is to provide the operator with a tabular set of guidelines within which he must control the system. Such a set of constraints is given in Table 1. These were prepared from the analysis of many hindered velocities measured over a wide range of settlement properties [11, 12]. The data were deivided into four ranges of SSVI as follows

Settlement	Range of SSVI (mlg^{-1})
excellent	60 – 79
good	80 – 99
poor	100 – 119
bad (bulked sludge)	120 – 139

Regression analysis, using the exponential expression given in equation (2), provided values for the arbitrary constants V_0 and k for each of the four ranges of settlement (Table 2). Two graphical examples representing analysis at underflow velocities of 0.25 and 0.75 mh^{-1} are given in Figs. 2 and 3 respectively. A comparison of the graphs shows how solids handling capacity increases with underflow rate, and it can be seen (Fig. 3) that, for the higher ranges of SSVI, there are no minima in the curves, only points of inflexion. Hence, if the tank was to be operated with an underflow rate of 0.75 mh^{-1}, it would no longer behave as a thickener for sludges SSVIs greater than about 100 mlg^{-1}. It would merely act as a clarifier, because the underflow velocity is above the critical value $V_0 \exp(-2)$, given by White [7]. In this case the maximum solids loading is given by solving equation (1) for the inlet concentration.

$$F = V_f C_f + \frac{Q_u}{A} C_f \tag{6}$$

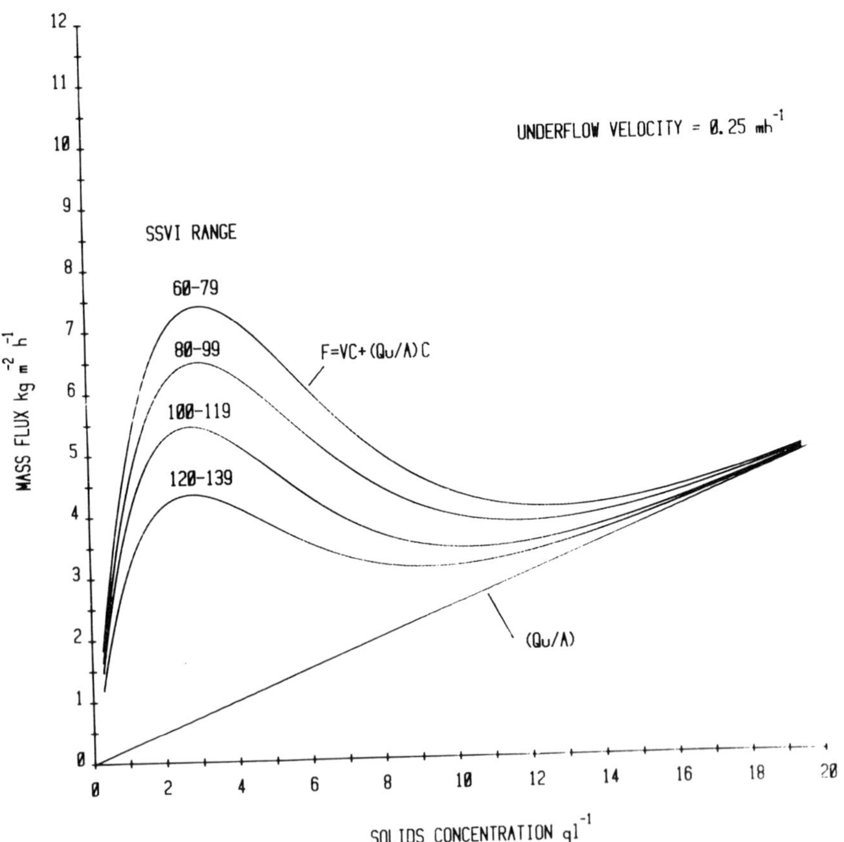

Fig. 2 – Total mass flux v. solids concentration.

Table 2

Curve fits of hindered settling velocity against concentration

SSVI range	Number of results	Solids concentration range (g/l)		Exponential $V = V_0 e^{-kC}$		
	n	Min.	Max.	V_0	K	r^2
60–79	83	0.65	9.80	6.500	0.3588	0.9607
80–99	310	1.23	11.0	5.826	0.3712	0.9591
100–119	148	1.55	9.62	5.300	0.4067	0.9374
120–139	168	1.53	8.31	4.234	0.4222	0.9618

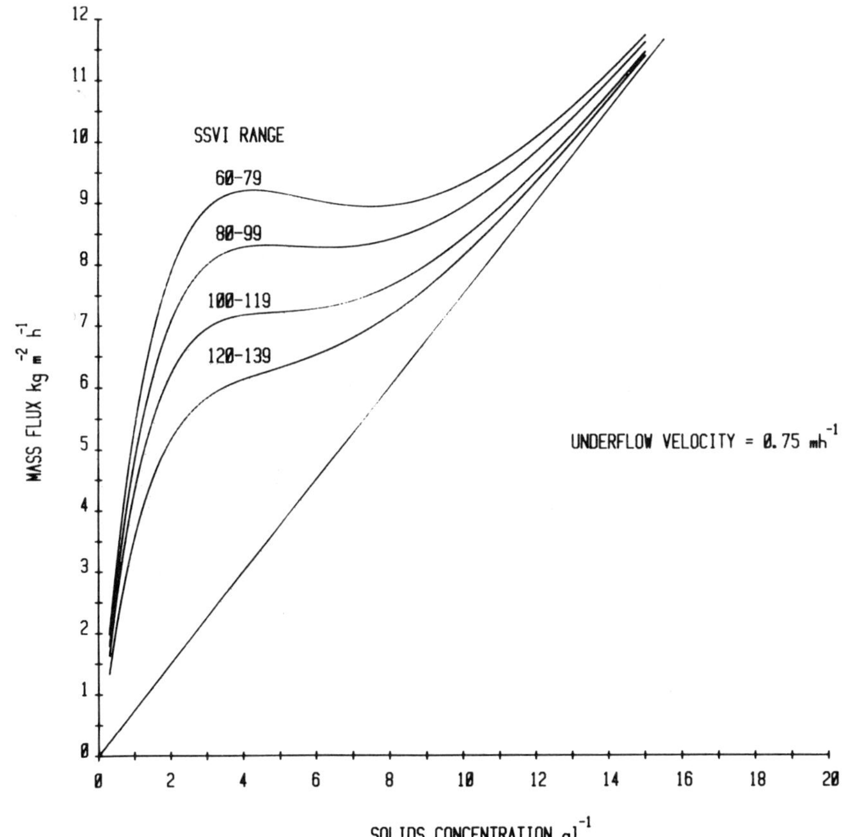

Fig. 3 – Total mass flux v. solids concentration.

The guidelines provided in Table 1 give the maximum value of MLSS (C_f) which can be maintained in the aeration tank to prevent the applied solids load, F_{app}, from exceeding the solids handling capacity of the tank. If there is a danger of F_{app} overloading the tank for a given set of conditions the operator can either increase solids handling capacity by increasing the return rate (i.e. underflow velocity) or he can reduce F_{app} by wasting sludge and reducing the MLSS concentration.

Sufficient studies on the use of a mass flux approach with SSVI measurements have now been made to establish the usefulness of the technique in the operation of final settlement tanks. The technique has also been used as a basis for design.

4. MASS FLUX THEORY AND DESIGN OF SETTLEMENT TANKS

In the design of final settlement tanks, mass flux theory can only assist in the selection of surface area and underflow rates. For the final design, engineering judgement must also be used. In common with other design philosophies mass flux theory requires estimates of sludge settleability but, unlike other philosophies, it is possible to predict tank performance from the estimated sludge settleability, if this settleability is given in terms of SSVI. In addition, mass flux theory predicts conditions when the tank will be overloaded with solids, the most common cause of unsatisfactory performance. Therefore, an approach to design, using mass flux in combination with SSVI, represents a significant improvement over the traditional approaches.

While this approach provides a rational basis for the selection of surface area and underflow rates, in its present form it has other limitations. Mass flux does not always appear to predict the concentration of solids in the underflow [9]. In studies by Johnstone *et al.* [11, 12] the underflow concentration was always 30–40% less than that predicted by theory. This was believed to be due to the shallowness of the tank preventing the solids from settling to a zone of compression. Considerable concentration gradients were found in the sludge layer, indicating that the sludge was in a hindered settling zone with minimal compression forces. This suggested that tank depth is important in establishing solids concentration in the underflow. The effect of depth on thickening is a feature of tank performance not considered by mass flux theory, although one attempt [9] has been made to account for it.

It is also important to provide sufficient depth to accommodate fluctuations in the level of the sludge/water interface caused by variation in flow and solids loading, i.e. storage. This is another aspect of tank performance not considered by mass flux theory nor by most other design philosophies.

The selection of inlet/outlet conditions, scum baffles, sludge hoppers, sidewall depths etc., still requires the use of empirical guidelines based upon experience. Such guidelines have been given by Sidwick [13].

Predicting effluent quality from activated sludge final tanks is extremely difficult and mass flux is of little help. It is difficult to estimate the settling velocity of the free-floating particles left after the main blanket has separated. In practice enhanced removal can be obtained by natural flocculation in the upward flowing liquid, and if the sludge blanket is above the level of the inlet, it will filter out solids. On the other hand an effluent poorer than that expected may be obtained if there is too much turbulence in a tank due to poor design of influent or effluent channels.

A comprehensive investigation by Chapman [14] studied the effect on effluent quality of seven variables, MLSS, sidewater depth, influent flowrate, depth of stilling chamber, rake speed, underflow rate, and airflow rate. He found only the first three variables were significant and derived a best fit equation.

$$C_e = -180.6 + 4.03\, C_f + 133.24\, Q_f/A + SWD\,(90.16 - 62.54\, Q_f/A) \qquad (7)$$

Even this analysis does not take account of the filtering effect of the blanket, the settleability of the sludge particles, or the effects of weir shape, and therefore there are still no hard and fast rules for design.

5. CONCLUSIONS

1. Mass flux theory can be used to predict (within 20%) the solids-handling capacity of a settlement tank treating activated sludge.
2. SSVI can be used as reliable measure of sludge settleability.
3. Together, mass flux and SSVI can be used as a basis for operational control of settlement tanks.
4. The combination of mass flux and SSVI can be used successfully in design to select surface areas and underflow rates.
5. The theory appears to be unable to predict the concentration of solids in the underflow.
6. Design of other features of settlement tanks for activated sludge such as inlet and outlet conditions etc. still requires engineering judgement on using well-tried guidelines.

ACKNOWLEDGEMENTS

The author would like to thank Mr J. L. Dakers, Thames Water, Western Division, for his permission to publish this paper. The views expressed are those of the author and are not necessarily those of Thames Water.

ABBREVIATIONS

SVI	Sludge volume index	$ml g^{-1}$
SSVI	Stirred specific volume index at solids concentration of $3.5\ g l^{-1}$	$ml g^{-1}$

MLSS	Mixed liquor suspended solids	
C	Solids concentration	gl^{-1} or kgm^{-3}
C_f	Concentration of feed solids (MLSS)	gl^{-1} or kgm^{-3}
C_u	Concentration of underflow solids	gl^{-1} or kgm^{-3}
C_e	Concentration of effluent solids	mgl^{-1}
F	Total mass flux of solids	$kg\ m^{-2} h^{-1}$
F_{max}	Solids handling capacity of settlement tank	$kg\ m^{-2} h^{-1}$
F_{app}	Applied solids loading	$kg\ m^{-2} h^{-1}$
V	Hindered settling velocity	mh^{-1}
V_0	Arbitrary constant	
k	Arbitrary constant	
Q_0	Flow rate of sewage	$m^3 h^{-1}$
Q_u	Flow rate of return sludge	$m^3 h^{-1}$
$Q_f = Q_0 + Q_u$		$m^3 h^{-1}$
A	Surface area of settlement tank	m^2
$\frac{Q_0}{A}$	Surface loading	mh^{-1}
$\frac{Q_u}{A}$	Underflow velocity	mh^{-1}
DWF	Dry weather flow	
SWD	Side wall depth	m

REFERENCES

[1] Dick, R. I., 'Folklore in the design of final settling tanks', *J. Wat. Pollut. Control Fed.*, **48**, No. 4, 633 (1976).

[2] Dick, R. I. and Ewing, B. B. 'Evaluation of activated sludge thickening theories', *J. Sanit. Enging Div. Am. Soc. civ. Engrs.*, **93**, SA4, 9–29 (1967).

[3] Dick, R. I., 'Gravity thickening of waste sludges', *Filtration and Separation*, **9**, 177 (1972).

[4] Dick, R. I. and Young, K. W., 'Analysis of thickening performance of final settling tanks', Paper presented at 27th Ind. Waste Conf., Purdue Univ., May, 1972.

[5] Dick, R. I., 'The role of activated sludge final settling tanks', *J. Sanit. Enging Div. Am. Soc. civ. Engrs.*, SA2 423 (1970).

[6] White, M. J. D., Design and control of secondary settlement tanks', *Wat. Pollut. Control*, **75**, 459 (1976).

[7] White, M. J. D., 'Settling of activated sludge', WRC TEchnical Report TR11, May (1976).

[8] Handley, J., 'Sedimentation: an introduction to solids flux theory'. *Wat. Pollut. Control*, 230 (1974).

[9] Hibberd, R. L. and Jones, W. F., 'The design and operation of activated sludge final settling tanks', *Wat. Pollut. Control*, 73.

[10] Vesilind, P. A., 'The influence of stirring in the thickening of biological sludge', Ph. D. Thesis, University of North Carolina (1968).

[11] Johnstone, D. W. M., Rachwal, A. J., Hanbury, M. J. and Critchard, D. J., 'Design and operation of final settlement tanks; use of stirred specific volume index and mass flux theory', *Trib. Cebedeau,* **33**, No. 443, 411 (1980).

[12] Johnstone, D. W. M., Rachwal, A. J., Hanbury, M. J. and Critchard, D. J., 'Design and operation of final settlement tanks', Paper presented at seminar on 'Future Trends in Sewage Treatment', Inst. Mun. Engrs., London, December (1981).

[13] Sidwick, J. M., 'Rationalisation of dimensions and shapes for sewage treatment works construction', Report No. 82, Construction Industry Research and Information Association (CIRIA) (1979).

[14] Chapman, D. T., 'The influence of process variables on secondary clarification', Paper presented at 55th Water Pollution Control Federation Conference St. Louis, Missouri, September (1982).

CHAPTER 5

Hydrocyclones in Chemical Engineering

L. SVAROVSKY, Schools of Chemical Engineering and Powder Technology, University of Bradford, Bradford, West Yorks, BD7 1DP, UK

---

1. INTRODUCTION

The separation action of hydrocyclones is based on the effect of centrifugal forces. In contrast to sedimenting centrifuges, however, hydocyclones do not have any rotating parts and the necessary vortex action is produced by pumping the fluid tangentially into a cono-cylindrical body (see textbooks such as ref. [1] or [2] for full description). The cylindrical part is closed on top by a cover, through which the liquid overflow pipe (often called 'vortex finder') protrudes some distance into the cyclone body. The underflow, which carries most of the solids, leaves through the opening in the apex of the cone. There have been many different designs in use in some specialized fields but for separation of granular solids, the above-described conventional design is still the best.

The idea is almost 100 years old (patented for the first time in 1891) but hydrocyclones did not find significant application in industry until after the Second World War. First in mineral processing and mining, but lately also increasingly in the chemical industry, petrochemicals, power generation, textile industry, metal working industry and many others. Hydrocyclones are now well established and their application is still growing.

The diameters of individual cyclones range from 10 mm to 2.5 m, cut sizes for most solids range from 2 to 250 μm, flowrates (capacities) of single units range from 0.1 to 7200 m^3 h^{-1}. The operating pressure drops vary from 0.34 to 6 bar, with smaller units usually operated at higher pressures than the large ones.

This paper gives, in summary form, some results and observations regarding the use of hydrocyclones in chemical engineering, based on experience from many years of hydrocyclone testing, consultancy and research.

2. NOTES ON INSTALLATION, DESIGN AND OPERATION

There is obviously a strong vortex flow inside a hydrocyclone and a study of the flow patterns has been made by many workers. From the point of view of a user, however, there are only a few important observations regarding the flow. Firstly and most importantly, for the solid particles to separate under the effect of the centrifugal force, there has to be a finite density difference between the solid and the liquid. Secondly, the flow must be steady, free of any disturbances or fluctuations. Thirdly, the vortex has a depression in the centre and any dispersed or dissolved gas in the feed tends to migrate to the centre and form a gas core. This is quite a desirable phenomenon as long as the gas core is straight. It gets distorted by sharp bends immediately above the vortex finder and these have to be avoided. The fourth important observation is that there are very steep velocity profiles in the flow, leading to high shearing forces which tend to break any loose flocs, agglomerates or droplets in the flow. In general, therefore, it would be a waste of money to use any flocculants to improve the separation performance of hydrocyclones. In fact, in classification duties, this is a positive advantage because the classification is then according to the size of individual particles rather than that of agglomerates.

The above stated generalizations represent the conventional view, however, and have been recently somewhat eroded by evidence of some flocs surviving the shear in the flow. If this happens, the effect manifests itself by a hump on the grade efficiency curve as shown in Fig. 1 for barium sulphate in water obtained

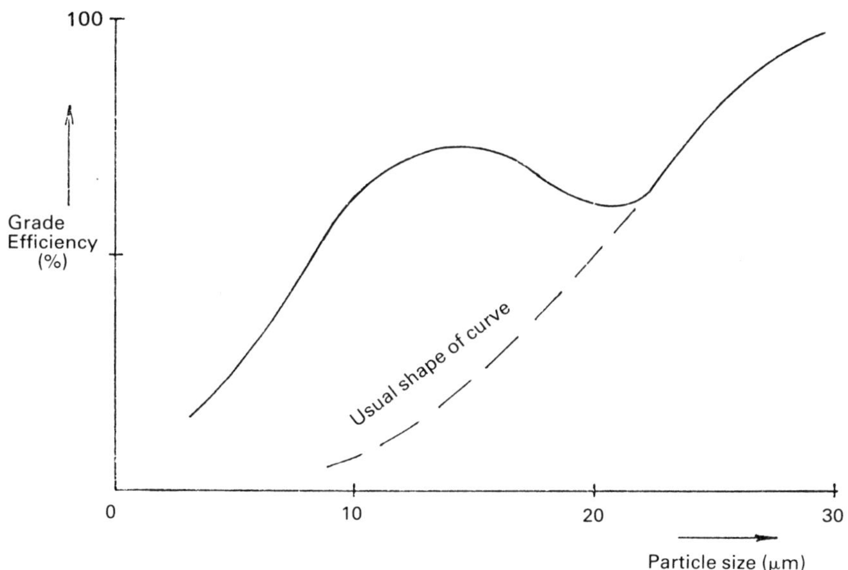

Fig. 1 – Grade efficiency curve obtained with barium sulphate, showing flocculation.

with a 125 mm diameter hydrocyclone. The reason for the hump is that the fine particles are separated as agglomerates at high efficiency but the particle size measurement method (sedimentation in this case) breaks the agglomerates into single particles. The recent trend towards separation of small particles in small diameter cyclones brings about increased likelihood of agglomerates surviving the passage through the cyclone because fine particles have relatively more surface and are more difficult to break apart when agglomerated.

Lastly and contrary to common belief, the discharge of the separated particles into the underflow orifice is due to the flow itself, which pushes the particle layer at the wall down into the apex. Gravity therefore does not significantly affect the separation, except in very large cyclones which tend to be used for separation of large particles. It also follows that hydrocyclones do not necessarily have to be operated in a vertical position and can be inclined or, for the smaller ones, even inverted if necessary.

Materials of construction of the cyclone body obviously vary with application, but the most common are steel (rubber or ceramics lined if necessary), aluminium oxide, ceramics, polyurethane and other plastics. There is a slight advantage in making the inlet rectangular, with the longer side along the cylindrical wall, rather than a simple circular inlet. This is no complication from the manufacturer's point of view because many cyclones are cast nowadays.

There has been a lot published about the relative proportions of the cyclone dimensions and their effect on separation efficiency and pressure drop. A user need not be concerned with this aspect except for two observations. There is a rule by which every measure which increases resistance to flow improves separation efficiency and vice versa. This applies to all proportions of the cyclone body, within certain reasonable limits, except for the length of the cyclone. Thus, for example, a cyclone with relatively small inlet and outlet openings is expected to give higher mass recovery but will offer higher resistance to flow and therefore have lower capacity. There have been several 'optimum' or recommended designs published and, as those have been well tested and enough is known about their performance, they may be adopted if needed [1, 2].

One dimension of a cyclone should be made variable and that is the underflow orifice. Correct adjustment of this is vital for the best operation of the cyclone and the size of the opening cannot be reliably predicted. It is best adjusted after the start-up of the plant and also during the operation whenever some operating conditions change. Several possible designs are available for this: replaceable nozzles, mechanically adjustable orifices, pneumatically operated adjustable orifices or even self-adjusting devices, which maintain a constant underflow density or concentration.

There is a whole host of operating conditions that affect the performance of hydrocyclones. Perhaps the most important are the operating pressure drop and the feed concentration. With increasing pressure drop the efficiency of separation increases but the law of diminishing returns applies here. There is little point in increasing the pressure beyond 5 or 6 bar, the typical operating pressures for larger cyclones being between 1 and 2 bar. With increasing feed concentration the efficiency of separation rapidly falls off and hydrocyclones are therefore operated with dilute feeds whenever high total mass recoveries are sought.

The scale-up of hydrocyclones is based on the concept of cut size, defined as the particle size at 50% on the grade efficiency curve, because the shape of the grade efficiency curve remains similar for a given family of geometrically similar cyclones. The following is a short account of the underlying theory; the reader is referred to ref. [1], Chapter 6 for further details.

3. DIMENSIONLESS SCALE-UP

At low solids concentrations, the flow pattern in hydrocyclones is unaffected by the presence of particles, and particle–particle interaction is negligible. Only few particles report to the underflow and the underflow-to-throughput ratio R_f is low and can be assumed to have no effect on the cut size x_{50}. Dimensional analysis coupled with theories of separation in hydrocyclones gives two relationships between three dimensionless groups:

$$Stk_{50} \, . \, Eu \; = \; \text{const.} \tag{1}$$

and

$$Eu \; = \; K_p \, . \, (Re)^{n_p} \tag{2}$$

where const., K_p and n_p are empirical constants for a family of geometrically similar cyclones, and the Reynolds number is defined as

$$Re \; = \; vD\rho/\mu \tag{3}$$

The Euler number is a pressure loss factor based on the static pressure drop across the cyclone Δp:

$$Eu \; = \; \Delta p/(\rho v^2/2) \tag{4}$$

Stk_{50} is the Stokes number (for cut size x_{50}) defined as

$$Stk_{50} \; = \; x_{50}^2 \, (\rho_s - \rho)v/18\mu D \tag{5}$$

where ρ_s and ρ are densities of the solids and of the liquid respectively, μ is liquid viscosity and D is cyclone diameter.

All the above equations use the superficial velocity in the cyclone body as the characteristic velocity, i.e.

$$v \; = \; 4Q/(\pi D^2) \tag{6}$$

The values of Stk_{50} . Eu, K_p and n_p for different cyclone designs have been published elsewhere (ref. [1] Chapter 6). The product of Stk_{50} . Eu varies from 0.06 to 0.22, the exponent n_p is usually between 0.2 and 0.4.

Equations (1) and (2) allow reliable cyclone design and scale-up at low concentrations up to about 1% by volume.

At higher input concentrations, the underflow-to-throughput ratio R_f has to be increased (by opening up the underflow orifice) to allow a greater volume of the separated solids to be discharged. The underflow-to-throughput ratio R_f as well as the volumetric concentration of the solids in the feed c then have to be added to the dimensionless groups that affect the separation process.

Quantitative prediction of the cyclone performance under such conditions is uncertain and the following scale-up method is to be treated as only very approximate. The pressure drop across a cyclone usually shows a small decrease up to concentrations of 8 to 12% by volume, followed by a gradual increase at higher concentrations. Some empirical correlations exist to describe this effect but those require two or more constants to be determined for each material and cyclone. With respect to the influence of solids concentration on the cut size, several models have been proposed, some of which are no more than a linear regression analysis of tests with specific materials and cyclone designs. Some theoretical justification can be shown for the following correlation

$$Stk'_{50} = K_1(1 - R_f)\exp(K_2 c) \tag{7}$$

where Stk'_{50} is based on the reduced cut size x'_{50} when the effect of the flow splitting is taken into account (this is determined from the 'reduced' grade efficiency curve now well established in describing the separation performance of hydrocyclones (consult refs [1] or [2] for details), K_1 and K_2 are empirical constants. For limestone in water for example, the values of K_1 and K_2 were found [3] to be equal to 9.1×10^{-5} and 6.46 respectively, for concentrations above 8% by volume and an AKW Rw 2515 cyclone.

4. APPLICATIONS

Applications of hydrocyclones in chemical engineering fall into several broad categories, as in the following sections.

4.1 Clarification

The aim here is to produce clear overflow, which is the same as to maximize the mass recovery of solids from the feed. The operating variables for this process are: dilute feed, relatively open underflow orifice (i.e. dilute underflow, typically below 12% of solids by volume), high pressure drop. The design variables are: small diameter cyclones (there is a limit here due to liability to blocking) nested in parallel in multi-cyclone arrangements, high efficiency design (i.e. small inlet and overflow orifices, etc.). In single pass installations, hydrocyclones can give minimum cut sizes around 3 μm but this can be further improved by using multiple-pass or recirculating systems. A multiple pass arrangement is shown schematically in Fig. 2; hydrocyclones are particularly suited for this. If for example 50% of the cyclone overflow is returned into the holding tank and the cyclone alone gives 48% solids recovery, the total recovery of the whole plant will be 65%. An additional benefit from this arrangement is that it dilutes the feed for the cyclone which improves the cyclone efficiency.

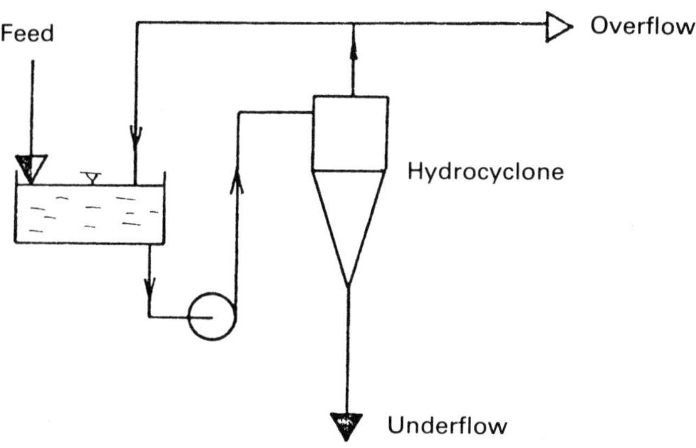

Fig. 2 – Multiple pass arrangement.

In order to maximize the clarification capability of hydrocyclones many manufacturers have recently made available multiple cyclone units, housing a number of 10 mm cyclones (from six to several hundreds) in parallel. As the underflow orifice must be greater than a certain minimum size (about 1 mm) to avoid blocking, the underflow-to-throughput ratio is usually quite large with 10 mm cyclones: 40% for the Dorr-Oliver Doxie 5, for example. The recently introduced Mozley 6 times 10 mm unit [4] shown in Fig. 3 gives R_f from 18 to 20%. As can be seen from Fig. 3, a Y-strainer has to be used in front of the unit in order to avoid blocking the underflow orifices with grit. Providing that this precaution is taken and that the slurry does not contain much coarse material, this unit gives trouble-free operation. Clean liquid tests on such a unit fitted with 2 mm overflow orifices give per single cyclone K_p = 121 and n_p = 0.347 in equation 2 (valid up to Re = 4700, i.e. pressure drop of about 5 bar with water).

Fig. 3 – A new unit of six cyclones of 10 mm dia. from Mozley, mounted in a test rig.

Some multiple cyclone units, such as for example the Doxie 5 mentioned above, have a strainer already built in. The applications include clarification of liquids as well as classification of solids at very low cut sizes of 2 or 3 μm for materials with specific gravity from 2 to 3. Recovery of catalyst in the oil and chemical industries, classification of abrasives or other fine particles in non-aqueous suspensions are examples of recent additions to the list of applications and the list is still growing.

4.2 Thickening

If hydrocyclones are to be used to produce thick underflows, the total mass recovery of the feed solids has to be sacrificed because throttling the underflow orifice inevitably leads to some loss of the solids to the overflow. A hydrocyclone as a single unit cannot therefore be used for both clarification and thickening at the same time. The underflow concentrations that can be achieved with hydrocyclones may be as high as 50% with some materials. Thus for example the AKW cyclone, 125 mm in diameter, shown in Fig. 4 installed in a test rig can give 45% by volume with chalk [5]. In this thickening performance hydrocyclones compare favourably with gravity thickeners and hydrocyclone systems are sometimes used to replace the much larger and more expensive gravity thickeners.

Fig. 4 – Large scale hydrocyclone test rig with a 125 mm dia. cyclone and a multi-cyclone (4 × 2″ dia.) installed.

The limiting factor in how far can the underflow orifice be closed is the danger of blocking. In terms of design variables, there is no particular need for small cyclone diameters; smaller units are in any case more liable to block. A variable underflow orifice is very useful here.

A typical example of hydrocyclone as a thickener is in pre-thickening of feed to vacuum filters, screens or dewatering centrifuges. Unless the feed is coarse, however, the overflow from the hydrocyclone used as a thickener has to be further clarified.

4.3 Thickening and clarification

If hydrocyclones are to perform both functions simultaneously, two or more units have to be connected in series, with one cyclone operated as a thickener and the others as clarifiers. In one arrangement, shown schematically in Fig. 5 as an example [2], the thickener is in the first stage, followed by two clarification stages, the underflows from which are returned to the feed. The overall recovery of the whole plant is better than the recovery of any of the individual cyclones used. The recycle should best be more dilute than the feed, thus diluting the feed to the thickening cyclone because divergent performance of the plant could be obtained otherwise.

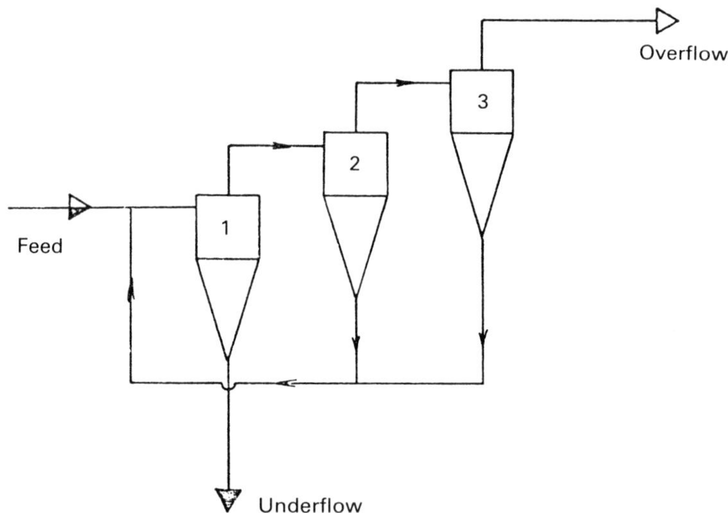

Fig. 5 – Arrangement of three hydrocyclones for thickening (No. 1) and clarification (Nos. 2 and 3).

If the feed is very dilute (less than say 1%) and the thickening cyclone might not produce sufficiently thick underflow in one stage, another arrangement may be used where the first stage is clarification, with the thickener treating the underflow of the first stage. Another clarifier is then used to clean the overflow from the thickener. The plant comes out smaller, achieves good thickening, but the recovery is not as good as in the arrangement in Fig. 5.

The cyclones in the above examples have to be properly sized, designed and operated to take best advantage of the arrangement. The diagram in Fig. 5 does not show pumps or sumps; it is advisable to use pumps between stages and discharge both the underflow and overflow of each unit into atmospheric pressure because this ensures correct hydraulic balance and steady flow in the cyclone. It is possible, however, to operate the outlets against back pressure but care has to be taken to balance the system and keep the flow steady.

4.4 Classification

This application of hydrocyclones is for solid–solid separation by particle size. As the grade efficiency of a cyclone increases with particle size, it can be used to split the feed solids into fine and coarse fractions. This may be a process requirement, by which coarse and fine solids are separated to follow different routes in the plant, such as for example in closed circuit wet grinding where the oversize particles are returned to the mill for further grinding.

Most frequently in this category, hydrocyclones are used either to remove coarse particles from the product (in a de-gritting or 'refining' operation) or to remove fine particles from the product (in a de-sliming or 'washing' operation). Hydrocyclones give a relatively poor sharpness of cut in one pass and consequently, in order to minimize the amount of misplaced material, the classification may be done in two or more stages. Figure 6 shows how the classification can be sharpened by using another stage on both the overflow and the underflow. If the cut size of all of the three cyclones is the same, the process will cut at the same particle size as the front cyclone but the grade efficiency will be significantly steeper. Typically, a sharpness index (x_{75}/x_{25}) of 2.4 for a single cyclone can be reduced to 1.6 using this arrangement. Note from Fig. 6 that if the underflow from the previous stages is to be further classified, it must be diluted in order to obtain good sharpness of cut and the required cut size.

Hydrocyclones are also used as classifiers simply to improve the performance of other filtration or separation equipment. A good example is in applications where the cyclone separates the feed into coarse and fine particles, and the coarse material is fed onto the horizontal belt filter first, as a pre-coat, with fines to follow. This will give good overflow clarity but in some cases at the expense of the final moisture content, because non-homogeneous cakes are known to give somewhat poorer dewatering characteristics.

4.5 Counter-current washing

Hydrocyclones can also be used for washing of solids by arranging several stages in a counter-current arrangement similar to that used with gravity thickeners. Such systems are found in production of potato or corn starch and similar applications. Washing here usually means removal of one solid from another like in gluten–starch separation, although removal of solubles from the feed is achieved using identical arrangements known as counter-current decantation. Two or more cyclones are used in series through which the solids and liquid move in the opposite directions.

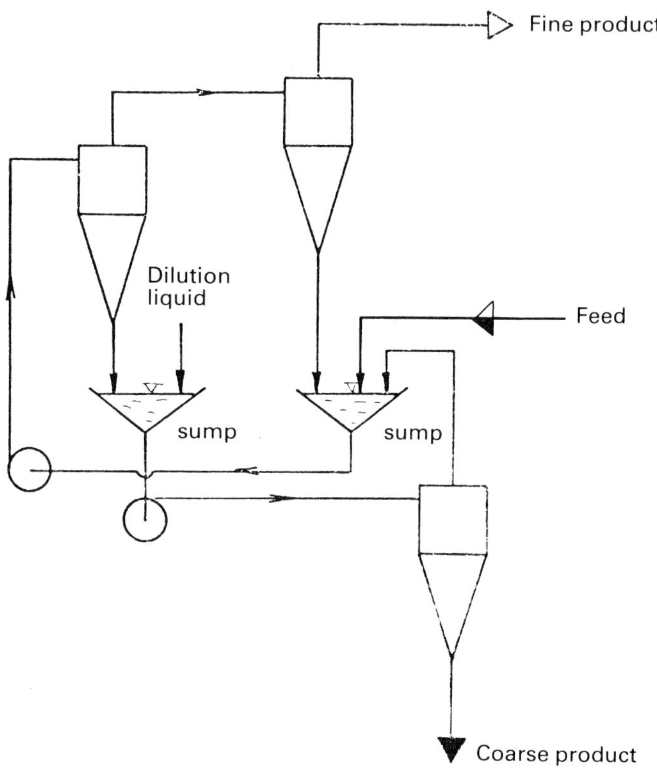

Fig. 6 – Re-classification arrangement designed to sharpen the cut.

4.6 Other two-phase separations

Specially adapted hydrocyclones (notably without the conical section) can be successfully applied to separation of two immiscible liquids. Recent developments in this field [5] show that such cyclones can separate oil from water, dewater light oils and produce highly concentrated samples of a lighter dispersed phase.

Another application is that of separation of gas bubbles from the feed liquid, such as in degassing of crude oil. Specially adapted cyclones [5] are thus used to replace large gravity separators where space is at a premium, such as on off-shore platforms.

5. CONCLUDING REMARKS

The conventional hydrocyclone has been subjected to considerable development to improve its characteristics. Sharpness of cut for example can be improved by injection of clean liquid into the flow near the apex of the cone. Another way is to operate the cyclone upside down so that the solids are separated into a container above the cyclone and subsequently fall back into the cyclone to be washed again. There is also some evidence that a flat-bottom hydrocyclone can give better sharpness of cut [5].

Cleaning efficiency and blocking characteristics of cyclones applied to pulps can be improved by using spiral cones [5]. These have a single or double spiral cast into the cone, with negative angle of the sides which represents no or negligible extra cost with plastic cones. Our own test work on one such unit shows that this invention does not offer any advantage in conventional separation of granular solids.

Various devices exist to counter the problems of blocking of the underflow orifice. Variable orifice anti-clogging reject chambers are available from one manufacturer while another offers manual or automatic orifice clearing devices without interrupting the feed supply. Automatic clearing is triggered off by a signal from a thermistor-based device which monitors the existence of underflow from each individual cyclone [5].

Significant further development of the conventional hydrocyclone is unlikely. It is expected, however, that hydrocyclones will spread into more industries and find wider use and new applications, as the know-how about their correct installation and operation becomes more wide-spread. Most hydrocyclone manufacturers have collected a wealth of experience and can offer useful advice. Many industries which are not in equipment manufacture make hydrocyclones for their own use and there is no reason why this also should not become more common in future.

6. REFERENCES

[1] Svarovsky, L., (ed.), *Solid–Liquid Separation,* Butterworths, London (1977).
[2] Bradley, D., *The Hydrocyclone,* Pergamon Press, Oxford (1965).

[3] Marasinghe, B. S., 'The effect of feed concentration on the performance of hydrocyclones', *M. Phil. Thesis,* University of Bradford (1980).

[4] *News International,* Issue 3 (1982), Richard Mozley Ltd., Woodlane, Falmouth, Cornwall TR11 4RF.

[5] *Proceedings of the International Conference on Hydrocyclones,* 1–3 October 1980, BHRA Fluid Engineering, Cranfield, Bedford.

CHAPTER 6

Effect of Chemicals on the Dewatering of Clay Suspensions from Sand Washing Operations

G. J. SPARROW, CSIRO Division of Mineral Chemistry, PO Box 124, Port Melbourne, Victoria 3207, Australia

---

ABSTRACT

A suspension of fine kaolin and quartz, generated as tailing by a sand washing operation, was dewatered to a solid by allowing it to settle in a dam, and then ponding the thickened material in shallow depths onto sand drainage beds.

In order to minimize the area required for the dewatering, the tailing should be ponded at the highest possible solids content. Attempts were made to increase the degree of compaction in the dam by dispersing the clay in the suspension. The pH of the suspension was raised from its natural value of pH 3–4 by the addition of sodium hydroxide, sodium carbonate or lime. It was possible to disperse the clay and obtain more compact sediments when the sand was washed with tap water. However, in the saline mill water, the clay could not be dispersed and so it was not possible to obtain more compact sediments. Addition of flocculants to the tailing in mill water increased the zone settling rate of the suspension but the resulting sediments had a lower solids content than the untreated material.

The addition of flocculants to the thickened tailing increased the rate at which water was lost by drainage when it was ponded onto a sand bed. However, the reduction in the time required to dewater the tailing to a solid was insufficient to make the use of flocculants an economic proposition.

1. INTRODUCTION

An aqueous suspension of finely divided solids is generated as tailing by many mining operations. The tailing is usually discharged into a containment area where the solids are allowed to settle and compact. When water is removed by decantation in the early stages, and by evaporation in the later stages, the tailing may be dewatered to a load-bearing solid.

The tailing generated during the washing of sand for the building and construction industries is a suspension of fine quartz and clay. The tailing is usually discharged into a dam or area from which the sand has been extracted and the fine solids are allowed to settle. However, when clay minerals are a major component of the tailing, a load-bearing solid may not be obtained even after many years because the clay does not settle and compact to any great degree. When an area has been filled and the surface water removed, a thin crust forms over the surface as water is lost by evaporation. However, the bulk of the tailing beneath the crust remains fluid.

In the mining industry where similar clay suspensions are generated as tailing (e.g. in the coal washing industry), mechanical equipment such as vacuum and pressure filters, and centrifuges have been used to dewater the tailing [1]. Similar equipment has been used in tests with tailing from sand washing operations [2] but, due to the low value of the product, the use of mechanical equipment to dewater the tailing is unlikely to be an economic proposition in this industry. With this in mind, we have used sand drainage beds to dewater thickened tailing from the dams since this dewatering method would not be expected to require equipment other than that already available at the washing plant.

Sand beds have been used extensively to dewater waste water and sewage sludge [3, 4]. Red mud, the tailing from the Bayer process, has been dewatered on drainage beds [5], and tests with dredge materials [6], and coal washery wastes [7] have been reported. However, this approach to dewatering tailing from sand washing operations has not been widely used in Australia. In this paper, the results of our work with sand washing tailing will be summarized, with particular emphasis on the use of chemicals to improve the degree of compaction achieved in the dam, and to increase the rate of dewatering on the sand drainage beds.

2. SAMPLES

The samples of tailing used in this work were taken from a sand washing plant in Melbourne, Victoria. X-ray diffraction indicated that the solids in the tailing were kaolin ($Al_2(Si_2O_5)(OH)_4$) with a little quartz (SiO_2) and traces of gibbsite ($Al_2O_3 . 3H_2O$) and pyrite (FeS_2). The tailing also contained 2–4 wt-% organic matter with spectral properties similar to humic acid. A particle size analysis indicated that the solids were very fine with 92 wt-% less than 2 μm and 84 wt-% less than 1 μm.

The mill water was very saline with a total dissolved solids content (TDS) of more than 2000 p.p.m. A typical analysis of the mill water indicated that it could be considered to be 1.4×10^{-2} M NaCl, 1×10^{-3} M Na_2SO_4, 6×10^{-3} M $MgSO_4$, and 2×10^{-3} M $CaSO_4$ (TDS 2260 p.p.m.). As a result of the organic material, and more particularly oxidation of the pyrite in the sand deposit, the pH of the mill water was in the range 3–4.

Tailing was taken from the pond at 18.2 wt-% solids and diluted with mill water to 5.1 wt-% solids to determine its dewatering characteristics and to 1.9 or 2.0 wt-% solids for the sedimentation tests.

Tests were also carried out with a suspension obtained by shaking the dry sand with Melbourne tap water (TDS 100 p.p.m.), allowing the coarse sand to settle, and then decanting the fine clay. A well aggregated suspension with pH 3.2 was obtained.

Sedimentation tests were carried out with 100 ml samples in 28-mm diameter cylinders. The pH was adjusted to the desired value, the cylinder was up-ended 10 times and the solids were allowed to settle undisturbed. The zone settling rate of the mud line was measured and, after 7 days, the volume and solids content of the sediment were determined. Samples were flocculated by first allowing the solids to settle for 24 h. The flocculant (as a 400 p.p.m. solution in distilled water) was then added to the supernatant water and contacted with the solids by up-ending the cylinder 10 times.

3. RESULTS AND DISCUSSION

The tailing generated by the sand washing operation was a dilute suspension of only 1–5 wt-% solids. In the saline, acidic mill water, the solids were well aggregated and settled slowly to yield a voluminous sediment with a low solids content. However, in order to minimize the area required for sand beds to dewater the tailing, the tailing should be ponded onto the beds at the maximum solids content possible to reduce the volume of water that must be removed. Accordingly, attempts were made in the laboratory to disperse the clay in the suspensions and so improve the degree of compaction achieved in the dam.

The kaolin particles in the suspensions have a plate-like shape. It has been shown [8] that the faces of the platelets carry an excess negative charge while the edges may have an excess positive or negative charge, depending on the properties of the suspension. Since increasing the pH of the suspension increases the negative character of the edges, attempts were made to disperse the clay by raising the pH of the suspension so that both the edges and faces carried an excess negative charge.

3.1 Effect of pH on the sedimentation properties

As the pH of a 2.0 wt-% suspension in tap water was raised by the addition of sodium hydroxide, the zone settling rate of the suspension decreased from 154 mm h^{-1} to 48 mm h^{-1} at pH 6, but above pH 7 the solids were dispersed and did not settle with a mud line (Fig. 1). There was little change in the solids content of the sediments as the pH was raised to pH 6 but above pH 7 it increased from 10–15 wt-% solids to over 30 wt-% solids (Fig. 2).

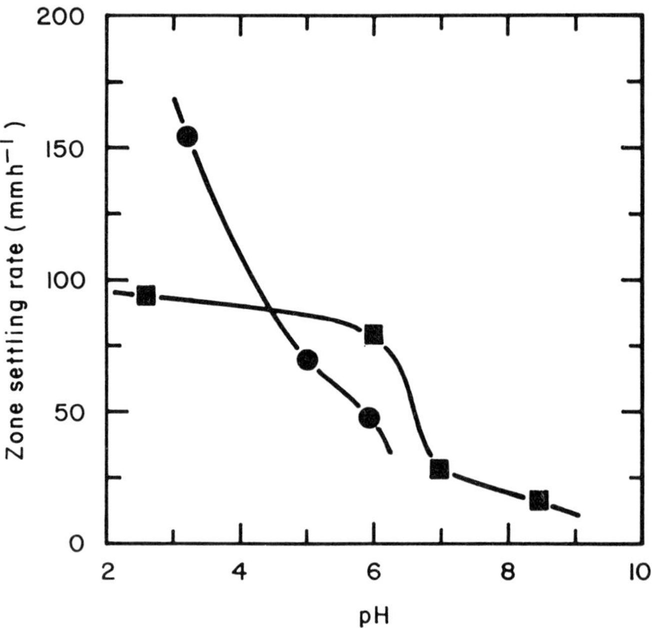

Fig. 1 – Effect of pH on the zone settling rate of the tailing in tap water (•) and mill water (▪).

While it was possible to achieve a significant increase in the degree of compaction of the tailing for a suspension in tap water, this was not possible with tailing in the mill water. There was a reduction in the zone settling rate between pH 6 and 7, indicating that the addition of alkali had altered the aggregated state of the solids (Fig. 1). However, the clay did not disperse even when the pH was raised to 9–10. There was no increase in solids content of the tailing associated with the reduction in zone settling rate. In fact, it decreased from 14 to 10 wt-% solids (Fig. 2). Similar results were obtained with additions of sodium carbonate and lime.

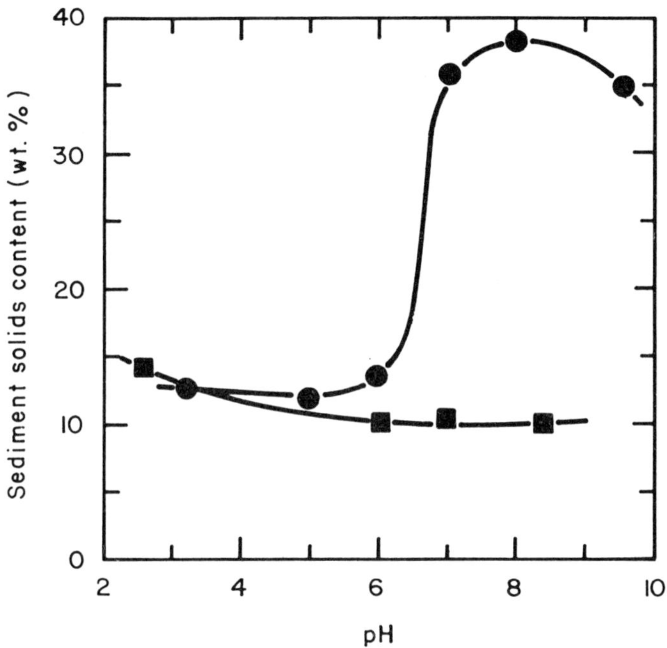

Fig. 2 – Effect of pH on the solids content of the sediment in tap water (•) and mill water (▪).

Only by the addition of a polyphosphate (sodium hexametaphosphate at a final concentration of 2.5 g/l) at pH 10 was it possible to disperse the clay in mill water. However, while the tailing could be dispersed under these conditions they were not considered to be a practical way of obtaining a more compact sediment because of the high alkali additions required and the possibility of eutrophication of the dam as phosphate levels in the water increased. (Alkali would need to be added continuously to neutralize the acid in the freshly washed sand.) Also, it is necessary to return clean water from the dam to the washing plant to wash further sand. When the clay was dispersed with alkali and polyphosphate, the supernatant water was not clear after 7 days, and so a flocculation or coagulation step would be required in order to obtain clean water. When the tailing in tap water was dispersed by raising the pH, the dispersed solids did settle over a period of 7 days.

3.2 Effect of flocculants on the sedimentation properties

A range of natural gums and starches and synthetic polyacrylamide-based flocculants were added to a 1.9 wt-% suspension in mill water and the zone settling rate and sediment volume were measured [9, 10].

A significant increase in zone settling rate was obtained with the polyacrylamides. The largest increases were obtained with non-ionic and cationic derivatives (Table 1). Addition of Superfloc C100 and N100 (4 p.p.m.) increased the zone settling rates by a factor of 40 and 10 respectively, while a 300-fold increase required additions of 12 p.p.m. C100, and 16 p.p.m. N100. These high additions are probably due to the very fine particle size of the clay and the presence of humic material in the tailing which has been shown to react with polyelectrolyte flocculants [11]. Since the tailing was allowed to settle undisturbed, there was no appreciable reduction in the volume of the sediment, and no corresponding increase in the solids content, when the flocculants were added. In most cases a less compact sediment was obtained (Table 1).

Table 1

Effect of flocculants on the sedimentation properties of the tailing

	Zone settling rate (m h^{-1}) at flocculant dose of (p.p.m.)						Sediment solids content† (wt-%) at flocculant dose of (p.p.m.)					
Flocculant	0	4	8	12	16	20	0	4	8	12	16	20
Superfloc A130	0.09	0.10	0.11				14.1	14.0	13.7			
A110			0.44	0.56		0.56			11.8			
N100	0.09	0.86	6.7	12.6	30.0	60.0	14.1	11.3	10.7	9.8	10.1	9.4
C100	0.10	3.6	10.8	30.0	48.0		12.5‡	10.1‡	9.5‡		9.5‡	
C110	0.09	0.57	2.4	6.7	10.8		13.8	11.0	10.2			

† Determined after 7 days.
‡ Determined after 1 day.

3.3 Dewatering thickened tailing on a sand bed

The results in Figs. 1 and 2 and Table 1 indicated that no significant increase in the solids content of the settled tailing at the washing plant would be achieved by the addition of alkalis and flocculants. Consequently, all the dewatering tests on a sand bed were carried out with untreated thickened tailing as recovered from the dam (at between 15 and 30 wt-% solids).

When the thickened tailing was ponded onto a sand drainage bed, water was lost by drainage through the sand and evaporation from the surface. In the initial stages of the dewatering, the solids content of tailing at the surface and at the bottom increased while the material in the middle remained virtually unchanged [12]. However, as the dewatering proceeded cracks began to form and the tailing was gradually dewatered as they penetrated into the solids. In order to be handled as a solid, the tailing from the sand washing operation in Melbourne needed to be dewatered to 55–60 wt-% solids and this was only possible if the initial depth of the tailing was not too great (say, <1 m) so that the cracks could penetrate to the sand bed. When the tailing is dewatered by evaporation only, the initial ponding depth is smaller because the cracks cannot penetrate to as great a depth into the wetter (at the bottom) tailing.

Laboratory [9] and field [12] tests have shown that the dewatering can be described by equation (1),

$$i = St^{1/2} + Et \tag{1}$$

where i is the cumulative water loss in time t, $St^{1/2}$ is the drainage component and Et is the evaporation component. S, the sorptivity, is a parameter which can be related to the hydraulic conductivity of the tailing, and E is the evaporation rate. Equation (1) can be used to predict the optimum ponding depth and so the area required to dewater the tailing from a particular operation. In order to do this it is necessary to determine the dewatering characteristics of the tailing, in particular, the moisture characteristic (the equilibrium moisture content–pressure relationship) and the sorptivity–pressure relationship. They were obtained from a series of constant-pressure filtration experiments in a stainless-steel pressure cell (capacity 200 ml, filtration area 6.08×10^{-3} m^2) fitted with a membrane filter (0.8 μm pore size) and a prefilter, both of which were presaturated with water. A sample of tailing (100 ml, 5.1 wt-% solids) was added to the cell and pressure applied from a constant-pressure gas supply. The pressure, maintained to ±5 mm water, was measured on a water or mercury manometer.

The cumulative outflow from the suspension was collected in the plane of the membrane filter. The outflow was rapid initially but became much slower with time until it gradually ceased. The plot of the cumulative outflow against square root of the time was a straight line, the slope of which, divided by the filtration area, is the sorptivity value, S, for the particular conditions. At the end of each filtration test, the cell was dismantled and the moisture ratio of the cake determined. The moisture ratio is the volume of water per unit volume of solid. The sorptivity values and moisture ratios from a series of experiments were plotted against the appropriate moisture potential to obtain the drainage characteristics of the tailing.

The moisture potential, Ψ, is defined [13] as the work done in removing unit weight of water from the suspension. The work is done against the forces arising from the interaction of water and the solid particles (i.e. against capillary and adsorptive forces). Defined in this way, $\Psi \leqslant 0$ because energy is released on wetting the solid. It is the major moisture-retaining potential which has to be overcome in order to dewater the tailing. In the constant-pressure filtration experiments, the moisture potential is the applied pressure except at pressures below 200 mm water where the overburden potential of the slurry must be added to the applied pressure to obtain the magnitude of the moisture potential. The overburden potential, Ω, is given by [13]

$$\Omega = \frac{\vartheta + \gamma_c}{\vartheta + 1} \,.\, T \tag{2}$$

where ϑ is the moisture ratio, γ_c the specific gravity of the solids and T is the depth of the slurry in the cell. When the tailing is ponded onto a sand bed, the moisture potential corresponds to the overburden potential for the depth of tailing.

In practice there is a limit to the amount of water that can be removed by drainage. Once the limiting value for drainage has been reached, subsequent dewatering is by evaporation only. To calculate the time required to dewater tailing for a chosen ponding depth, the limiting value for drainage was first determined from the moisture characteristic. The sorptivity value was then taken from the sorptivity–moisture potential curve and the dewatering time was calculated using equation (1). Details of the calculation have been given elsewhere [14].

3.4 Effect of flocculants on the dewatering rate

While the addition of chemicals cannot increase the rate at which water is removed by evaporation, the sorptivity values, and so the drainage rates, can be increased by the addition of polyelectrolyte flocculants.

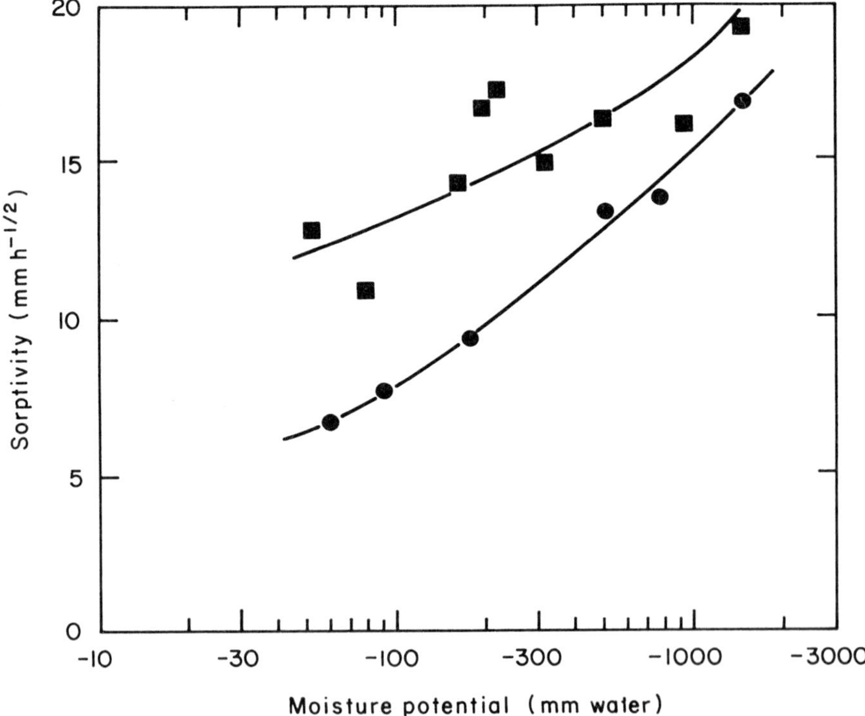

Fig. 3 – Sorptivity–moisture potential relationship for the unflocculated tailing (•) and tailing flocculated with 50 p.p.m. Superfloc N100 (▪). Initial solids content 5.1 wt-% solids.

The results in Fig. 3 show that the sorptivity values are increased by the addition of a flocculant to the tailing (Superfloc N100, 50 p.p.m.). An increase in sorptivity values will mean that the loss of water from the tailing by drainage will be faster. However, the moisture characteristic of the tailing is also affected by the addition of flocculant (Fig. 4). The curve for the flocculated tailing is above that of the untreated material which means that for the same moisture potential the flocculated tailing will retain more water at equilibrium. Thus, while the rate at which water is lost by drainage is increased when a flocculant is added, the amount of water removed is less. In practice, this will mean that a greater proportion of water must be removed by evaporation from the flocculated tailing.

Equation (1) was used to determine the effect of an increase in sorptivity and equilibrium moisture ratio on the time required to dewater the tailing to a solid state. The results in Figs. 3 and 4 indicate that sorptivity values may increase by up to 100% when the tailing is flocculated while the equilibrium

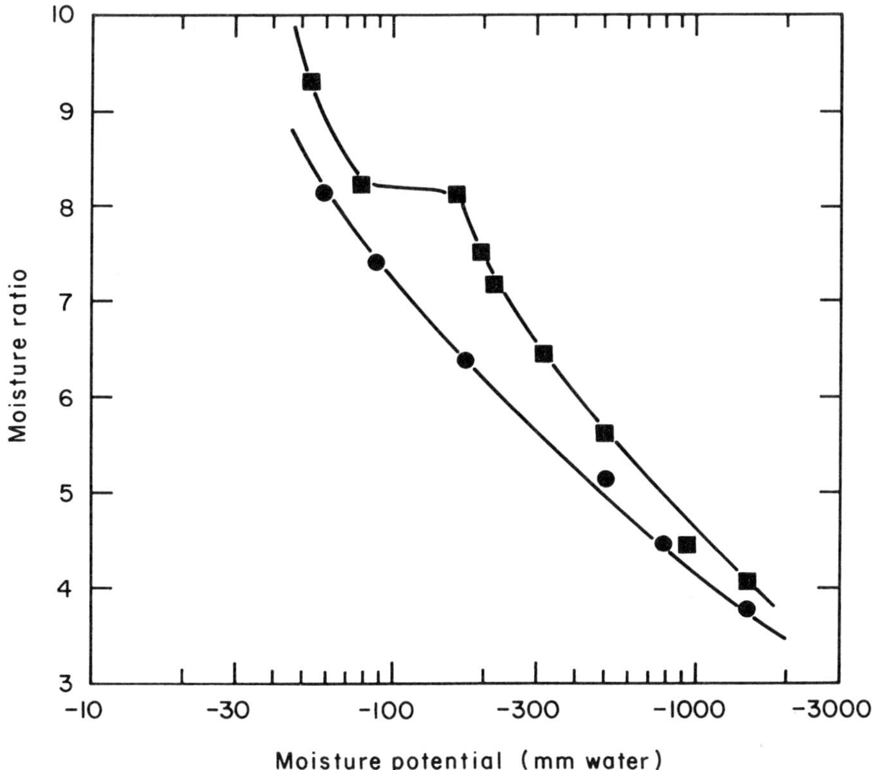

Fig. 4 – Moisture characteristic curve of the unflocculated tailing (•) and tailing flocculated with 50 p.p.m. Superfloc N100 (▪).

moisture ratio increases by approximately 10%. (The 'kink' in the moisture characteristic curve of the flocculated tailing at low moisture potentials may be due to interactions between the clay platelets forming a structure which resists compression.)

Calculations were carried out for a 10% increase in the moisture ratio and up to a 100% increase in the sorptivity values. The results in Table 2 show that the time to dewater the thickened tailing from 18 to 60 wt-% solids is shorter when the tailing is flocculated and the greatest reduction in dewatering time in a particular month occurs when the increase in the sorptivity value is greatest. An increase of 100% in the sorptivity value is obtained at low moisture potentials and in practice this means that the ponding depth must be very small (<200 mm). In the summer months (December–February) when evaporation rates are high and the maximum amount of dewatering takes place, it is better in practice to use a large ponding depth rather than a series of shallow depths, and so the maximum advantage resulting from the large increase in sorptivity values when the tailing is flocculated cannot be completely realized. As a result it is unlikely that the actual reduction in dewatering time as a result of flocculating the tailing will be great enough to compensate for the expense of adding the flocculant.

Table 2

Effect of increases in the moisture ratio and sorptivity values on the time to dewater the tailing from 18 to 60 wt-% solids

	Increase in moisture ratio	(%)	0	10	10	10	10
	Increase in sorptivity	(%)	0	10	25	50	100
July:	ponding depth – 100 mm						
	dewatering time	(days)	33.3	33.8	32.5	30.7	28.6
	change in dewatering time	(%)		+1.5	−2.4	−7.8	−14
October:	ponding depth – 200 mm						
	dewatering time	(days)	29.7	29.7	28.5	27.0	24.3
	change in dewatering time	(%)		0	−4.0	−9.1	−18
January:	ponding depth – 350 mm						
	dewatering time	(days)	31.1	30.8	29.7	28.0	
	change in dewatering time	(%)		−1.0	−4.5	−10	

Total evaporation for the months of July, October and January is 42.9, 110.5 and 220.7 mm, respectively.

4. CONCLUSIONS

The results of this work have shown that while it may be possible to control conditions in the laboratory to improve the dewatering of the tailing, this is not always possible for chemical or economic reasons at the sand washing plant.

Samples of the tailing in tap water were dispersed and yielded more compact sediments when the pH of the suspension was raised above pH 7.. However, in mill water it was not possible to disperse the clay due to the high salt content of the water.

Also, it was shown that the addition of a flocculant could increase the rate at which water was lost by drainage from the tailing when it was ponded onto a sand bed. As a result it was possible to reduce the time required to dewater the tailing to a solid state. However, the reduction in time was not great eneough to compensate for the expense of adding the flocculant.

Current practice is to dewater the tailing as recovered from the dam without any chemical treatment to improve the degree of compaction in the dam or increase the rate at which water is lost by drainage.

The approach of dewatering the tailing by ponding the thickened material in shallow depths onto the ground is being used at several sand washing plants in Victoria.

5. REFERENCES

[1] Wakeman, R. J., Mehrotra, V. P., and Sastry, K. V. S., 'Mechanical dewatering of fine coal and refuse slurries', *Bulk Solids Handling,* **1**, 281 (1981).

[2] Sparrow, G. J., 'Dewatering tailing from sand-washing operations by evaporation and drainage', *Quarry Manage. and Prod.,* **9**, 623 (1982).

[3] Beardsley, J. A., 'Sludge drying beds are practical', *Water Sewage Works,* **123**, July 82–84 and August 42–44 (1976).

[4] Novak, J. T., and Montgomery, G. E., 'Chemical sludge dewatering on sand beds', *J. Environ. Eng. Div., Am. Soc. Civ. Eng.,* **101**, February 1–14 (1975).

[5] Vogt, M. F., and Stein, D. L., 'Dewatering large volume aqueous slurries; Sand bed filtration of bauxite residue', in *Light Metals 1976,* Vol. II, pp. 117–132, S. R. Leavitt (ed.), American Institute Mining, Metallurgical and Petroleum Engineers, New York, (1976).

[6] Haliburton, T. A., 'Development of alternatives for dewatering dredged material', in *Proceedings,* Conference on Geotechnical Practice for Disposal of Solid Waste Materials, Ann Arbor, 13–15 June 1977, pp. 615–631, American Society of Civil Engineers, New York (1977).

[7] Backer, R. R., and Busch, R. A., 'Fine coal-refuse slurry dewatering', United States Bureau of Mines, *Report of Investigations* RI 8581 (1981).

[8] van Olphen, H., *An Introduction to Clay Colloid Chemistry,* 2nd ed., Wiley, New York (1977).

[9] Sparrow, G. J., and Ihle, S. W., 'Dewatering clay slimes. An approach to the problem', CSIRO, Minerals Research Laboratories, *Report* No. IR 126 (1978).

[10] Ihle, S. W., and Sparrow, G. J., 'The effect of flocculants on the dewatering time of a clay suspension', *Proc. Australas. Inst. Min. Metall.,* No. 274, 37–40 (1980).

[11] Narkis, N., Rebhun, M., and Sperber, H., 'Flocculation of clay suspensions in the presence of humic and fulvic acids', *Israel J. Chem.,* **6**, 295–305 (1968).

[12] Ihle, S. W., Johnson, B. B., Raven, T. A., and Sparrow, G. J., 'Field test of an equation that describes the dewatering of clay suspensions on a sand bed', *Int. J. Mineral Proc.*, accepted for publication.

[13] Phillip, J. R., 'Moisture equilibrium in the vertical in swelling soils. I. Basic theory', *Aust. J. Soil Res.*, 7, 99–120 (1969).

[14] Sparrow, G. J., 'Calculation of maximum ponding depth and corresponding area required to dewater a clay tailing', CSIRO, Division of Mineral Chemistry, *Report* No. MCC 290 (1981).

CHAPTER 7

Thickening and Dewatering of Hydroxide Sludges

J. H. WARDEN, WRC Processes, Elder Way, Stevenage, Hertfordshire SG1 1TH, UK

---

1. INTRODUCTION

Although hydroxides are precipitated as valuable products in some processing plants, the term 'hydroxide sludge' is usually understood as referring to a mixed waste in which hydroxides are important and perhaps dominant constituents. The thickening and dewatering of these sludges can be a problem. By their nature the solids are often very difficult to separate from water and the problem is often compounded by the variability of both concentration and composition of the solids.

This paper recounts some of the work carried out at the Water Research Centre on the coagulant hydroxide sludges produced in potable water treatment. As a result of this work there is a new appreciation of the potential of polyelectrolytes for treating sludges which have a natural tendency to flocculate. Thickened sludges can be produced from a continuous thickener which contain 'no free water'; that is to say, all the water which is removable under the force of gravity has been removed, and standing will not cause any further separation. This paper touches only lightly on the design of the process equipment but it must be emphasized that correctly designed plant is essential in achieving the best results. The practical aspects of this work have been published elsewhere [1].

2. HOW WATERWORKS COAGULANT SLUDGES ARISE

About two-thirds of UK potable water is drawn from surface sources and is treated with a coagulant. On most works the coagulant is aluminium sulphate, but ferric salts are also used, coagulant solution being dosed into the raw water. Aluminium or ferric hydroxide is precipitated, as the case may be. In the precipitation, suspended and some dissolved impurities, bacteria and viruses are co-precipitated, adsorbed or entrapped as part of the hydroxide floc. Where the raw water quality is high and the coagulant dose is low, this floc is collected in sand filters, being removed periodically by backwashing. A typical filter will generate about 100 m^3 of washwater with an average solids content of 0.02% to 0.05% every 12 to 72 hours, according to loading. Where the sludge load in the coagulated water is too high for efficient removal by filtration alone, a preliminary sedimentation stage is used. Most sedimentation tanks operate continuously and produce a flow of sludge amounting to about 1 or 2% of the throughput, often released in short surges at 15–20 minute intervals at a solids concentration of between 0.1% and 0.5%.

The proportion of coagulant hydroxide in a sludge may vary from about 45% of dry weight for a reservoir water to about 11% where the water is taken directly from a lowland river. The physical nature of the hydroxide component will depend upon the other materials present, notably organic compounds, particularly in the case of ferric sludges. Coagulant sludges are quite strongly flocculent. If a sample is taken in a cyclinder, shaken vigorously and then allowed to stand, the initially uniform suspension will be seen to flocculate as it settles. The flocs are weak and easily broken by agitation but they will form again under quiescent conditions. It is this natural tendency to flocculate which characterizes the coagulant sludge.

Aluminium and ferric salts have been chosen as coagulants because they produce these loose flocculent hydrophilic precipitates, ideal for clarification. These are the very properties which make them difficult to thicken and dewater.

3. LABORATORY STUDIES

In the early part of the work attempts were made to measure the settling properties of polyelectrolyte dosed sludges and to construct mass flux curves as described by Talmage and Fitch [2] and Yoshioka [3]. Samples of sludge were dosed with polymer and, after mixing, were observed settling in a clear perspex cylinder 9.2 cm in diameter, calibrated to a depth of 50 cm. It had a working volume of about 3.3 litres and was fitted with a stirrer comprising two vertical 5 mm diameter bars mounted at pitch radii of 3.5 cm and 7.5 cm, rotating at a speed of 1 rev/min. The function of the stirrer was to provide standardized settling conditions by making vertical channels behind the bars for the release of separated water and to prevent flocculated solids from bridging across the cylinder.

For each settling test a measured amount of sludge was put into the cylinder and the dose of polymer solution was put into a 10-litre bucket. Then the sludge was poured into the bucket to mix with the polymer and back to the cylinder. To ensure complete mixing the dosed sludge was poured into the bucket and back to the cylinder several times before the settling rate was measured.

This test was found to have several shortcomings. First, if the sludge concentration was less than about 0.06% by weight (and washwater concentrations are often as low as 0.03%) the flocs settled separately and a distinct interface did not develop between the sludge and the supernatant water. Second, although the settling rate for a given sludge was supposed to be a function of concentration alone, it was found that after polymer was dosed, the settling rate was affected by the number of times it was poured from bucket to cylinder, decreasing with the amount of pouring. This is illustrated in Fig. 1. It appeared that the polymer reacted with the sludge solids in a much shorter time than it took to pour the sample of sludge from bucket to cylinder and that the subsequent handling only degraded the floc. This made it impossible to obtain mass flux curves.

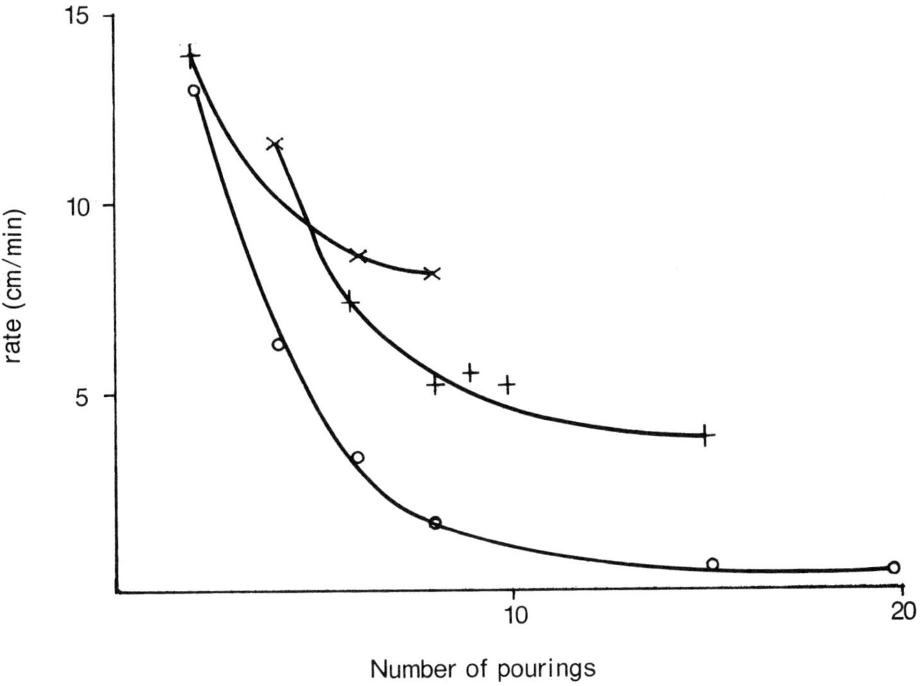

Fig. 1 – Sludge Settling Rate versus Number of Pourings.

Lastly, the bottom bar of the stirrer operated about 1 cm above the bottom of the cylinder. This residual unstirred volume was a significant proportion of the final settled volume when a sludge had concentrated from, say, 0.1% solids to 3%.

In fact the final settled sludge concentration obtained in stirred cylinders was usually much lower than that obtained from a pilot scale continuous thickener treating the same sludge, and there was no correlation between the results. The idea of using laboratory tests to predict thickener performance was therefore abandoned and all the experimental work was done in pilot plant sited at waterworks and using the sludges from the processes as they were produced.

4. PILOT PLANT

The continuous thickener consisted of a cylindrical steel tank 1070 mm in diameter with a peripheral overflow weir, radial feed launder and a central feedwell of about 300 mm dia. and 230 mm deep. The tank stood about 1980 mm high to the overflow. Three legs were fastened to the top section which was spanned by a bridge carrying a right-angle reduction gear and drive motor for the scraper. The top section was 610 mm deep to a flanged joint and during the research several shapes and depths of bottom were connected to this flange, and several types of scraper were tried.

Feed sludge was drawn continuously or semi-continuously from the treatment plant into agitated buffer storage. From here it was possible continuously to pump a homogeneous sludge which reflected changes in the works' operating conditions accurately without preventing the thickener establishing and maintaining a dynamic steady state. Polymer solution was injected continuously into the thickener feed pipe.

When the thickener was in continuous operation there was a sharp interface between settling sludge and clarified water. A photoelectric level probe was used, set at a depth of about 200 mm, to control desludging. The presence of sludge was proved when the light beam had been extinguished for a continuous period of about 2 minutes. Then the positive displacement desludging pump ran for a preset time to withdraw a batch of thickened sludge, about 10 litres. A counter noted the number of desludges, between 5 and 100 per day, according to the load. This simple automatic control allowed the pilot plant to run continuously with only a daily check by the operator on his sampling visit.

Because the pilot plant was treating a real sludge which could be expected to vary with changes in raw water and coagulant use, the experimental evidence was obtained by running two thickeners side by side, one being operated under steady conditions as a control. An experiment started with both tanks empty. A daily check was made of all the operating parameters and samples were taken of feeds and thickened sludges for determination of solids contents. It took between one and three days for the thickeners to settle to steady state and an experiment lasted until it was reasonably sure that the results obtained truly reflected the operating conditions. Under good conditions each experiment took a little less than one week.

5. FLOCCULATION

Initial performance of the pilot plant at low flow rate was poor compared with the results obtained in laboratory settling tests but, contrary to expectations, thickening became much better as the feed rate was increased as shown in Table 1.

Table 1

Effect of feed rate on thickening

Experiment	Duration (days)	Thickener No. 1 Rate (l/min)	Thickener No. 1 sludge (%)	Thickener No. 2 Rate (l/min)	Thickener No. 2 sludge (%)
E1	5	5	4.5	15	5.4
E2	5	5	4.4	20	5.4
E3	3	5	3.5	25	5.0
E4	3	5	3.4	30	5.8
E5	4	5	3.3	35	5.9

During the 3 weeks of these experiments the sludge drawn from the control thickener became progressively thinner while that from the other became just perceptibly thicker.

It had been expected that as the feed rate increased, solids residence time in the thickener would decrease to the detriment of thickening. This result was quite unexpected and was found to be related to turbulence in the feed pipe, and particularly at the polymer dosing point, illustrated in Fig. 2. Sludge was pumped to the thickener through a 25-mm bore platic pipe and polymer was injected from a small-bore branch which was cemented into the pipe just upstream of a flanged joint. An orifice plate was mounted between the flanges to induce some turbulence to help mixing of the polymer.

A series of experiments was carried out to investigate the effect of orifice size relative to sludge flow rate and also the effects of turbulence after dosing. It appeared that good conditions for thickening were achieved when there was intense turbulence at the orifice plate. However, the levels of turbulence prevailing at right-angle bends in the pipe downstream were sufficient to have a deleterious effect on thickening.

By placing the orifice at the end of a section of clear-walled flexible plastic tube, the progress of flocculation could be observed. The first visible flocs were seen within 2 seconds of dosing and floc development in the pipe appeared to be completed after about 30 seconds. This experiment was made at two sites with alum sludges, one at 0.1% solids and the other at 0.03%. This experience seems to relate directly with the bucket pouring work and explains why laboratory tests give inconclusive results. It is impossible to mix polymer intimately with sludge in a beaker in a fraction of a second and then immediately to reduce shear to the low level which prevailed in the gently curved plastic tube.

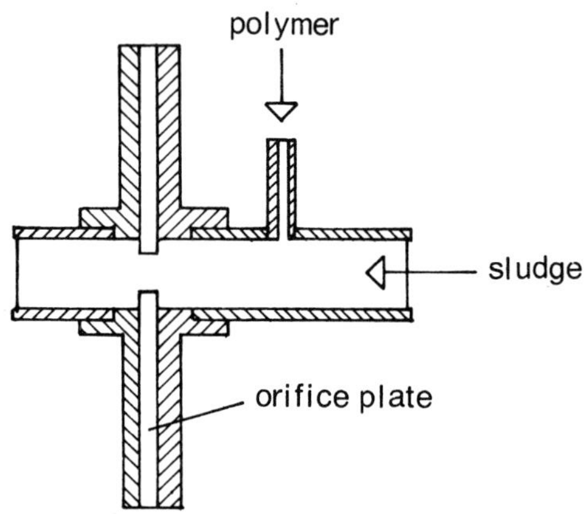

Fig. 2 – Detail of polymer dosing point.

Pressure drop measurements were made over a range of flow rates with all the orifice plates so that Reynolds numbers and velocity gradients could be calculated for the various sections of the feed pipe. Correlation with thickening was disappointing, mainly because of the random variations in thickening results. For example, Table 2 shows the solids content of sludge sampled on successive days from one thickener under constant operating conditions.

Table 2

Random variations in sludge thickening

Day	15	16	17	18	19	20	21	22	23	24
Sludge solids %	5.1	6.1	5.5	6.0	5.8	4.4	4.5	4.3	4.3	4.1

By describing thickening as 'poor', 'fair', 'good' or 'very good' a diagram was drawn, Fig. 3, which shows poor thickening at very low and very high values of applied shear and good thickening in an intermediate range. The diagram has the same form whether the abcissa is plotted as velocity gradient, Reynolds number or work done per unit volume (which is headloss). Every plot has one 'rogue' point. As a rule of thumb which has proved effective in subsequent work, it may be said that thickening is good if the average speed through the orifice is 5 m/s for an alum sludge and 11 m/s for a ferric sludge.

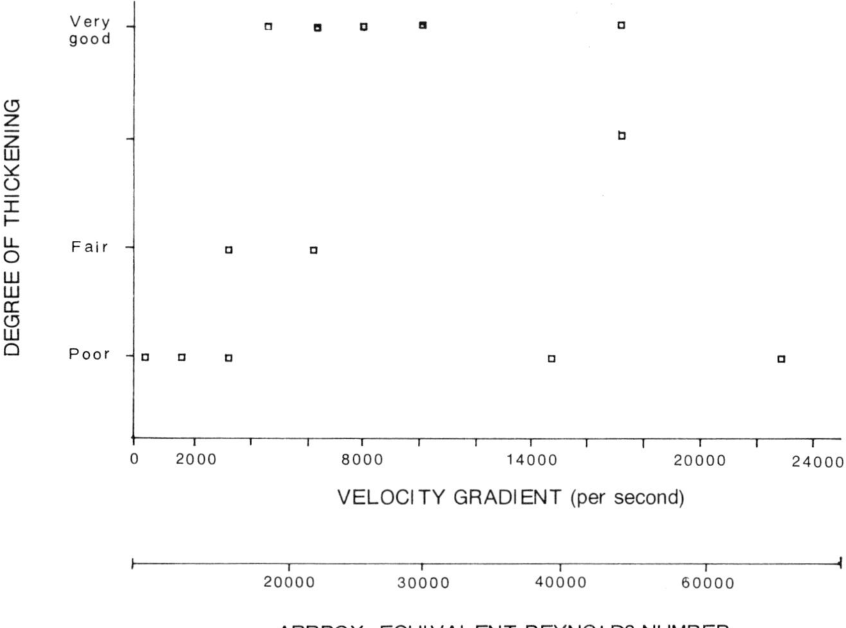

Fig. 3 – Degree of thickening related to turbulence at dosing point.

6. SELECTION OF POLYMER

All the early work was done using a moderately anionic waterworks grade polyacrylamide polymer of high molecular weight. In a series of experiments this was used as a standard in comparison with four other products, all waterworks grades from the same manufacturer, all dissolved to 0.05% concentration and dosed at 1.8 mg/l or about 2 kg/tonne dry solids.

A moderately cationic polymer of high molecular weight gave good thickening though not as good as the control, while a material of similar charge but lower molecular weight gave poor results. This suggested that high molecular weight was more important than ionic character. A non-ionic product of moderate molecular weight gave a bulky floc which did not thicken well. A material of the same anionic nature as the control but of higher molecular weight gave improved thickening, which improved further when it was dosed at a lower concentration.

For a period, one thickener was operated with no polymer dose. It was found that the feed rate had to be reduced by a factor of four to obtain a satisfactory interface between settling sludge and the supernatant water, and the sludge concentration was only one third of that achieved with polymer dosing.

7. EFFECTS OF THICKENER DESIGN

It was found in the pilot plant work that the design and method of operation of the thickener could have an important effect on the separation results obtained and on the perception of the processes of thickening and separation.

7.1 Overflow quality

When the pilot thickener was operated with a sludge level below the bottom of the feedwell, the fresh feed could be seen settling through clear water with small unattached flocs moving away and upwards with the flow. There was always free floc in the overflow. If the sludge were allowed to accumulate, however, as soon as it came up to the bottom of the feedwell the overflow suddenly became very much clearer as the small flocs were trapped in a form of floc blanket. After this was noted the sludge level was always kept high and the photoelectric control maintained this level automatically. The application of this

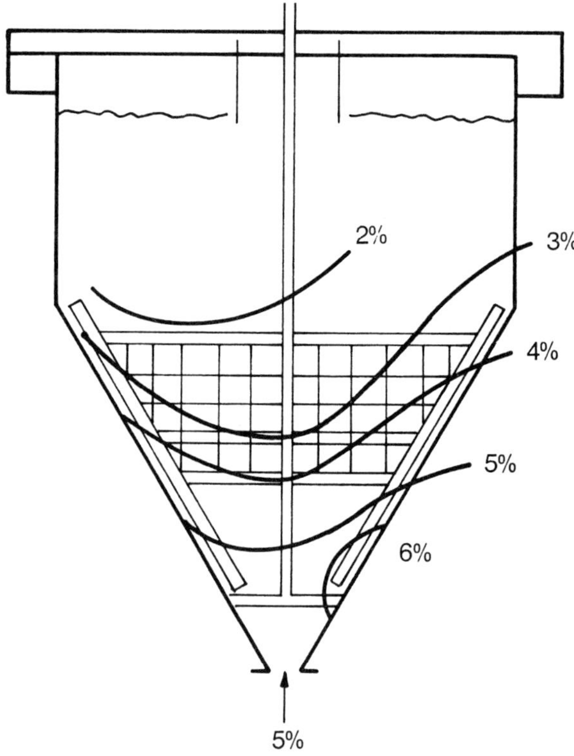

Fig. 4 – Concentration contours.

simple piece of automation made a significant improvement in the quality and consistency of the experimental results together with a significant reduction in the amount of attention the plant needed.

7.2 Thickening

The first pilot thickeners had conical bottoms with 60° included angle. Samples taken from points throughout the thickener volume indicated that sludge near the axis travelled quickly downwards with relatively little thickening while the best thickening occurred at the wall where the sludge came under the influence of the scraper, Fig. 4.

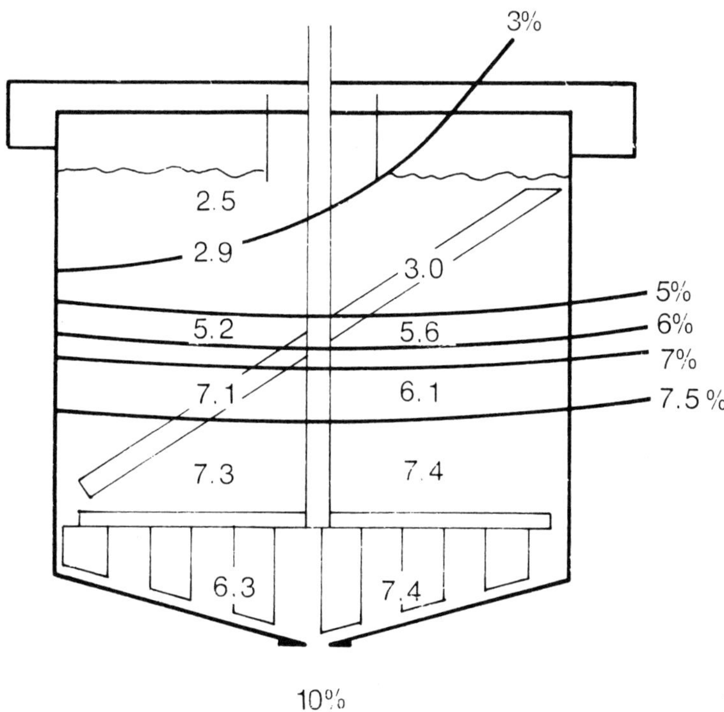

Fig. 5 – Concentration contours.

A flat-bottomed tank, shallower than the cone (but with the same volume to equalize residence time) gave equally good thickening. This tank had a conventional form of rake to draw sludge to the central outlet. Examination of concentration contours such as those of Fig. 5 suggests that while it is transporting thickened sludge to the outlet, the rake itself is making an important contribution to thickening. The concentrations of 5% and above could not have been achieved simply by gravity settlement and compression by the weight of overlying solids, because the 5% contour was only 30 cm below the water surface. It is believed that the rake transport capacity was much greater than the average sludge withdrawal rate and material was recycled many times through the rake before being discharged.

It was usual at the end of an experiment to empty the thickener by running the desludging pump continuously, leaving the rake running to clear the tank floor. Usually the pump would deliver sludge for a few minutes and then the supernatant water would break through so that the thickest sludge was discharged last. In one experiment the rake had been run at a higher speed than usual and at the end of this run all the sludge was withdrawn while the water stayed on top. It was concluded that on this occasion the rake had been able to draw sludge to the outlet at a rate greater than the pump's capacity and so the pump had no opportunity to draw water. As a result of this experience a new design was developed and published [4], the special feature being that the rake is built to deliver a specified flow to the thickener outlet. A number of rakes have been built to this design and one of these was found in a performance trial to exceed the design delivery by a satisfactory margin.

8. PREDICTION OF THICKENER PERFORMANCE

One of the main problems in thickener design has always been to size the unit correctly for the design loading. Despite the great efforts that have been put into theoretical analysis of sludge treatment and thickener behaviour, it is still considered necessary to carry out laboratory tests on representative samples if a thickener is to be sized correctly.

The WRC work showed that laboratory tests on polymer flocculated sludges could not be relied on to give reliable design data and so effort was concentrated on the alternative approach of building a statistical model from the results of pilot plant trials carried out on a wide range of sludges. The water authority engineer now has simple formulae [1] by which he can decide the dimensions of a thickener and predict the solids concentration of the underflow, knowing nothing more than the colour and turbidity of the raw water to be treated. Thickeners designed by this procedure are small and all those built so far are performing well.

9. DEWATERING THICKENED SLUDGES

Dewatering is usually regarded as the means by which a sludge is converted from a viscous liquid to something that may be described, however loosely, as solid. The purpose of dewatering hydroxide sludges is almost always to prepare them for disposal. The degree of dryness required depends on the means and place of disposal. The water authority engineer is advised [1] to take note of the source, amount and quality of sludge discharged from the treatment processes, to investigate the location and requirements of the possible disposal sites, and then to select the least-cost method of translating the waste from one to the other. Thickening alone may suffice in some cases and dewatering without thickening in others. Dewatering processes to be considered include drying beds, vacuum filters, centrifuges and filter presses.

A drying bed to dewater a thickened sludge from which all the 'free water' has already been removed does not need to have drainage except to remove rainwater and almost any open area of ground has potential. However since the main operating cost is in sludge lifting, the area must be suitable for this; so it should be reasonably flat, hard and slightly sloped to allow rain to run off.

Rotary vacuum filters are only used on waterworks for dewatering calcium carbonate sludges from softening plants and, even if there is also coagulation on the works, the sludges are not really classifiable as coagulant sludges. An interesting innovation, however, is a system that may be described as a vacuum assisted drying bed. One has recently been installed at a waterworks in Devon. Here the sludge is not thickened, but is dosed with polymer as it is delivered on to the bed. When the bed is filled to a depth of about 400 mm drainage begins, at first under gravity. As the drainage rate falls a small vacuum pump is switched on to evacuate the space under the bed so that drainage and compression of the cake is achieved by gravity plus atmospheric pressure. It is claimed that this process is much more rapid than conventional drying beds so that much smaller areas are needed.

Centrifuges are used on some waterworks for dewatering coagulant sludges. The polymer dose required is somewhat higher than is required for gravity thickening but the degree of concentration is much greater. Polymer thickening before centrifuging has not been tried, but gravity thickening without polymer would be logical.

A filter press produces a dryer cake than any process other than extended air drying. Although its output is usually measured in terms of the solids throughput, it is in fact determined by the amount of water to be removed. The WRC work has shown that polymer-thickened sludges can be filter pressed quickly and efficiently without additional conditioning. In the new generation of plants now being built at waterworks, filter presses are only about half as big as hitherto, resulting in considerable savings.

10. REFERENCES

[1] Sludge Treatment Plant for Waterworks, *WRC Technical Report* TR 189 (1983).
[2] Talmage, W. P. and Fitch, E. B. *Ind. Eng. Chem.*, **47**, No. 1, 40 (1955).
[3] Yoshioka, N., 25th Annual Congress, *Soc. Chem. Eng.*, Japan, 230 (1961).
[4] Warden, J. H., *Filtration and Separation,* **18**, 113 (1981).

CHAPTER 8

A Preliminary Investigation of Gravity Drainage from Particulate Beds

R. J. WAKEMAN and A. VINCE, Department of Chemical Engineering, University of Exeter, Exeter, Devon, UK

1. INTRODUCTION

The draining of liquid from porous media, or dewatering of the media, is important to many industrial operations for a variety of reasons. In many instances the liquid is allowed to drain from the voids under the influence of gravity, particularly when it is impracticable to use pressure or centrifugal forces to aid the rate of drainage. It is expected that an understanding of this problem will also help to develop the fundamentals of filter cake dewatering.

During dewatering it is important to be able to estimate, and preferably predict, the rate of liquid removal and the distribution of liquid inside the porous medium. To facilitate this a fundamental understanding of the basic physical mechanisms and phenomena of the microprocesses taking place in the voids of the medium is essential; from the practical point of view it is not possible to observe these processes taking place in an individual void, but it is possible to make suitable measurements of liquid distributions in the bed and infer mechanisms and phenomena by the application of mathematical models. Reported in this paper are some results from a preliminary investigation of local saturation measurements and predictions of some pertinent factors affecting gravitational drainage based on a relatively simple model of the process.

2. PREVIOUS WORK

A survey of the literature has shown that most attempts to deal with this problem have been largely empirical, and were aimed at the calculation of dewatering rates in ignorance of the liquid distribution inside the bed. Correlations derived empirically are applicable only to the specific system from which they were developed, and hence have limited use. It has been demonstrated that data scatter obtained, even when correlating residual moisture levels, is not understood [1]. Of the mathematical models proposed for the more general problem, including liquid displacement by applying vacuum or pressure in a displacing gas phase, three approaches are worthy of note and can be categorized as:

(i) Capillary models, based on the existence of mutually independent cylindrical capillaries extending throughout the length of the medium. In some cases a distribution of capillary sizes has been inferred from a capillary pressure curve [2], giving curved relative permeability plots, which are more realistic than the linear plots predicted on the basis of monosized capillaries. These models always tend to overestimate the rate of dewatering.

(ii) Network models, in which connectivity of the pores is recognized; the flow channels are still assumed to be capillaries and the pores not only have a distribution of sizes but it is also possible for any pore to feed fluid to more than one other [3–8]. The network models are generated by considering the pores in the medium to be a bundle of capillaries with a distribution of radii, and which pass throughout the length of the medium. If the medium is cut perpendicular to the capillaries into a series of slices, each slice is then rotated to a random extent, and then the medium is reconstructed from the slices, there is a chance that the rejoining will introduce an element of pore connectivity at the jointed faces where some pores will adjoin more than one other pore of a different size. The predicted relative permeability plots from such models agree quite well with experimental data and exhibit the correct qualitative shapes, although it is still necessary to measure a capillary pressure curve for the medium. To remove this problem the general shape of capillary pressure curves has been investigated, and the curves expressed by a general mathematical expression which appears to fit data from a wide range of sources and for a large number of different media [4, 8].

(iii) Film flow models. In gravitational and centrifugal drainage it has been supposed, with some justification, that the bulk liquid level falls in the medium to leave a quantity of liquid retained over the surface of the particles [9]. This liquid then drains filmwise over the particle surfaces at a much slower rate than the bulk of the liquid. The film drainage continues until continuity of the film is broken at some point, whence drainage ceases. This departure from the relative permeability concept was later refined [10] by consideration of the retarding effects of film flow due to changes of the capillary drain height.

Of the above, capillary and film-flow models tend to be deterministic, whereas network models utilize a stochastic description of the bed structure.

3. EXPERIMENTAL MEASUREMENTS

The experimental equipment is shown schematically in Fig. 1; the data logger and digital voltmeter were controlled by a microcomputer, and the draining bed formed one arm of a Wheatstone bridge.

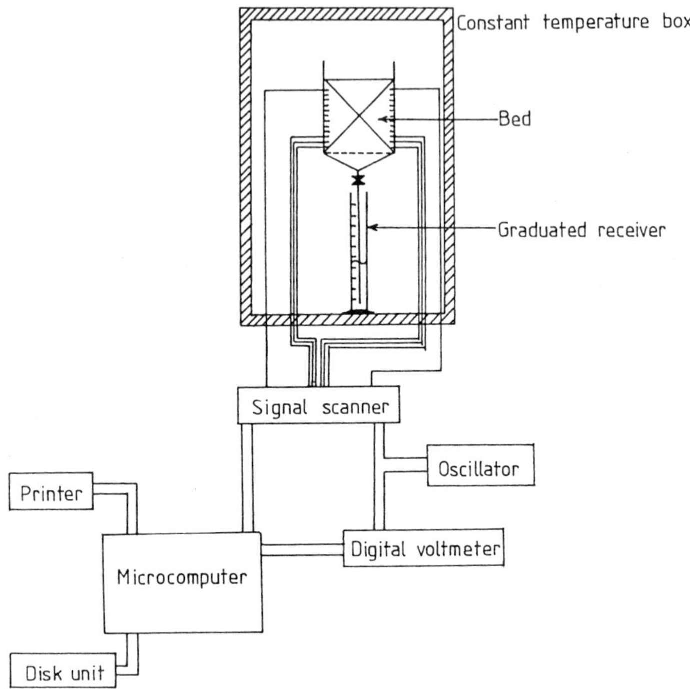

Fig. 1 – Schematic diagram of the experimental equipment.

The bed was held in a 10-cm diameter cylinder which was fitted with pairs of silver-plated pin electrodes located diametrally and at 1-cm intervals axially The impedance between each pair of electrodes was measured indirectly via the bridge circuit; impedance changes between an electrode pair due to the bed dewatering caused a change of the off-balance voltage over the bridge, which was measured by the DVM and recorded by the microcomputer. Dewatering kinetics were deduced by scanning the pin pairs periodically, and from gravimetric measurements the overall liquid content of the bed was also determined at the time of each scan.

The system was calibrated by draining thin beds ($\leqslant$ 2 cm thickness) so as to obtain an approximately uniform saturation distribution; the off-balance voltage measured across the bed at the axial centre corresponding to this resulting saturation was then determined. The saturation–porosity product could then be plotted against the ratio of the voltage for a saturated bed to that for a drained bed; calibrations were found to be independent of particle size, but dependent on the mean porosity of the bed. All of the dewatering runs were carried out using tap water stored at the temperature of the experiments (25 ± 0.5 °C). A calibration curve for ballotini is shown in Fig. 2.

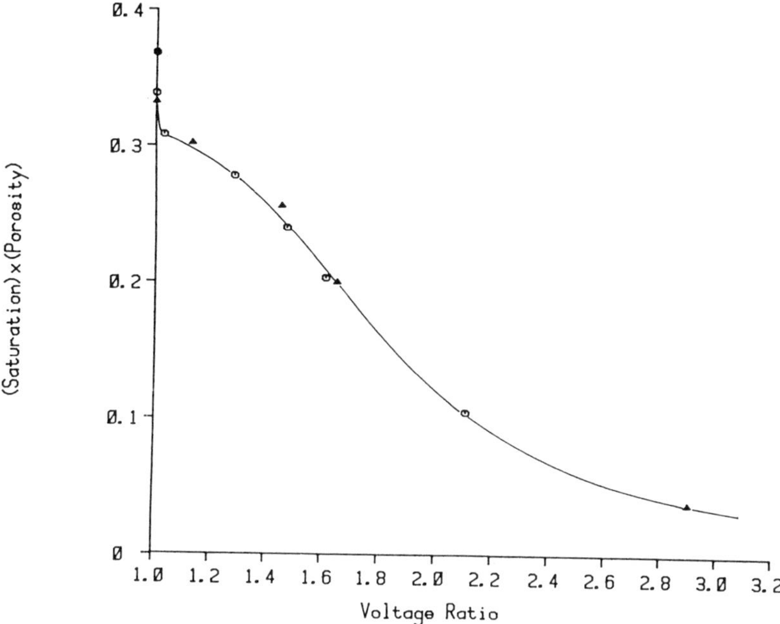

Fig. 2 – Calibration curve for ballotini beds, $\epsilon = 0.369 \pm 0.001$.

4. SOME EXPERIMENTAL AND CALCULATED RESULTS

Figures 3 and 4 show the time variations of local saturation within beds of sieved fractions of glass beads. As the bed drains the saturation profile varies gradually, to eventually reach a final equilibrium state. Internal saturation profiles show that there is a very rapid saturation reduction throughout the depth of each bed; close to the outflow face ($x = 0$) the saturation falls to about 80–85%, whilst at the free surface of the bed the saturation gradient is very steep and the saturation appears to approach a constant value at the free surface. This was predicted previously [5], and these experiments confirm that an irreducible saturation is reached instantaneously at the bed surface from which liquor is draining. The relative permeability of the bed to the liquid is therefore zero at this surface; it was previously argued [5] that this would be the case as there was no liquid flow into the voids at this surface, and that a zero relative permeability can only occur at the irreducible saturation.

It is also apparent from these experimental results that there are principal flow pores which drain very rapidly, reducing the saturation from 100% to 80–85%. This indicates that a funicular saturation state is reached almost instantaneously throughout most of the bed volume, whilst a pendular saturation state may exist at the free bed surface (provided that the capillary number at the

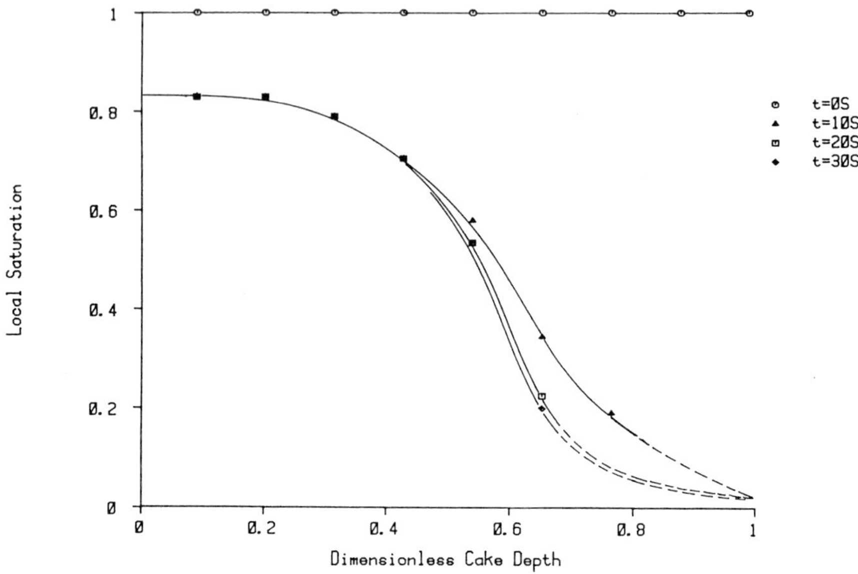

Fig. 3 – Experimental saturation profiles for 1100 μm beads [$L = 0.098$ m, $k = 1.64 \times 10^{-9}$ m^2, $\epsilon = 0.3681$.

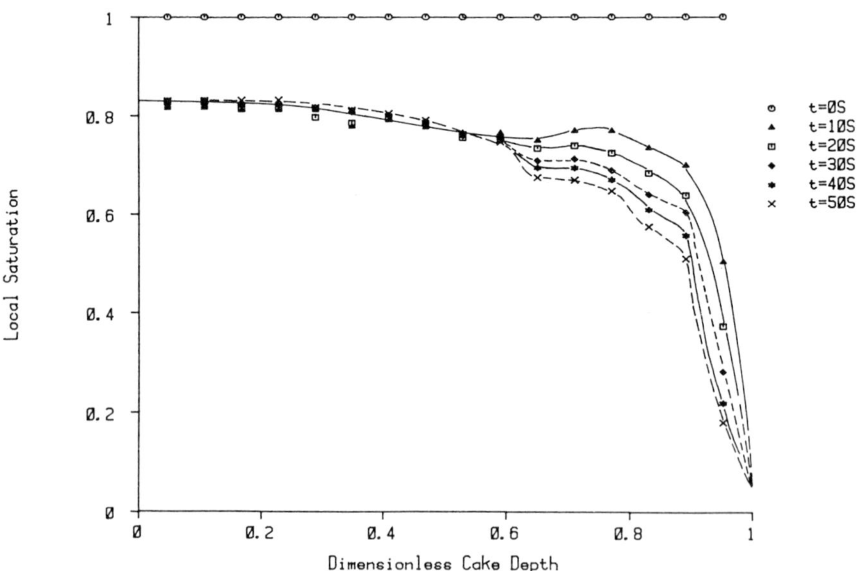

Fig. 4 – Experimental saturation profiles for 486 μm beads [$L = 0.168$ m, $k = 8.95 \times 10^{-10}$ m^2, $\epsilon = 0.37$].

free surface is sufficiently large). One important practical implication of saturation state distributions of this sort in filtration is that cake cracking will first occur at the surface of the filter cake, and propagate inwards to approach the cloth-supported surface. This is due to the fact, in general, that lower cake strengths can be associated with lower saturation states.

During gravity drainage air is sucked into the pores as the liquid leaves, and to a limited extent there is simultaneous flow of air and liquid within a pore as further liquid drains from liquid films remaining over particle surfaces or from connected pores. However, air flow effectively ceases once continuous air passages are formed and liquid flow is the most pertinent from a practical point of view; the liquid flux relative to the solids is described by Darcy's law:

$$q_l = \frac{-kk_{rl}}{\mu_l} \frac{\partial}{\partial x} (\rho_l g x - p_c) \tag{1}$$

where the capillary pressure p_c and the relative permeability k_{rl} are functions of the local saturation of the bed. For continuity of fluid flow any solution to the Darcy equation must also obey the liquid material balance:

$$\frac{\partial q_l}{\partial x} = -\epsilon \frac{\partial S_l}{\partial t} \tag{2}$$

when the bed has a uniform and constant porosity ϵ. For present purposes it will be sufficient to consider trends of saturation profiles generated by solutions to equations (1) and (2) and, for this, effects of capillary pressure will be ignored. This will therefore yield analogous solutions to those obtained by Baluais *et al.* [11]; it should be recognized that any ultimate solution must include capillary pressure terms in the equations, which will reduce the calculated rate of dewatering. Rates shown in this paper, and any other, which neglect capillary pressure will be too great, and only general forms of curves can be considered.

The relative permeability can be expressed in terms of the characteristics of the capillary pressure curve [5, 8] as:

$$k_{rl} = S_r^{(2+3\lambda)/\lambda} \tag{3}$$

where λ is the pore size distribution index and is the slope of the capillary pressure curve plotted as a reduced saturation, S_r, against the ratio of breakthrough to capillary pressures [4, 8]. The reduced saturation is defined by:

$$S_r = \frac{S_l - S_{l\infty}}{1 - S_{l\infty}} \tag{4}$$

where $S_{l\infty}$ is the irreducible liquid saturation and treats the irreducible saturation as though it were part of the solid structure of the bed. Utilizing equation (3) and the above assumptions, equation (1) can now be rewritten as:

$$q_1 = \frac{-k\rho_1 g}{\mu_1} S_r^{(2+3\lambda)/\lambda} . \quad (5)$$

Equations (2) and (5) can now be solved using (4) by dividing the bed into a series of layers, noting the following initial and boundary conditions:

$$\begin{aligned} &t = 0, \quad 0 \leqslant x \leqslant L, \quad S_r = 1 \\ &t > 0, \quad x = L, \quad q_1 = 0, \quad S_r = 0 . \end{aligned} \quad (6)$$

Calculations are started at the free surface of the bed ($x = L$) into which air is being sucked and completed when the bed outflow surface is reached. The computation sequence is outlined in Fig. 5, and some typical results for saturation

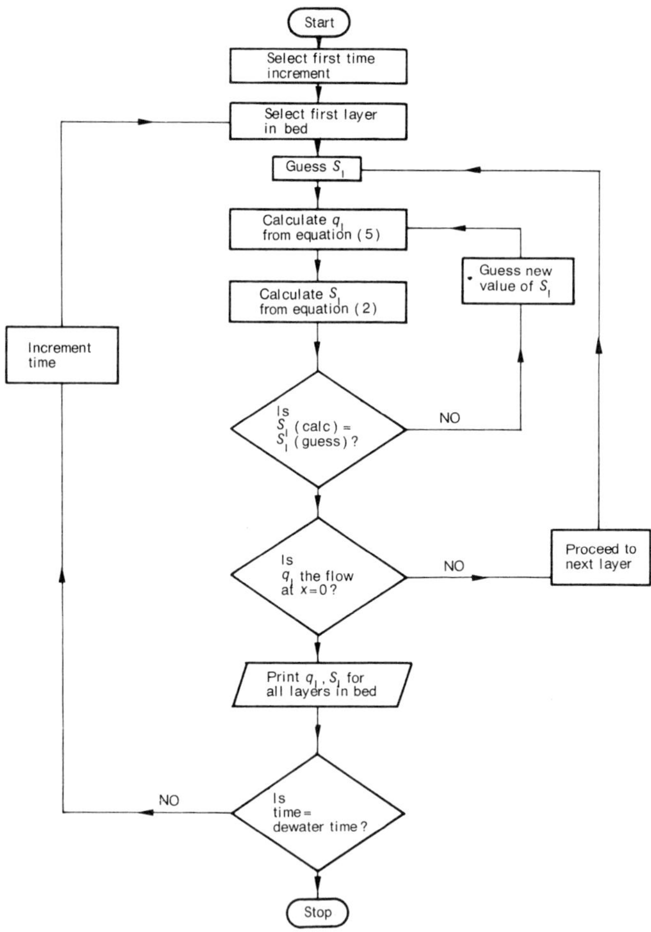

Fig. 5 – Flowchart for the calculation of saturation profiles.

profiles are shown in Fig. 6 for when the permeability of the bed controls the drainage process. The general shape of these curves is similar to those measured experimentally, excepting that too rapid a dewatering rate is indicated due to the omission of capillary pressure effects.

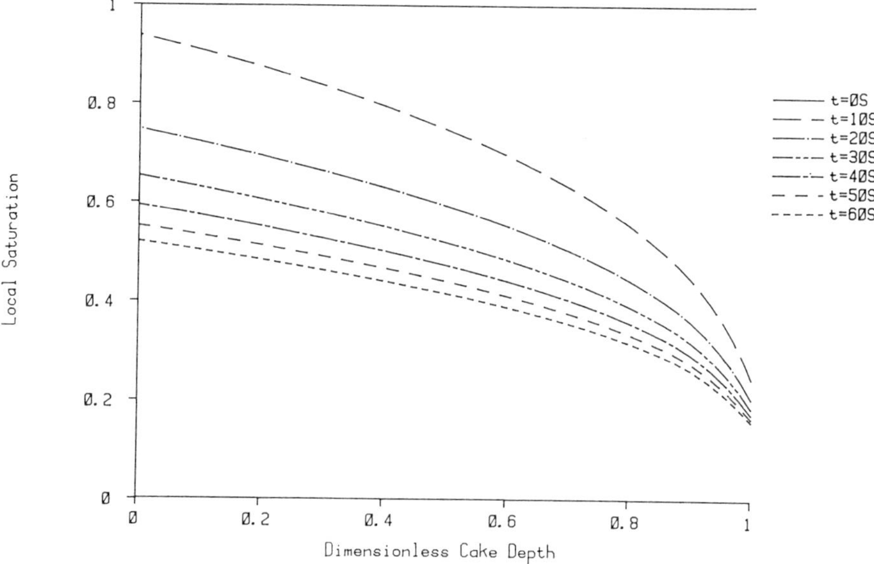

Fig. 6 – Computed saturation profiles when $k_{\text{cloth}} < k_{\text{bed}}$.

It is not uncommon in practice for the bed support to have a resistance to fluid flow of the same order or greater than that presented by the bed. The support resistance is envisaged to be in series with the resistance of each layer of bed, and so whichever is the greater might be expected to control the dewatering rate. However, as drainage proceeds it is quite possible for the rate control to transfer from the support to a layer in the bed at some time during the process, whence there is an abrupt change in the slope of the internal saturation profile. If the support permeability is denoted by k_m, then for layers in the bed where $(kk_{rl}) > k_m$ the dewatering rate is support controlled, but for layers where $(kk_{rl}) < k_m$ the rate is bed controlled. Since the saturation reduction proceeds from the free surface there can be a period when upper parts of the bed are controlled by local values of (kk_{rl}) whilst lower regions are controlled by k_m. As the saturation is further lowered k_{rl} is reduced and the entire bed is controlled by the local permeabilities (kk_{rl}). This sequence of events can be calculated from equations (1) to (6) and is shown in Fig. 7; the corresponding changes of the average bed saturation are shown in Fig. 8.

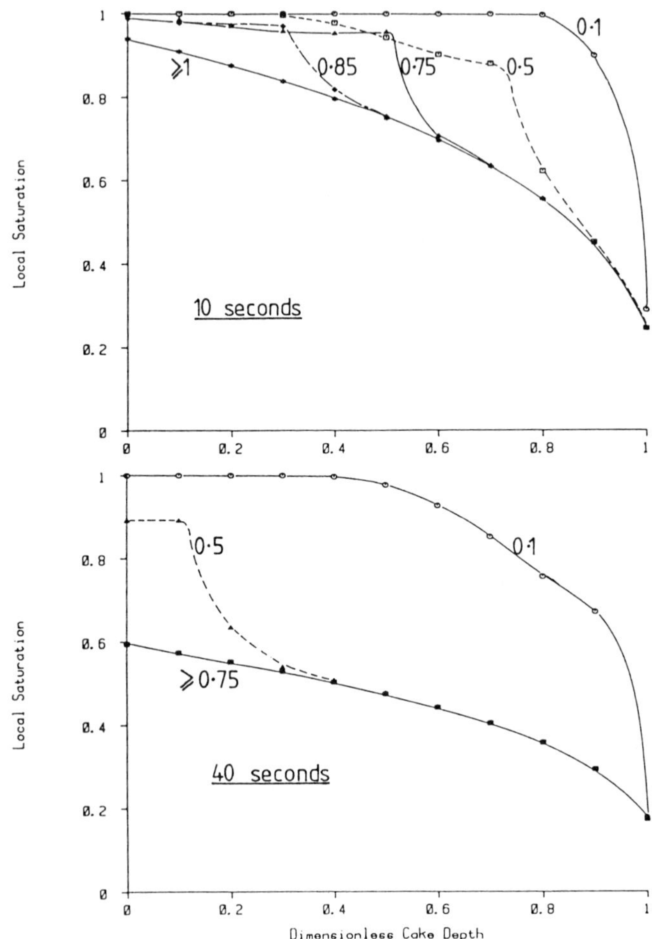

Fig. 7 – Effects of bed and support permeabilities on saturation profiles [the parameter is $k_{support}/k_{bed}$].

5. CONCLUDING REMARKS

This paper has described the application of conductivity measurements to the determination of liquid saturation in particulate beds, and gives a simplified analysis of the mechanism of the dewatering of homogeneous incompressible beds under gravity. The purpose of the analysis was to demonstrate some prominent features of the process, in particular the initial rapid reduction of saturation and the effects of a supporting medium which has a permeability lower than that of the bed. The structure of the bed is described by the pore size distribution index λ.

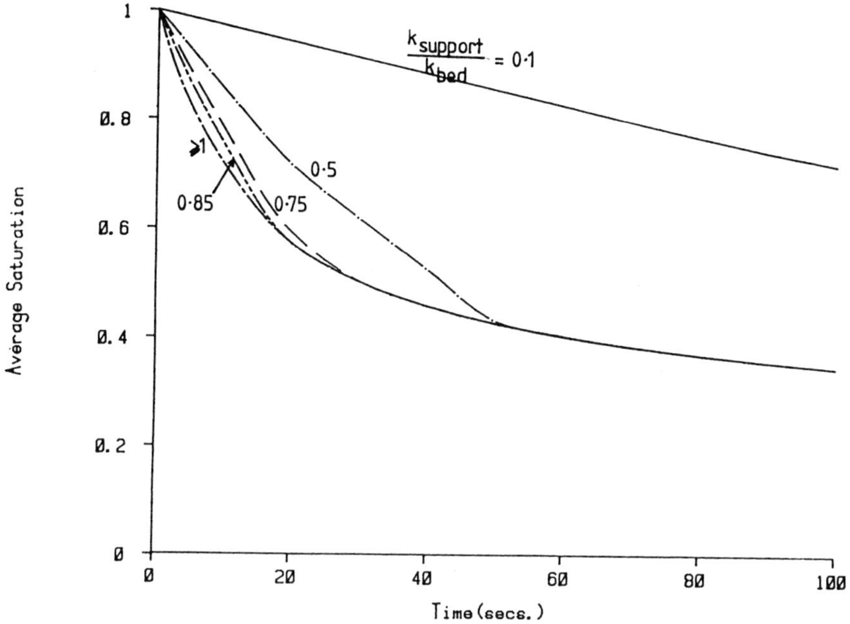

Fig. 8 – Effect of bed and support permeabilities on dewatering curves ($k_{\text{bed}} = 2 \times 10^{-10}$ m²).

This work is part of an on-going research programme to study the dewatering characteristics of porous media under gravitational, centrifugal and pressure gradients. The experimental determinations are being further extended and more rigorous mathematical statements and solutions of the problem are being sought, with an aim of providing better insight into the mechanisms of drainage of porous media. The data scatter associated with dewatering curve plots is generally unexplained; within this paper one reason for such scatter is alluded to but it is expected that further reasons will emerge as this research programme proceeds and an improved understanding of the interactions of all parameters is gained.

NOMENCLATURE

g	Acceleration due to gravity, m s^{-2}
k	Permeability, m^2
k_m	Permeability of the bed support medium, m^2
k_{rl}	Relative permeability
L	Depth of medium, m
p_c	Capillary pressure, N m^{-2}

q_l	Liquid flux, $m^3\ m^{-2}\ s^{-1}$
S_l	Saturation of liquid, fractional volume of voids occupied by liquid
$S_{l\infty}$	Irreducible liquid saturation
S_r	Reduced saturation
t	Time, s
x	Distance, m
ϵ	Porosity, volume of voids per unit volume of bed
λ	Pore size distribution index
μ_l	Liquid viscosity, $N\ s\ m^{-2}$
ρ_l	Liquid density, $kg\ m^{-3}$

ACKNOWLEDGEMENT

The authors of this paper wish to record their gratitude for a Science and Engineering Research Council grant to support a project of which this work formed a part.

REFERENCES

[1] Wakeman, R. J. and Rushton, A., *J. Powder & Bulk Solids Technol.*, **1**, 64 (1977).
[2] Gray, V. R., *J. Inst. Fuel*, **31**, 96 (1958).
[3] Lloyd, P. J. and Dodds, J. A., *Filtration and Separation*, **9**, 91 (1972).
[4] Wakeman, R. J., *Int. J. Miner. Process.*, **3**, 193 (1976).
[5] Wakeman, R. J., *Int. J. Miner. Process.*, **5**, 379 (1979).
[6] Wakeman, R. J., *Int. J. Miner. Process.*, **5**, 395 (1979).
[7] Mahgoub, I., Prost, C. and Dodds, J. A., *Powder Technol.*, **31**, 143 (1982).
[8] Wakeman, R. J., *J. Separ. Proc. Technol.*, **3**, 32 (1982).
[9] Nenniger, E. and Storrow, J. A., *A.I.Ch.E.J.*, **4**, 305 (1958).
[10] Shirato, M., Murase, T., Mori, H., Mizuno, Y., Ogura, T., and Nakai, E., *Int. Chem. Engrg.*, **21**, 294 (1981).
[11] Baluais, G., Dodds, J. A., and Tondeur, D., *Proc. POWTECH 83 Conference* – 280th EFCE Event, pp. 101–121, I. Chem. E. Symposium Series No. 69 (1983).

CHAPTER 9

Control of Structure in Particulate Solids Suspensions

R. F. STEWART and **D. SUTTON**, Imperial Chemical Industries Inc., Corporate Colloid Science Group, PO Box 11, The Heath, Runcorn, Cheshire, WA7 4QE, UK

ABSTRACT

Control of structure and properties of particulate solids suspensions is of widespread importance for solid–liquid separation and handling operations in the chemical industry.

In this paper we describe studies in which the effects of changes in basic parameters, such as solids content, particle size, and flocculating polymer type, on aggregated suspension characteristics, are carefully examined. Both laboratory-made polymer latices, and practical industrial systems were employed for the work. Suspension structure was examined directly by a number of techniques, notably freeze-etch microscopy. After mapping out some general principles concerning structure and properties of particulate solids suspensions we also briefly point out some consequences for the selection and design of solid–liquid separation systems.

1. INTRODUCTION

In the separation of fine particulate solids from liquids the effectiveness of the process, and the cost incurred, are not only dependent upon the mechanical equipment selected but they are also very much determined by the properties of suspension itself. Indeed, in many instances the latter prove to be the paramount constraints upon process design options [1–4]. Problems involving the separation and handling of colloidal solids are particularly prevalent in the chemical industry, and within our own company the variety of such materials with which we deal is very diverse, ranging from single cell protein to clays and precipitated inorganics. Because of the (apparently increasing) economic importance of such operations it has become essential to improve their efficiency by better matching of equipment to suspension properties in the process train. This requires the ability to control suspension structure, and its derivative properties such as sedimentation, filtration and rheological behaviour.

To achieve an improved understanding of the effects of changes in process variables on the characteristics of aggregated particulate solids suspensions, we have undertaken an investigation of how the structure and properties of standard systems of flocculated colloidal materials vary with basic parameters such as solids loading, particle size and flocculating polymer type. In the main, well-characterized polymer latices have been employed for the work so as to allow (unambiguous) examination of one variable at a time. However, practical industrial materials have also been studied. Suspension structure has been determined directly by a variety of techniques of optical and electron microscopy whilst properties investigated in parallel include sedimentation and centrifugation behaviour, shear moduli and suspension rheology under continuous shear conditions.

From the experimental data we map out some general principles concerning the influence of changes in basic parameters on morphology and properties of colloidal systems which have been coagulated/flocculated with electrolytes or polymers. We also briefly point out consequences for the selection and design of solid–liquid separation systems.

2. MATERIALS AND METHODS

2.1 Materials

The principle dispersions used for the work were soap free, (anionic) charge-stabilized polystyrene latices prepared by methods following those suggested by Goodwin *et al.* [5]. In each latex the particles were of uniform size. Species with diameters ranging from 0.2–4.0 μm were produced by appropriate modification of reaction conditions, and after preparation the suspended particles were characterized in terms of their size, electrophoretic mobility, surface charge and stability to added electrolyte. For the majority of the studies reported with these materials the coagulant used was 0.1 M $BaCl_2$, but investigations were also performed with a number of polymer flocculants. In particular a cationic polyelectrolyte, polyethyleneimine (PEI), of medium molecular weight (~50,000) from BDH Limited, and a number of high molecular weight, polyacrylamide-based flocculants from Research Department, Allied Colloids Limited, were employed.

The other systems used for the work were laboratory and commercial suspensions of precipitated calcium carbonate. No added coagulants or flocculants were required in this case as the materials were already flocculated under ambient conditions.

2.2 Structural Methods

In dilute systems individual floc sizes and shapes were found either from optical micrographs of materials suspended in a haemocytometer cell, or from electron micrographs of flocs freeze-dried on electron microscope grids. In concentrated systems, which were the main subject of this work, structure was characterized by stereological analysis (see Underwood [6]) of freeze-etch micrographs of the materials. In all cases a statistically significant amount of data was rapidly collected from the micrographs by means of 'Magiscan' image analysis equipment (from Joyce-Loebl Limited). The great advantage of freeze-etch microscopy was that the technique allowed *direct* examination of structure up to solids contents of >50% volume fraction without dilution, drying or other gross perturbation. The method, which involves crash-freezing of a small sample of suspension, followed by fracture, replication and replica TEM examination, has been employed by Menold *et al.* [7] to examine colloidal suspensions, though not to quantify floc structure. Examples of the types of micrograph produced by the technique are shown in Figs. 1 and 2. Great care was taken to verify that artefacts were not being introduced into the results by, for example, the freezing or fracturing procedures. Details of this analysis are given elsewhere [8]. Further description of the freeze-etch method, as developed by us, has been provided in a recent paper [9].

2.3 Mechanical Properties

A variety of mechanical property measurements were made on the various aggregated suspensions studied in this work. These included:

Sedimentation Volume and Rate which were measured in 10-ml sample tubes which were 10 cm in height. In the present work the sedimentation rate was taken to be ths speed at which the interface between supernatant and the consolidating sediment changed. For the majority of sedimentation volume measurements an initial standard concentration of 5% volume fraction of solids was used.

Shear Modulus Measurements which were made with a Rank Brothers Limited Pulse-Shearometer.

Continuous Shear Rheological Properties which were determined using a Deer PDR 81 rheometer fitted with double concentric cylinder platens. For more concentrated samples a Haake RV3 viscometer was employed.

Centrifugation Characteristics which were measured by use of MSE 'Chilspin' and HS25 laboratory centrifuges. In each case from the graph of equilibrium sediment height versus centrifuge speed, a compressive, or 'network', modulus ($K = -V\,(\mathrm{d}P/\mathrm{d}V)$, where V = volume of material, P = applied pressure) could be estimated (see Buscall [10]):

$$K = -H_{eq}\,(\Delta\rho . g . H_0 . \phi_0)/2 \frac{\mathrm{d}H_{eq}}{\mathrm{d}n} \qquad (1)$$

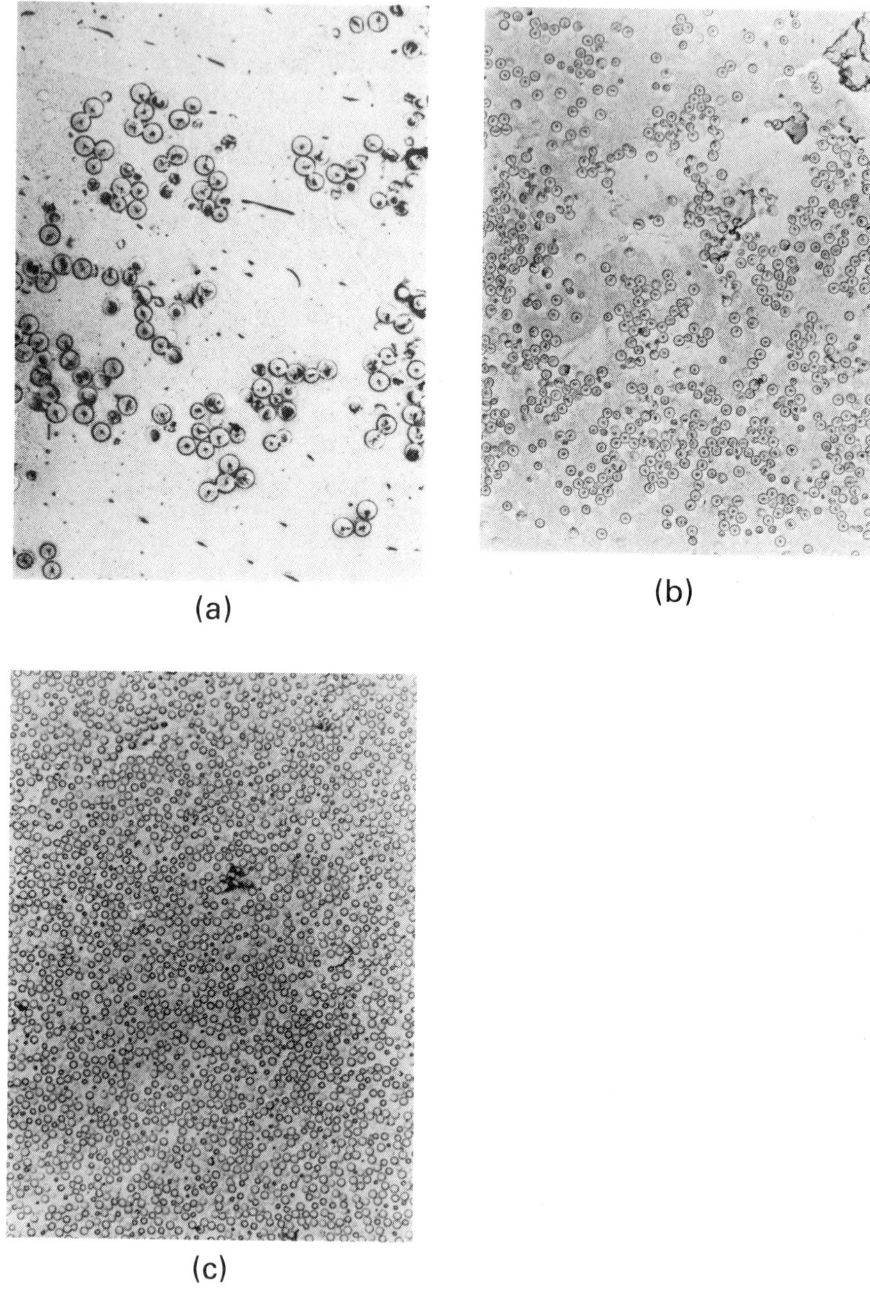

Fig. 1 – Freeze fracture replicas of coagulated polystyrene latices. Solids volume fractions are (a) 0.05, (b) 0.20 and (c) 0.30.

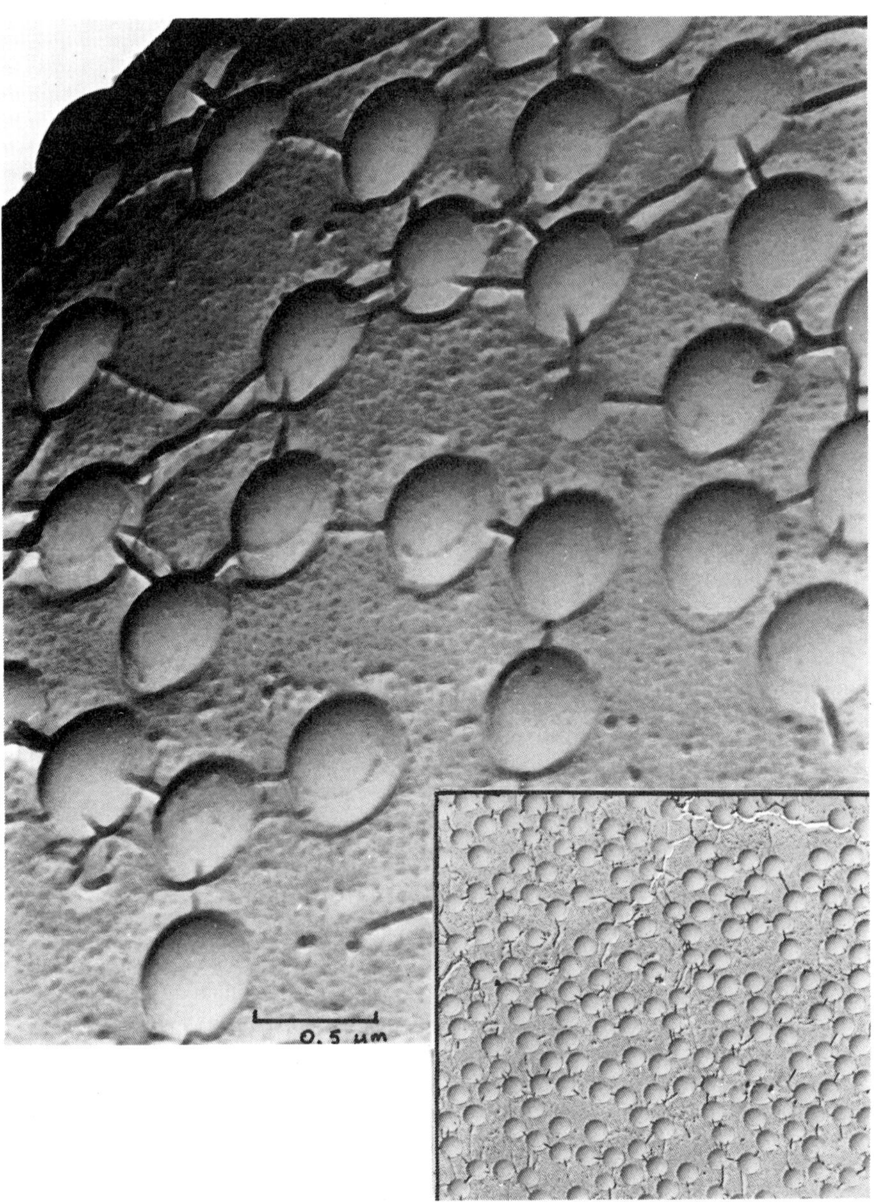

Fig. 2 – Polystyrene latex flocculated by the addition of high molecular weight anionic polyacrylamide. Inset shows larger region.

where

H_{eq} = equilibrium sediment height

H_0 = initial sediment height

ϕ_0 = initial sediment solids content

$\Delta\rho$ = density difference between solid particles and the medium

g = gravitational acceleration

(ng) = mean centrifugal acceleration factor.

For a variety of suspensions it has been shown that the network modulus derived as above, the shear modulus, and the compressive modulus determined by the compression cell method, are all essentially identical numerically [10–12].

3. RESULTS AND DISCUSSION

The effectiveness of the separation and handling of fine, aggregated solids is governed by a wide range of bulk properties of the material, these in turn being

Table 1

Lists of the more important bulk properties of flocculated systems, and of the major factors which influence these properties

Bulk Properties which are influenced by →		*System Parameters*
Sediment volume and porosity		Electrolyte concentration
Sedimentation rate		Ion type
Sediment compressibility	Network	Particle size
– Centrifuge response	Modulus	Particle shape
Zero- and low shear rheology		Solids content
(e.g. zero-shear viscosity)		Temperature
Medium and high-shear rheology		Shear rate/history
Turbidity		Mixing conditions
Filterability		Interparticle adhesion
Filter-cake compressibility		Added polymer characteristics
Filter-cake moisture content		Added polymer concentration
Size/shape/porosity of individual flocs†		
Strength of individual flocs†		

† Though these characteristics, of course, partly determine the properties above them in the list, in certain cases they may be important in their own right, e.g. where it is necessary to float the flocs or to have a dust-free product.

dependent upon microscopic parameters such as particle size and interparticle adhesion. In Table 1 we display the more important bulk properties and system parameters. Over the last few years, we have extensively examined the effects of changes in system parameters on structure and properties. However, space precludes us from discussing other than a few of the most important points here.

3.1 Effects of Solids Content and Particle Size in Coagulated Systems

The properties of simple coagulated materials, in which direct particle–particle adhesion is promoted through addition of electrolyte, provide a baseline from which one may usefully assess the characteristics of other types of aggregated suspensions. Of the many parameters (Table 1) which can influence the behaviour of these systems solids loading and primary particle size are generally the most important. In Fig. 3 we show how the network modulus changes with these

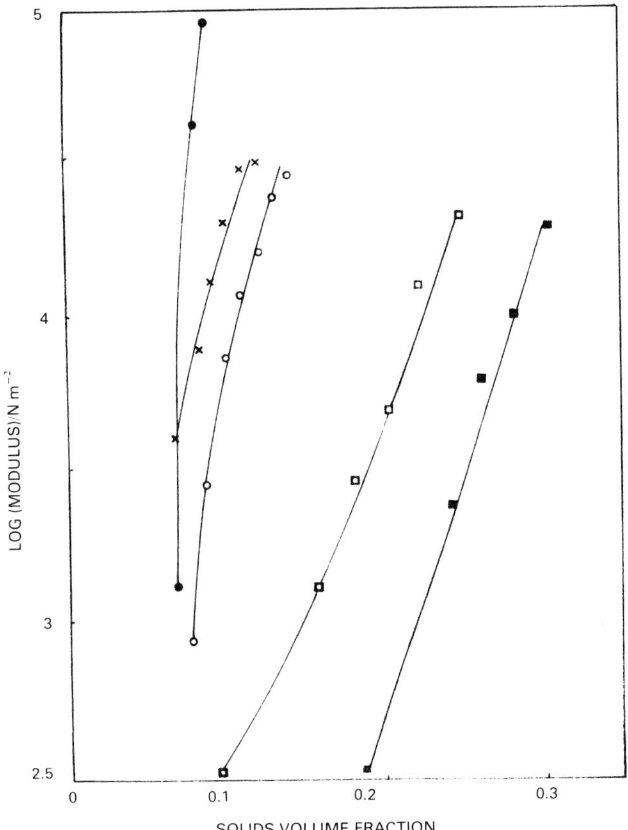

Fig. 3 – Network modulus vs. solids content for various sized latices in 0.1M $BaCl_2$. • = 0.17, × = 0.33, ○ = 0.52, □ = 0.96, ■ = 2.06 μm.

two variables for a series of coagulated polystyrene latices. Three points are immediately obvious from the results:

(i) In each case there is a quite specific critical concentration at which the suspension starts to show some solid-like behaviour as the (randomly aggregated) flocs begin to interact strongly to give a space-filling network (the system can be loosely considered as 'one large floc' at this 'gel point'). Interestingly, as particle size decreases there appears to be a convergence to a limiting solids content of ~8% volume fraction. It is likely that this is a fundamental consequence of the structural requirement of a space-filling network of (spherical) particles; our observations on limiting concentra-

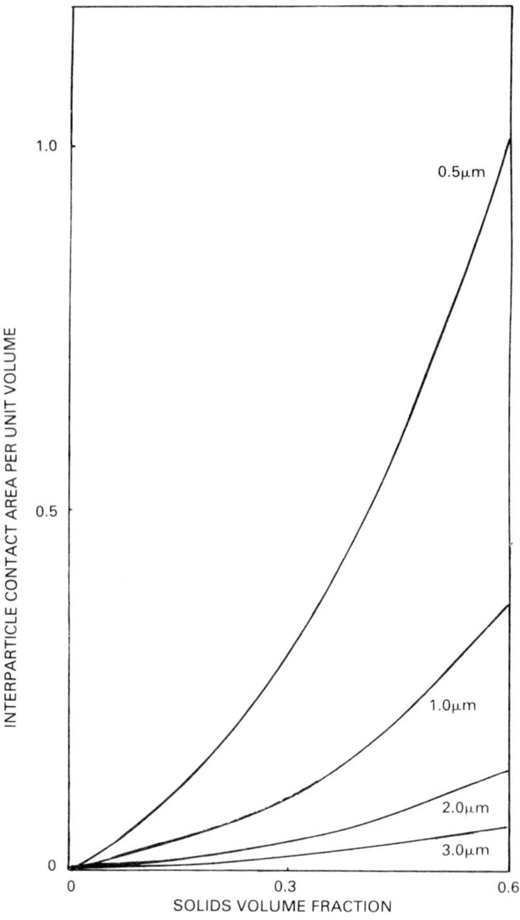

Fig. 4 – Interparticle contact area per unit volume vs. solids content for various sized latices coagulated with 0.1M $BaCl_2$.

tion are in harmony with that obtained theoretically from simulation of the build-up of a randomly aggregated system [13].

(ii) There is a clear particle size effect with the curves for larger primary species being translated to higher solids contents. The effect may be attributed to the rapid increase in the number of particle–particle contacts per unit volume of sediment ($\propto r^{-3}$), greatly outweighing the decrease in the strength of each individual contact ($\propto r$) as one moves to smaller sized species. This point is demonstrated in Fig. 4 in which interparticle contact area per unit volume (as *directly* estimated from the microscope data) is plotted for various latices.

(iii) In all cases there is an extremely rapid increase in $K(\phi)$ with ϕ. This cannot simply be explained in terms of an increase in the average primary particle coordination number with ϕ (Fig. 5). Instead what appears to

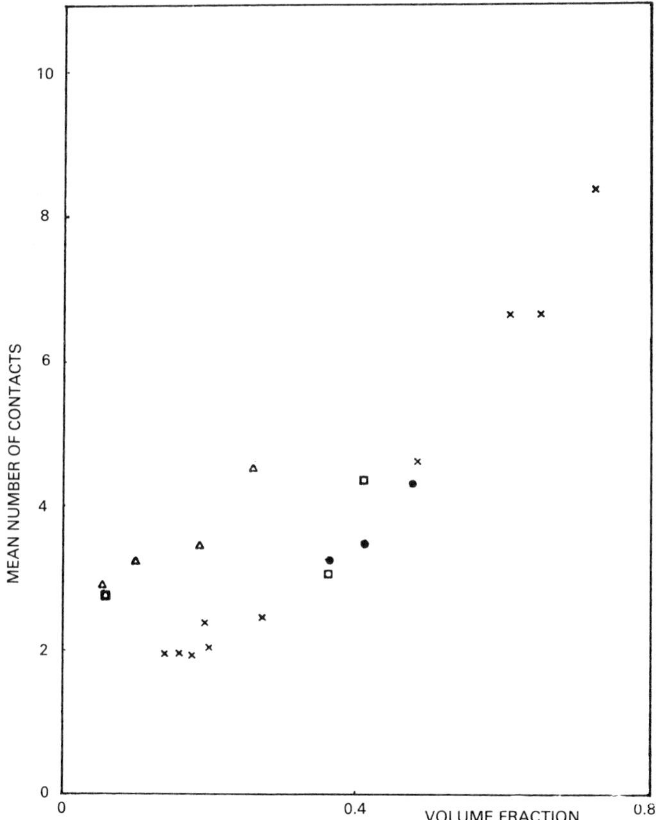

Fig. 5 – Mean number of contacts per particle vs. solids volume fraction. △, □, ● results for 0.5, 2.0 and 3.0 μm latices; × M. J. Vold simulation results, *J. Phys. Chem.*, **64**, 1616 (1960).

matter is the progressive engagement of floc 'tentacles' (with consequent elimination of weak, low coordination linkages) as solids content increases (see the freeze-etch micrographs in Fig. 1). Above the gel point (*circa* 10% volume fraction solids for the example shown in Fig. 1) particle rearrangement and cake densification became rapidly much more difficult as, with increasing solids content, even modest changes in the network require progressively greater cooperation between structural elements.

The above-noted rapidity of the rise in K with ϕ is of no little relevance to dewatering; once the critical solids concentration had been passed substantial increases in pressure are required to effect quite small increases in solids content, the phenomenon being particularly marked for smaller primary particles. This point will be returned to in section 4.

The structural changes imposed by variation of solids content or particle size also have a dominant influence on other mechanical properties of suspensions. For example, very rapid increases in viscosities measured at finite shear rates are observed above the gel point (Fig. 6). Sediment volumes, in addition, reflect the effect of particle size and other variables on sediment structure and strength. As is shown in Fig. 7, smaller particle sizes tend to be associated with progressively more voluminous materials.

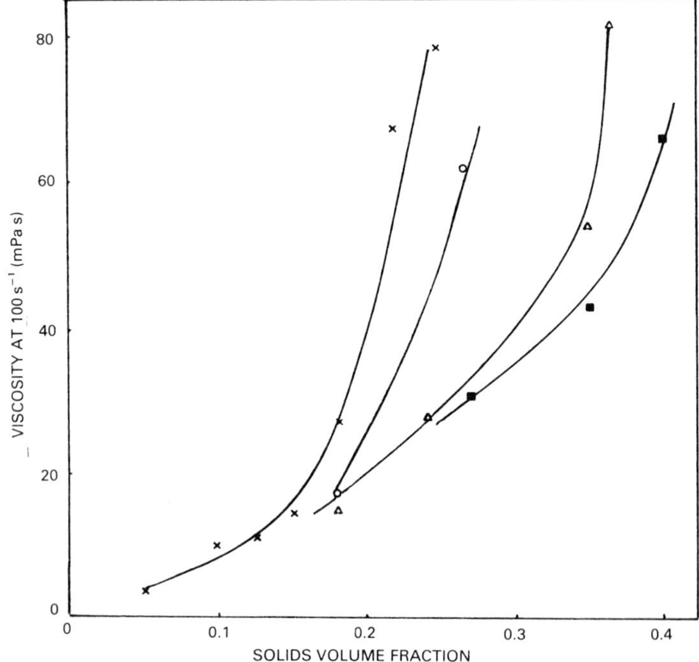

Fig. 6 – Viscosity at $100 s^{-1}$ vs. solids volume fraction for various sized latices. × = 0.5, ○ = 1.0, △ = 2.0, ■ = 3.0 μm.

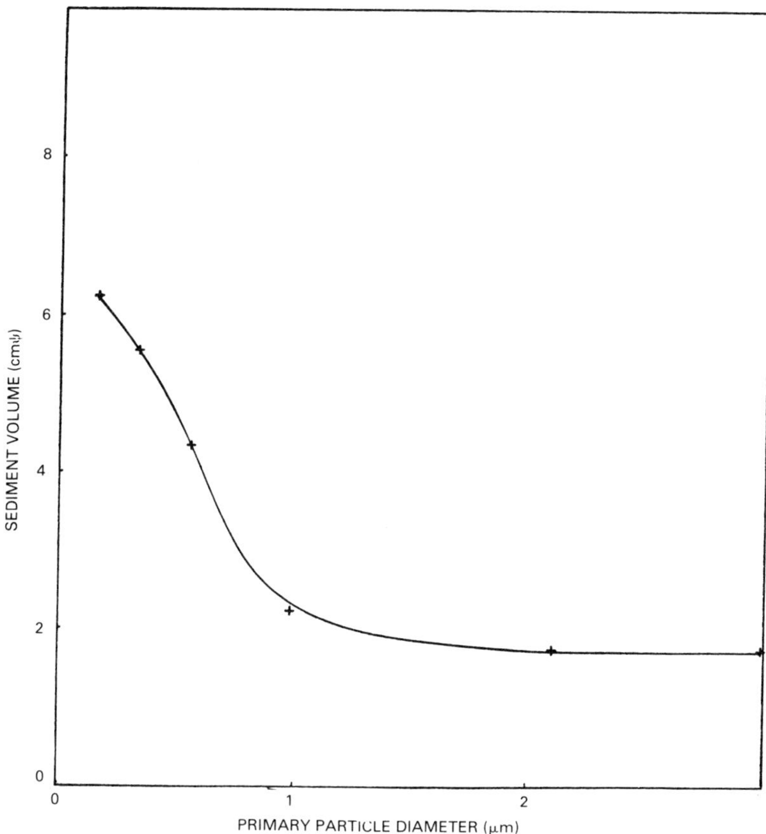

Fig. 7 – Sediment volume vs. primary particle diameter for latices coagulated in 0.1M $BaCl_2$.

3.2 Polymer Flocculated Materials

The properties of particulate suspensions in which aggregation is promoted by added polymer flocculants are of great importance owing to the widespread use of such agents in clarification and dewatering. In principle, too, they offer further degrees of control over the suspension characteristic owing to the possibilities they afford of differing *flocculation mechanisms* and differing *floc structures* from those observed in simple coagulated systems. We will comment briefly on the characteristics of two types of polymer flocculated systems:

With our latex systems flocculated with a medium molecular weight (~500,000) cationic polymer which may be expected to induce aggregation by a 'charge-patch' mechanism [14, 15] we have observed suspension structures which are only modestly perturbed from those observed on electrolyte coagulation of the same materials. The particles, which mutually adhere by direct particle–particle contact, are *nearly* randomly aggregated into small floccules which then (as before) strongly engage in the concentrated system. The only reflection of the differing flocculation mechanism is that analysis reveals that the floc units tend to be a little more compact than in electrolyte-coagulated case. Not surprisingly suspension characteristics differ only in degree rather than in kind for the two types of aggregation mechanism. Though the use of, say, a polyethyleneimine flocculant slightly increases floc and network strength compared with the electrolyte-coagulated material (Fig. 8), parameters such as solids content and particle size are still the dominant influences on the suspension properties.

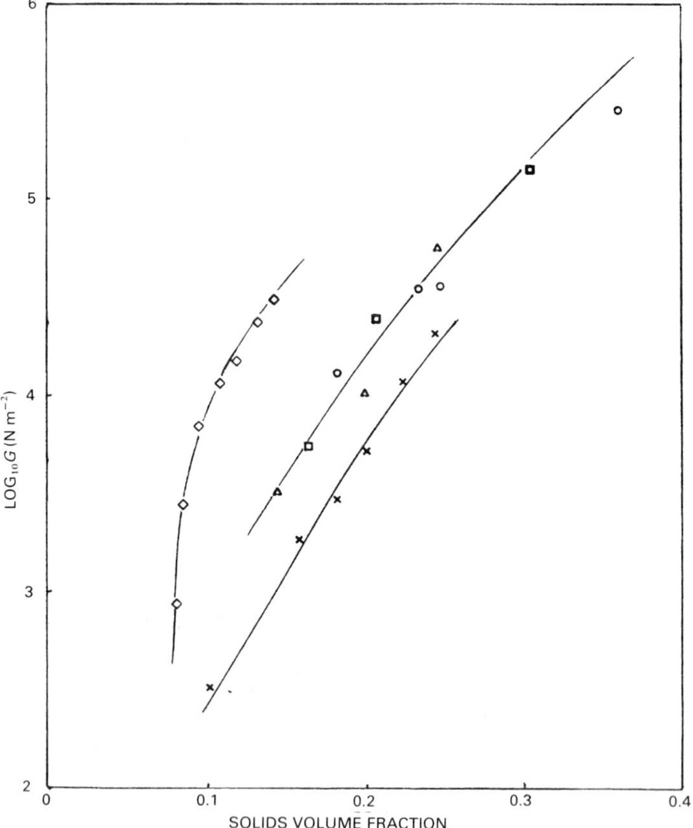

Fig. 8 – Modulus vs. solids volume fraction for PEI flocculated latex. (○ = optimum polymer concentration; △ = 40% underdose; □ = 40% overdose. Also shown results for 0.5 (◇) and 1.0 μm (X) latex coagulated with $BaCl_2$.

In contrast, use of high molecular weight, non-ionic or anionic flocculants with our latex systems gives rise to flocs in which there is no direct interparticle contact. Instead the particles are joined by flexible polymer linkages (Fig. 2)†, the properties of the material reflecting this type of morphology is that the aggregates rapidly settle to high solids content sediments: network or shear moduli are negligible except when the particles are near close packed; viscosities, though higher than for comparable disperse materials, are much lower than those observed for coagulated systems in which strong, open networks are formed (Fig. 9).

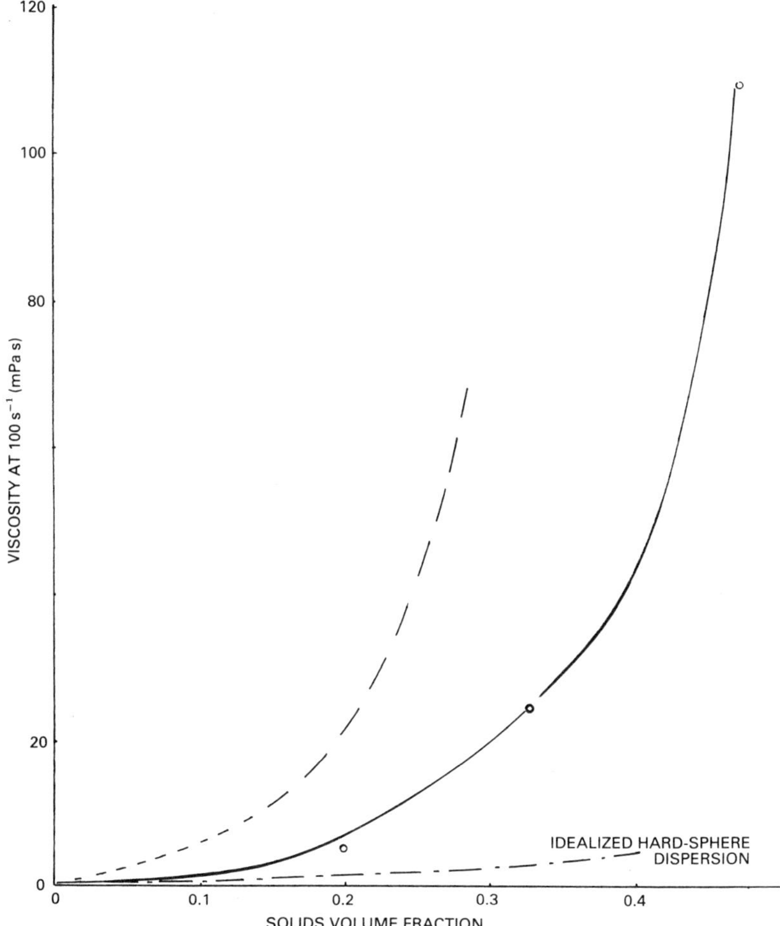

Fig. 9 – Viscosity at $100 s^{-1}$ vs. solid volume fraction for a 1.0 μm latex flocculated with high molecular weight anionic polymer (o —— o) and 0.1M $BaCl_2$ (— —).

† NB. The linkages in the micrograph are *not*, of course, claimed to be individual polymer linkages. Instead they probably represent the gel-like region formed by partly coiled flocculant molecules.

4. INDUSTRIAL APPLICATION

Efficient process design involving the separation and handling of aggregated suspensions necessitates two requirements:

(a) General rules concerning the manner in which properties change with basic parameters, and
(b) Rapid means of characterising materials and thence estimating the likely limits their properties place upon process options.

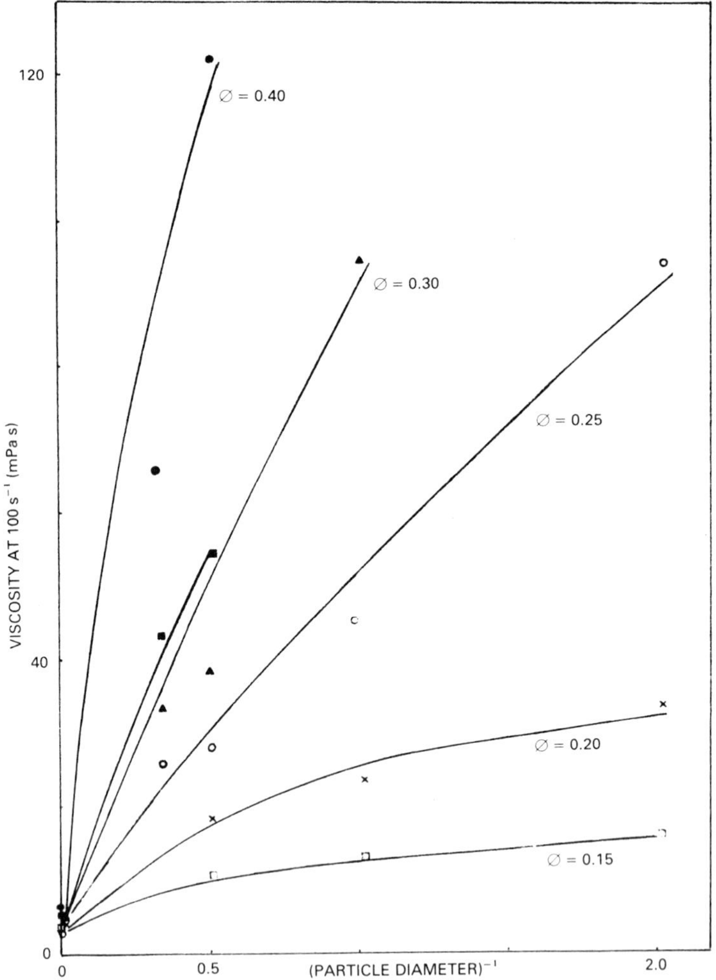

Fig. 10 – Viscosity at $100s^{-1}$ vs. (particle diameter)$^{-1}$ for the coagulated latices at various volume fractions.

Quantitative prediction of the type embodied by criterion (b) not only allows comparative estimation of the efficacy of differing types of apparatus (e.g. centrifuge, pressure filter) in, say, dewatering particular suspensions but it also enables the effects of different pressure regimes or chemical additives to be screened rapidly in the laboratory, minimizing the need for expensive pilot or plant trials.

Concerning general trends, many points leap out even from the limited data presented here – for example the rapidly increasing difficulty which must be expected in dewatering as primary particle size diminishes. It is also important to note the increasing sensitivity of suspension viscosity to, say, a small proportion of fines as ϕ becomes larger (Fig. 10). If high solids, aggregated slurries are to be pumped, care has to be taken to avoid inadvertant changes in fines content, otherwise results can be potentially disastrous! Examples are given elsewhere [16] of the use of general structural principles to aid process design in the particular case of bacterial separations.

Quantitative prediction of the effects of different separation regimes can often be obtained readily from consideration of the mechanical properties of the particulate network [10–12]. For example, as first shown by Buscall [10], network moduli obtained from laboratory centrifuge data can be inverted:

$$P_{eq} = \int_0^{\phi_{eq}} \frac{K(\phi)}{\phi} \, d\phi \qquad (2)$$

to yield a dewatered solids/pressure correlation. Illustrations of the types of relationship which can be obtained are shown in Fig. 11 for the use of a number of precipitated $CaCO_3$ magmas. This example shows the earlier mentioned effects of particle size and (less importantly) interparticle adhesion on dewatering. It also demonstrates the diminishing returns (in terms of solids content yielded by increased pressure) observed when dealing with very cohesive materials in which structural strength builds up rapidly with ϕ (Table 2).

Table 2
Degree of dewatering of two $CaCO_3$ magmas under different pressures as determined from equation (2)

	Moisture by weight in wet cake (%)	
Pressure (bar)	*Magma A*	*Magma C*
0.25	74.0	59.0
0.5	71.0	57.0
1.0	69.2	55.0
5.0	61.8	48.7
10.0	58.0	45.1
50.0	48.2	36.2
100.0	42.5	30.0

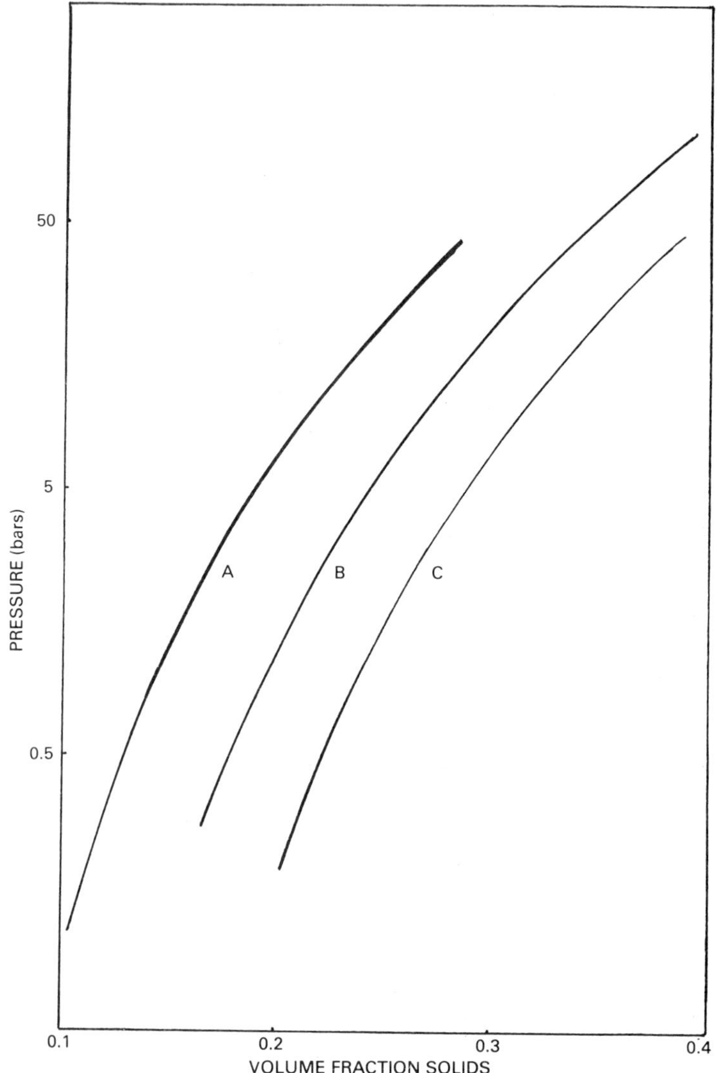

Fig. 11 – Pressure/solids content relationships for three precipitated $CaCO_3$ magmas. B and C are larger sized materials, B having a hydrophobic coating which increases interparticle adhesion in aqueous suspensions.

5. CONCLUSIONS

Effective separation and handling of fine, aggregated, particulate solids requires careful matching of process design to suspension properties. From consideration of the effects of changes in basic variables on suspension structure, unifying principles concerning material characteristics can be determined. These enable much more reliable qualitative and quantitative assessment of the results of process parameter changes on operational efficiency.

ACKNOWLEDGEMENTS

We are grateful for the assistance of Mrs E. Jones and Mr J. McMahon with work described in this paper. We should also like to thank Dr R. Buscall, of ICI Corporate Colloid Science Group, for helpful discussions.

REFERENCES

[1] Novak, J. T., *AIChE Symp. Ser.*, **73**, No. 162, 62 (1977).
[2] Novak, J. T. and Montgomery, G. E., *J. Env. Eng. Div. ASCE,* **101**, 1 (1975).
[3] Novak, J. T., Becker, H., and Zuraw, A., *J. Env. Enq. Div. ASCE,* **103**, 815 (1977).
[4] Schwartzberg, H. G., Rosenau, J. R., and Richardson, G. *AIChE Symp. Ser.*, **73**, No. 163, 177 (1977).
[5] Goodwin, J. W., Hearn, J., Ho, C. C., and Ottewill, R. H., *Coll. Polym. Sci.*, **252**, 464 (1974).
[6] Underwood, E. E., *Quantitative Stereology,* Addison Wesley, New York (1970).
[7] Menold, R., Luttge, B., and Kaiser, W., *Adv. Coll. Interface Sci.,* **5**, 281 (1970).
[8] Jones, E., Stewart, R. F., and Sutton, D., *The Morphology and Properties of Flocculated Suspensions, I, Materials and Methods,* (submitted for publication).
[9] Luckham, P. F., Vincent, B., McMahon, J., and Tadros, Th. F., *Colloids and Surfaces,* **6**, 83 (1983).
[10] Buscall, R., *Colloids and Surfaces,* **5**, 269 (1982).
[11] Buscall, R., *Proceedings 3rd Australian National Rheology Conference,* May 1983 (1983).
[12] Stewart, R. F., and Buscall, R., Paper given at the Symposium, *Emulsion and Colloid Science: The Requirements for Successful Industrial Application,* I. Chem. E., Chester, 22/23 February 1983 (1983).
[13] Vold, M. J., *J. Coll. Sci.,* **14**, 168 (1959).
[14] Kasper, D. R., Ph.D. Thesis, California Institute of Technology (1971).
[15] Gregory, J., *J. Coll. Interface Sci.,* **42**, 448 (1973).

[16] Stageman, J. F., and Stewart, R. F., Paper presented at the SCI Symposium *Flocculation and Separation of Biological Cells,* London, 27th September 1983 (1983).

CHAPTER 10

Effects of Organics in Natural Organo-clay Dispersions on Flocculation

H. SPERBER†, **N. NARKIS** and **M. REBHUN,** Technion, Israel Institute of Technology, Haifa, Israel

ABSTRACT

Difficulties and high flocculant demand were encountered in flocculation of turbid waters from 'playa' lakes in northern Texas. The high content – 7.3% – of sorbed organic matter was considered the cause of the high flocculant demand – above 10 mg/l of cationic polymer. To study the role of sorbed organics, the sediments from 'playas' were treated with various solvents and methods to extract and eliminate the organics from the clay minerals. The sediments were then dispersed in water and flocculated.

Extraction with ethanol reduced the organic content to 6.3% with little effect on flocculation, aqueous extraction reduced the organics to 5.2% and the polymer dose to 2 mg/l, NaOH extraction reduced the polymer dose to 1.5 mg/l. Elimination of organics from clays by mild oxidation with H_2O_2 reduced organic content to 1.3% and polymer dose demand to 0.5 mg/l close to the demand of pure clay dispersion of the same solid concentration.

Addition of the extracted organics to pure clay dispersions accordingly increased the cationic polymer demand.

1. INTRODUCTION

Previous publications by the authors on coagulation–flocculation in presence of organic substances in water, were on model aquatic–humic and reference clay–humic systems [1, 2]. The mechanism of coagulation of such systems was elucidated and stoichiometric relationships in coagulation of such systems were shown.

The present paper deals with coagulation–flocculation of naturally occurring organo-clay systems – NOC. This paper intends to show that the mechanisms

† Deceased.

and relationships found in model systems apply to, and are valid also in, NOC systems. Turbid surface waters with a relatively high organic content from 'playa' lakes in northwest Texas, USA served as the NOC system in this research.

For meaningful results the study had to be performed under controlled and reproducible conditions. Therefore, a large quantity of 'playa' water was evaporated *in situ* upon exposure to natural surroundings, an air-dried sediment was obtained which was used for preparation of working suspensions with reproducible colloidal properties similar to fresh natural turbid water of the 'playa'.

The study programme included:

(1) Flocculation characteristics of NOC dispersions and comparison with dispersions of similar reference clay minerals but without organic matter.
(2) Extraction of organic substances from the NOC sediment by using various solvents or methods. Flocculation studies of the sediment dispersions after removal of organics.
(3) Addition of the organic extracts to pure reference clay dispersions and the study of their flocculation characteristics.

2. PROCEDURES

2.1 Flocculation

All flocculation experiments were performed on freshly prepared dispersions. The ionic composition was always adjusted to the ionic composition of the natural 'playa' lake water. This preparation procedure ensured reproducible colloidal and flocculation properties. Flocculation was by a jar test method, electrophoretic mobilities were measured by Zeta-meter and turbidities were determined by photometer. All procedures were as described in Refs [3] and [4]. Cationic polymer was Purifloc C-3 a polyethyleneimine of m.wt. 30,000; alum and polymer solutions were applied as described before [1, 2].

2.2 Sample Preparation

Water extraction of organic matter from the original sediment was a follows: A weighed sample of NOC sediment was subjected to a series of 'washings' with distilled water and centrifugations. The procedure was repeated until a colourless centrifugate was obtained. All extracts were collected for examination and for preparation of reference clay – organic mixtures. These mixtures were prepared by dispersing reference Kaolinite K-4 in a solution of the extract corresponding to its concentration and ionic composition of the natural water. Alkaline extraction was performed in the same way but using a 1% solution of sodium hydroxide.

The fractionation of humic and fulvic acids in the alkaline extracts was by procedures described by Bear [5].

Oxidation of organic matter was by a mild oxidation method, using hydrogen peroxide as described by Jackson [6].

2.3 Characteristics of the 'playa' NOC sediment

Chemical, X-ray and cation exchange capacity analyses of the sediment showed that 70% of the mineral part are clay minerals, with poorly developed crystallisation. Kaolinite and illite could be clearly defined and indications of montmorilonite were evident. The total organic content of the sediment was 7% as organic carbon, the humic substances content was 4.5% as organic carbon, with a humic to fulvic ratio of 1:3.

3. RESULTS

3.1 Flocculation of the NOC dispersion

Flocculation characteristics of the NOC dispersion by the cationic polymer and corresponding mobility data are shown in Fig. 1. For comparison, flocculation

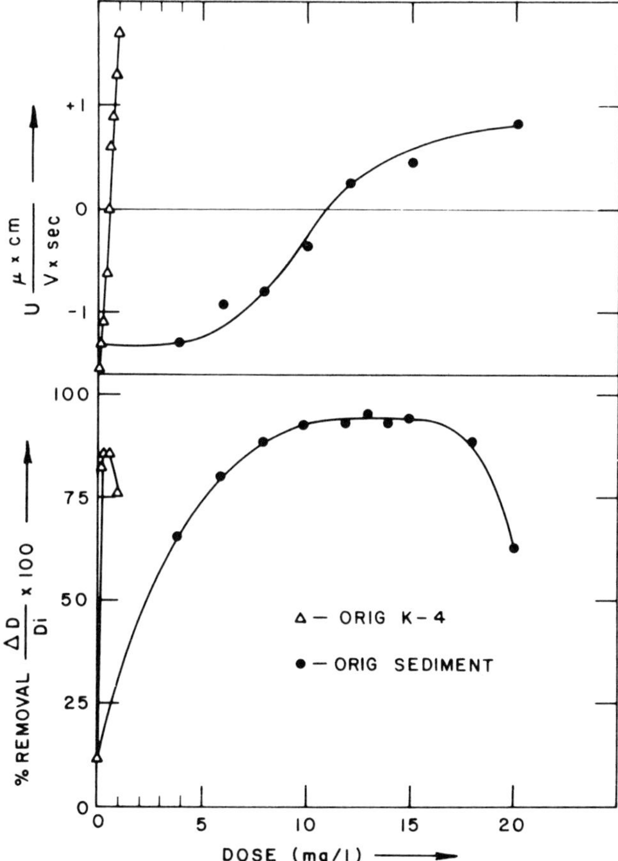

Fig. 1 – Flocculation and electrophoretic mobility curves of natural organo clay – NOC – dispersion and reference clay K-4 (both 400 mg/l) with cationic polymer.

of dispersion of a pure reference clay mineral K-4 is also shown in the same figure. The polymer dose required for clarification of NOC was 11 mg/l, many times higher than the dose required for the reference clay dispersion 0.3 mg/l. Both curves exhibit the typical pattern of cationic polymer flocculation. The mobility curves show that the optimum flocculation dose for NOC coincides with the isoelectric point, while good flocculation of pure reference clay mineral was obtained at relatively high mobility values (−1.2 μ/sec/V/cm) close to the mobility of the untreated suspension.

Flocculation of the NOC sediment dispersions at different solids concentrations, showed a linear relationship between optimal dose required and NOC solids concentrations. The specific cationic polymer demand was 25 mg/g for NOC *vs* 1 mg/g for reference kaolinite K-4.

Flocculation of NOC dispersions with alum is shown in Fig. 2, the dose requirement was 100 mg/l *vs* 12 mg/l required for reference clay suspension of the same concentration. A linear relationship was found between NOC concentration and alum dose required.

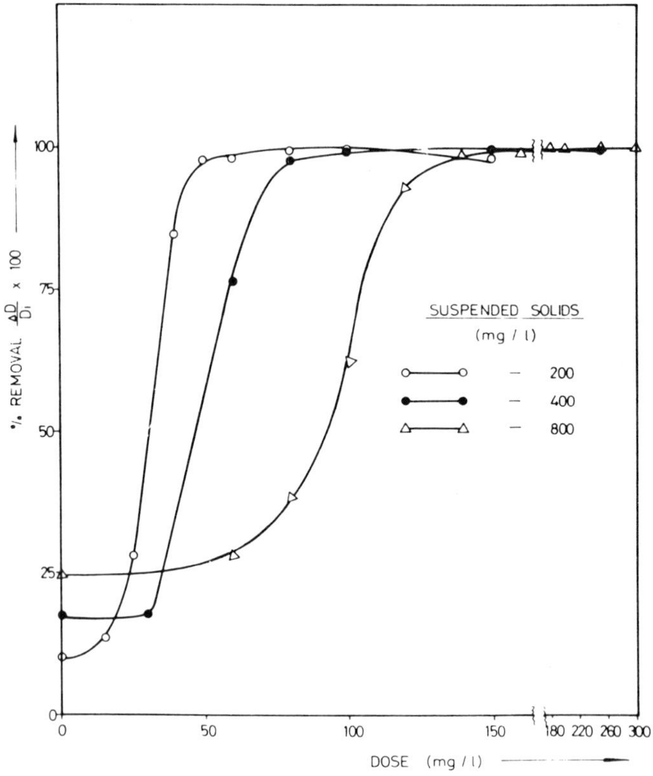

Fig. 2 – Flocculation of NOC dispersions of different suspended solids concentration with alum.

The above findings show clearly two major flocculation characteristics of NOC:

High flocculant demand – manyfold higher than for any reference clay mineral;
Coincidence of optimum dose with the isoelectric point.

These characteristics are typical for flocculation of systems containing humic substances [1–3]. The hypothesis was that also in the 'playa' water the organics (humic substances) are responsible for this flocculation behaviour. This hypothesis was checked by extraction–purification of NOC sediment from the organics and flocculation studies of the purified NOC.

3.2 Purified NOC Sediment after Extraction

The organic content of the NOC sediment after the different extraction procedures can be seen in Table 1. Little reduction of organics was effected by ethanol extraction, while high removals of organics from NOC sediments were obtained by water and NaOH extraction and the highest degree of sediment purification was obtained through H_2O_2 oxidation.

The 'purified' NOC sediments were used to prepare suspensions for flocculation experiments. The extracts were used for preparation of organic–clay dispersions by dispersing pure reference clay minerals in solutions of the extracts under controlled conditions.

3.3 Flocculation of Dispersions of NOC Sediments after Different Degrees of Organics Extractions (Different Organics Content)

Flocculation of dispersions of NOC sediments, after different methods and degrees of purification from organic matter, are summarised in Fig. 3 with alum, and Fig. 4 with cationic polymer, in the form of flocculation curves and electrophoretic mobility. Original NOC sediment dispersions and pure reference clay mineral dispersions are also included. Initial dispersed solids concentration and ionic composition were identical for all samples.

Ethanol extraction of the sediment, did not affect significantly the flocculation behaviour. Water extraction of the organics from the sediment caused a dramatic change in flocculation behaviour. The flocculation curves of NOC sediments after water extraction compared to original sediment are also shown separately in Figs 5 and 6 for clearer comparison.

The alum dose requirement was 30 mg/l *vs* 100 mg/l for the original NOC sediment, and the cationic polymer dose required was 2 mg/l *vs* 11 mg/l for the original NOC. Also the degree of clarification at optimum dose of cationic polymer was higher for the purified sediment.

The sediment after NaOH extraction had a flocculation behaviour similar to that after water extraction with slightly lower focculant dose requirements. Oxidation of organics in the sediment had the strongest effect on flocculation, lowering the doses required to 0.5 mg/l of cationic polymer and 25 mg/l of alum, approaching the dose requirements of reference clays.

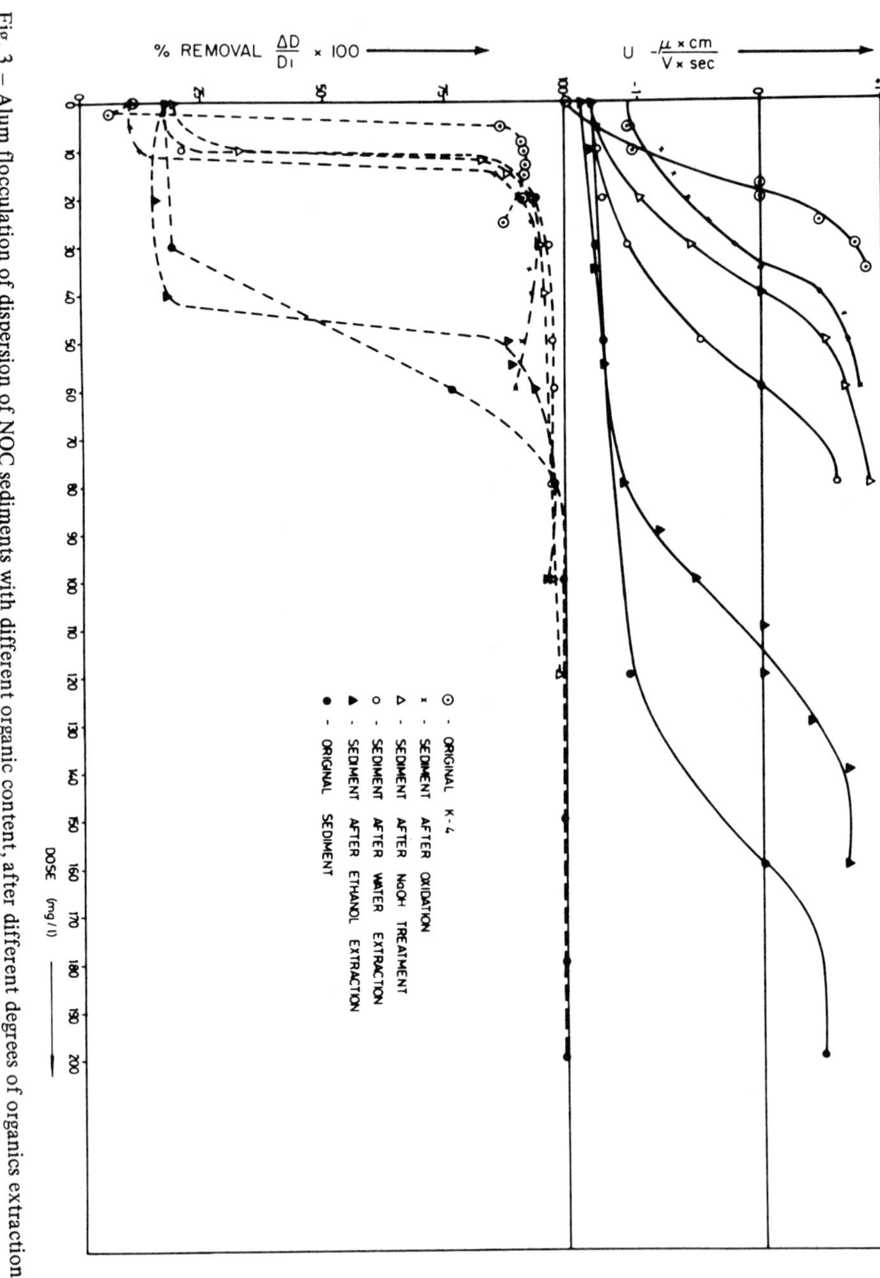

Fig. 3 – Alum flocculation of dispersion of NOC sediments with different organic content, after different degrees of organics extractions.

Table 1

Organic content and flocculation behaviour of suspensions prepared from NOC sediment 'purified' by different methods

Determined properties / Dispersed material	Organic carbon content (%)	Doses required for good flocculation (mg/l)		Mobility values (μ/sec/V/cm)			Dose required for reaching the isoelectric point (mg/l)	
				of the untreated susp.	At the dose required for good flocculation			
		Alum	Purifloc C-31		Alum	Purifloc C-31	Alum	Purifloc C-31
Original NOC sediment (non-purified)	7.3	100	11	–1.4	–1.1	0	160	11
Sediment after ethanol extraction	6.3	80	8	–1.47	–1.14	0	120	8
Sediment after water extraction	5.22	30	2	–1.24	–1.1	–0.4	60	2.8
Sediment after NaOH extraction	3.4	30	1.5	–1.38	–0.6	–0.22	40	1.8
Sediment after oxidation	1.3	25	0.5	–1.2	–0.4	–0.8	35	1.5
Reference clay K-4	0.14	12	0.2	–1.5	–0.9	–1.08	20	0.5

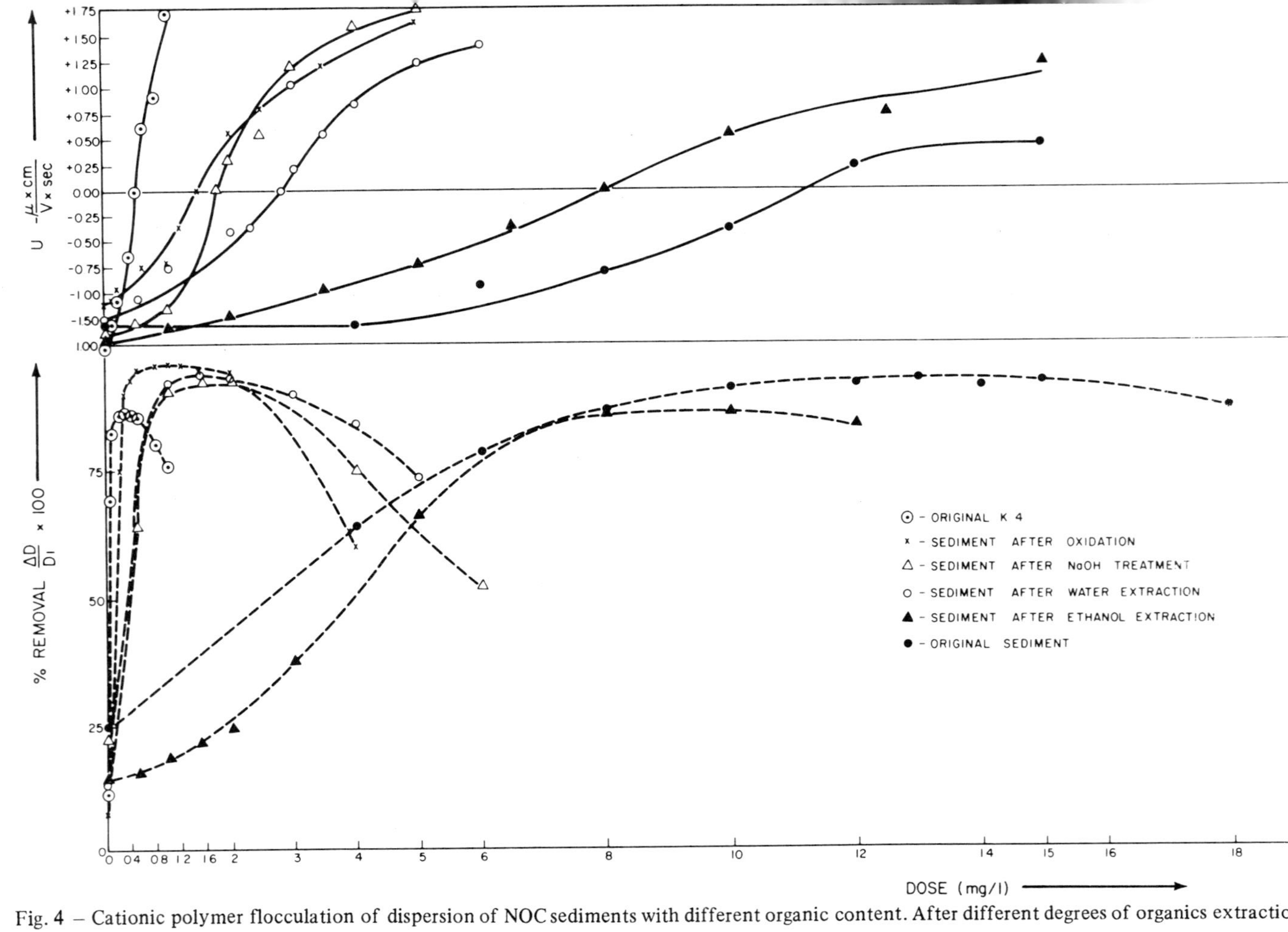

Fig. 4 – Cationic polymer flocculation of dispersion of NOC sediments with different organic content. After different degrees of organics extraction.

The extraction procedures affected also the relations of electrophoretic mobility to applied dose and to flocculation.

In case of the original NOC sediment and after ethanol extraction the isoelectric points coincided with the optimal flocculant doses, which were high for these dispersions. After water extraction, optimal flocculation was reached before the isoelectric point though not far from it, and the flocculant doses needed to reach the isoelectric point were many times lower than for the original NOC. After oxidation, optimal flocculation could be obtained at rather high mobility values (far from the isoelectric point) just slightly lower than the

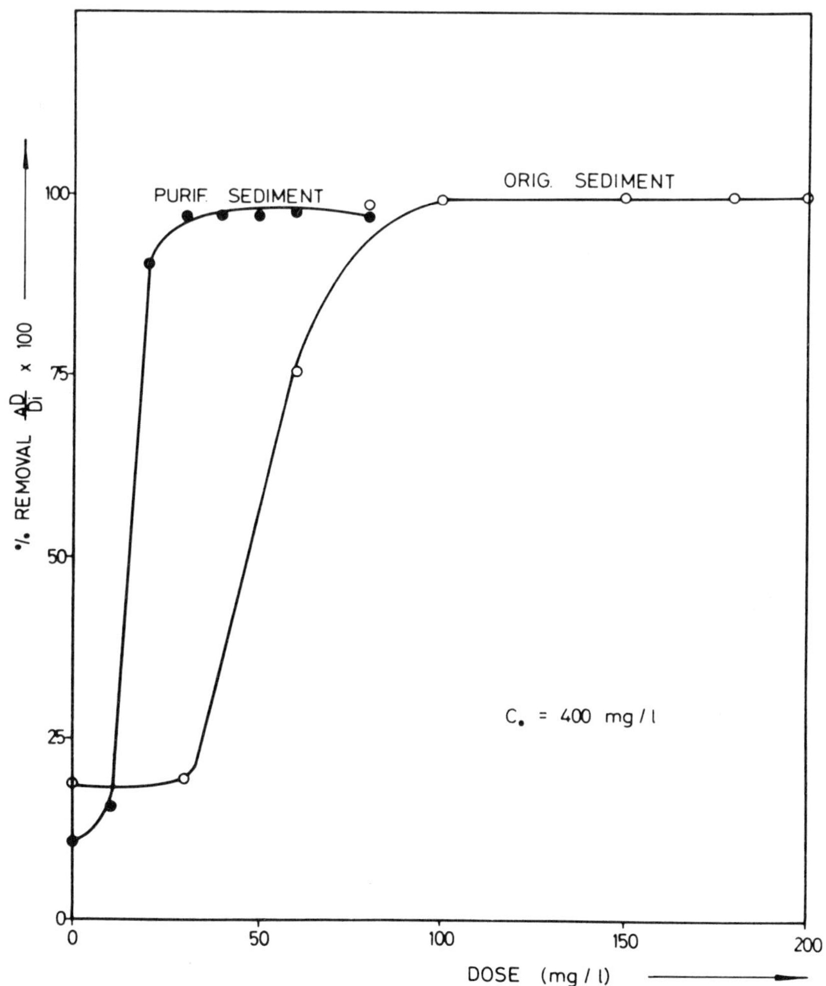

Fig. 5 – Alum flocculation of dispersions of original NOC sediment and of NOC after water extraction.

mobility of the dispersion in the absence of flocculant. Flocculation behaviour of the oxidized sediment approached closely the behaviour of reference clay dispersions.

Major flocculation characteristics of the dispersions are summarised in Table 1.

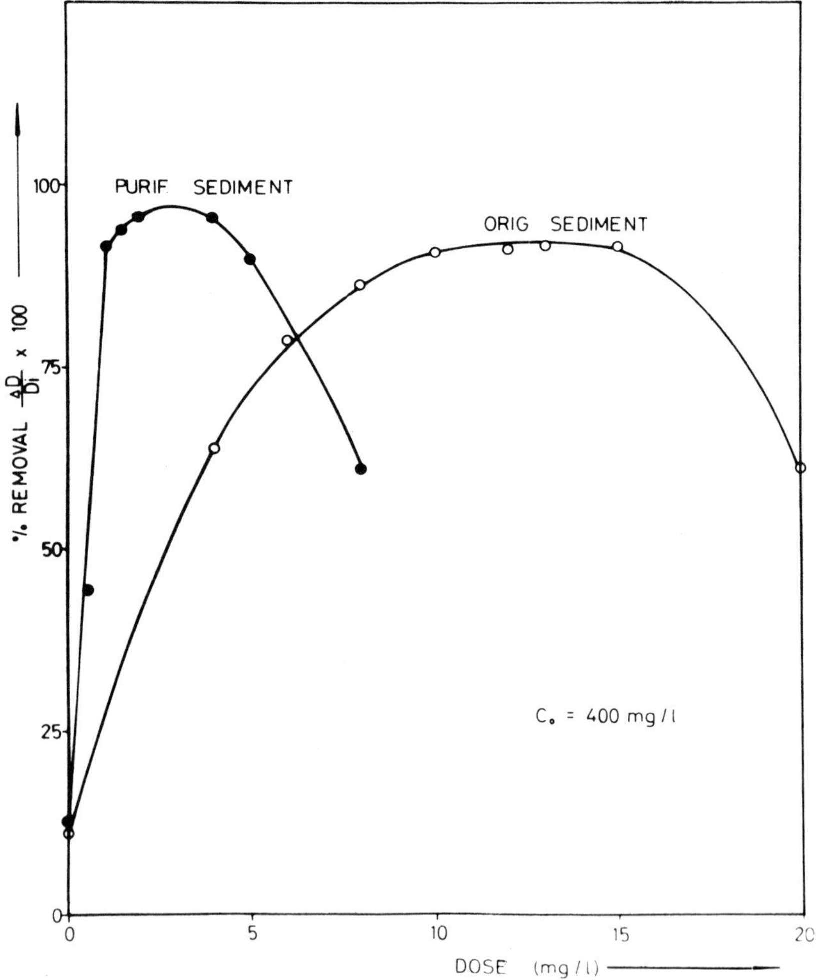

Fig. 6 – Cationic polymer flocculation of dispersions of original NOC sediment and NOC after water extraction.

3.4 Effect of Extracted Material from NOC on Flocculation of Reference Clay Suspensions

An additional way to examine the effect and role of organic substances present in NOC on flocculation processes was to prepare reference clay mineral dispersions in aqueous solutions of the organics extracted from NOC. The dispersions were at mineral and organic concentrations similar to the original NOC systems.

The effects of water-extracted organics on flocculation of reference clay mineral K-4 are shown in Figs 7 and 8 for alum and cationic polymer respectively. The extract increased the alum dose requirement from 12 mg/l to 50 mg/l and cationic polymer requirement from 0.2 mg/l to 2.5 mg/l and impaired the final clarity of flocculated water. The addition of alkaline extracts to K-4 dispersion had a similar effect.

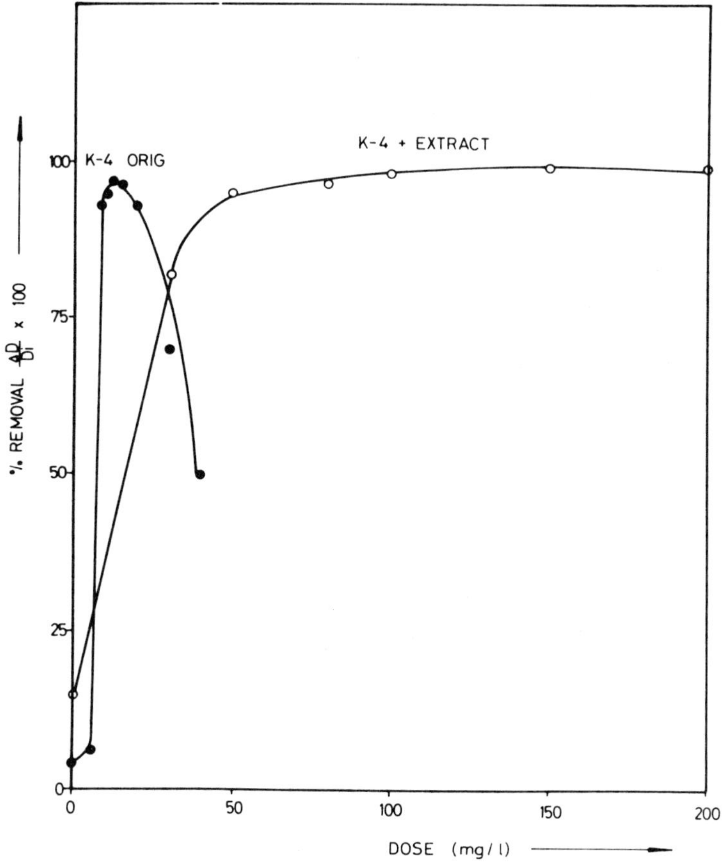

Fig. 7 – Effect of water extract from NOC on alum flocculation of reference clay K-4 dispersion.

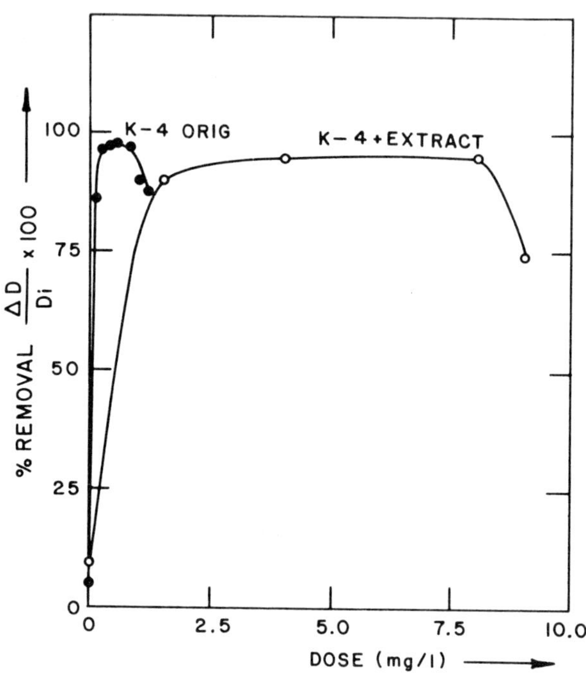

Fig. 8 – Effect of water extract from NOC on cationic polymer flocculation of reference clay K-4 dispersion.

4. DISCUSSION

Flocculation behaviour of turbid surface waters is often significantly different from flocculation of dispersions of reference clay minerals of the same concentration and same ionic composition of the aqueous solution. The flocculant dose is higher and difficulties in floc formation and clarification are encountered. These differences can be ascribed to different mineralogical structure and surface properties of the turbidity particles in natural waters as compared to reference clay mineral, to mineral impurities or to organic substances which affect and modify the process.

The flocculation behaviour of the natural colloidal system – the playa lake turbid waters – could also be ascribed to one of the above mentioned factors.

This study indicates that the organic substances forming a natural organo-clay system, and particularly the humic substances, are responsible for the interferences in flocculation and high flocculant demand.

Three basic features are typical for flocculation of humic substances: high flocculant demand, strict stoichiometry and coincidence of the isoelectric print with optimum flocculation [1–4]. All these features were found in the flocculation of the NOC dispersions of the playa sediment. Removal-purification of the organics from the NOC sediment by extraction or oxidation reduced the flocculant demand and the flocculation characteristics of the putrified sediment approached the characteristics of reference clay mineral dispersions.

The various purification methods applied brought about various degrees of removal of organic interfering matter and different response to flocculation was obtained. The different flocculation properties are represented graphically in the form of two families of flocculation curves and electrophoretic mobility curves. The position of each curve in relation to reference clay mineral curve indicates the degree of organic interference. The major flocculation parameters were also tabulated in order of decreasing organic content of sediments.

In general, good correspondence was found between the position of flocculation and mobility curves and the organic carbon content. The higher the reduction of the organic content the lower the flocculant dose required and the flocculation characteristics approach closer the reference clay mineral.

However, the amount of organic carbon in the NOC is not the only factor, the type of organics present is also important. The sediment after ethanol extraction has not changed significantly its flocculation characteristics compared to original NOC sediment. Ethanol does not extract humic substances efficiently; the humic substances are, however, the major organic group affecting flocculation. The water and alkaline extractions, which removed the humic substances, drastically changed the flocculation characteristics of sediment dispersions, though the remaining organic content of the sediment was still relatively high (5.3%). One of the most pronounced characteristics of flocculation of systems containing humic materials is the coincidence of the isoelectric point with optimal flocculation dose.

This is due to the fact that the removal of humics is by reaction and charge neutralization with the cationic flocculant and the completeness of this reaction (mobility = 0) is a prerequisite for optimum flocculation. With increased degree of purification of the NOC sediment from the organics the optimum flocculation preceded the isoelectric point. The highest degree of sediment purification from organics was obtained by oxidation. After oxidation, the flocculation and mobility relationships were very close to the relationships of clay mineral dispersions.

The humic substances are present in organo-clay systems in two forms: as free humates and adsorbed on clay mineral particles [1, 2]. The effects of the free and the adsorbed humates are different. The water and alkaline extractions removed mainly the free humates from the sediment, the oxidation procedure removed a significant amount of the adsorbed humates.

The humates extracted by water or alkaline solution – the free humates – exerted the major flocculant demand, while the residual humates that were removed by oxidation only – the adsorbed humate – exerted a relatively low demand. This corresponds well to the findings in model organo-clay systems [1]. The NOC after water and alkaline extraction can be regarded as an organo-clay complex, while the original NOC dispersion can be regarded as an organo-clay complex in humate solution. The flocculation characteristics of the respective NOC dispersions were similar to model organo-clay systems with respect to response-dose, and electrophoretic mobility relationships, indicating similar flocculation mechanisms.

The study confirmed the hypothesis that in natural turbid surface waters containing organic (humic) substances in significant concentration these organic substances are the major factor determining the flocculation behaviour.

ACKNOWLEDGMENT

The invaluable assistance of Mrs Regina Offer is greatly appreciated. This study was supported by USDA Grant No. FG-IS-148(1966-67).

This paper was written by M. Rebhun in memory of Henryk Sperber who contributed greatly to the progress of the research projects on flocculation.

REFERENCES

[1] Narkis, N, and Rebhun, M., The mechanism of flocculation processes in the presence of humic substances, *Jour. Amer. Water Works Assoc.*, **67**, 101–108 (1975).

[2] Narkis, N. and Rebhun, M., Stoichiometric relationships between humic and fulvic acids and flocculants, *Jour. Amer. Water Works Assoc.*, **69**, 325–328 (1977).

[3] Black, A. D. and Williams, D. G., Electrophoretic studies of coagulation for removal of organic color, *Jour. Amer. Water Works Assoc.*, **53**, 589–604 (1961).

[4] Hall, E.S. and Packham, R. F., Coagulation of organic color with hydrolysing coagulants, *Jour. Amer. Water Works Assoc.*, **57**, 1149–1166 (1965).

[5] Bear, F. W., *Chemistry of Soil*, Reinhold Publishing Co. (1960).

[6] Jackson, M. L., *Soil Chemical Analysis*, Constable Co. Ltd. (1958).

CHAPTER 11

The Development of Improved Methods for the Industrial Recovery of Protein Precipitates

N. DEVEREUX, M. HOARE, P. DUNNILL, Department of Chemical and Biochemical Engineering, University College London, Torrington Place, London WC1E 7JE, UK and **D. J. BELL,** D.S.I.R., Private Bag, Petone, New Zealand

ABSTRACT

Effective large-scale recovery of protein precipitates by centrifugation or ultrafiltration is governed by the characteristics of the precipitate particles and the precipitate sediment and the relationship of their properties to the shear fields associated with both recovery procedures.

A detailed study has been made of the characteristics of isoelectric soya precipitate particles including their size, density and sensitivity to shear fields. These properties have been related to the separation efficiency of disc and scroll industrial centrifuges and a hollow fibre ultrafiltration system.

The rheological properties of protein precipitate suspensions and sediments have been examined. The plasticity, shear modulus and apparent viscosity of the material are related to the size and shape of the precipitate aggregates and depend on the method of preparation. The shear modulus and plasticity of the sediment determines the dewatering capabilities of scroll discharge centrifuges and intermittent-discharge, disc centrifuges, while the operation of a hollow fibre ultrafiltration system is governed by the apparent viscosity of the retentate.

The relative merits of centrifugation and ultrafiltration for protein precipitate recovery are discussed against the background of these rheological factors.

1. INTRODUCTION

The design of protein precipitation and precipitate recovery processes [1] involves the specification of high-speed industrial centrifuges or membrane separation equipment for the recovery of the protein precipitate. Blinding of conventional filter septa and product contamination with filter aid material generally precludes the use of filtration for this operation.

The specification of centrifuge or membrane separation equipment is conditioned by the properties of the protein precipitate. In previous publications we have defined the effect of reactor design conditions on precipitate aggregate size [2, 3], shape [4], density [5] and resistance to shear break-up [6]. These properties may also be influenced by batch ageing of the precipitate which can improve resistance to shear break-up [6], and by pumping [3] and flow through the centrifuge feed zone [6] or membrane system [7] which may cause aggregate break-up.

The useful throughput of a centrifuge is governed principally by the small size of protein precipitate aggregates and the limited density difference between the aggregate and mother liquor, but the rheological and dewatering characteristics of the sediment also are of great importance. Ultrafiltration techniques represent an alternative to centrifugation in the recovery of proteins though reduction in flux at high protein concentration ($> 20\%$ w/v) limits the usefulness of this technique when applied to soluble proteins. This problem can be partly overcome by the ultrafiltration of precipitated proteins [7] and the approach has the advantage compared with centrifugation that it does not rely on density difference. However, as with centrifugation, the rheological properties of the concentrated protein precipitate are a key operating parameter and it is this aspect which is the main subject of the present paper.

2. PROTEIN PRECIPITATE PREPARATION

Soya protein total water extract [2, 4, 5] precipitated at its isoelectric point, pH 4.8, was the test material for studies of rheological characteristics and of precipitate recovery by centrifugation or ultrafiltration. Solutions containing 25 to 90 kg m^{-3} protein were contacted with the precipitating agent, sulphuric acid, in either batch-stirred reactors of 0.67 l, 2 l and 200 l, [3, 4, 5] or in a continuous tubular reactor, 15 mm i.d., 19 m long [2]. For rheological studies the precipitate suspensions were concentrated by gravity settling, up to 150 kg m^{-3} total protein, or for higher concentrations, by centrifugation ($\sim$ 450 RCF, 30 ml angle head bench centrifuge (Baird and Tatlock, Essex, England)). Details of precipitate preparation are given with accompanying results.

The precipitate particles prepared are aggregates of 1 μm diameter primary particles [5]. Batch, turbine-stirred, reactors which give a mean velocity gradient, $\bar{G}$, of $\sim$ 200 s^{-1} and a residence time, t, of $\sim$ 1800 s, yield compact and regular-shaped aggregates [4] with maximum resistance to shear break-up at $\bar{G}t > 10^5$ [6]. Protein precipitation in a continuous reactor [2] with $\bar{G} \sim 340$ s^{-1} and a mean residence time of $\sim$ 25 s, yields loose and irregular-shaped aggregates [4] with less resistance to shear break-up [6].

Precipitate particle size analysis was by Coulter counter models TA and TA II (Coulter Electronics, Harpenden, Herts.). Material prepared in the tubular reactor gave values of d_m of 18 ± 3 μm and d_{90} of 7 ± 2 μm. For the pilot-scale, batch reactor (200 l vessel, i.d. = height = 0.67 m, with an off-centred, 3-bladed, pitched paddle, diameter 0.19 m, speed 230 rpm) precipitate particle sizes of d_m = 8 ± 1 μm and d_{90} = 6 ± 1 μm were obtained. Particle sizes for the precipitate prepared using small-scale, batch reactors are given with the relevant results.

3. MEASUREMENT OF RHEOLOGICAL PROPERTIES

For creep flow rheology a constant torque rheometer (Deer, Rheometer Marketing Ltd., Leeds, England) was chosen to measure the strain–stress relationships for small samples of precipitate suspension (~ 0.1 ml). Cone and plate (MG 1601, 35 μm gap) or single concentric cylinder (MG 7101, 76 μm gap) geometries were used. Evaporation effects were minimized using a stainless steel hood. The rheometer was calibrated using silicone fluids, viscosity 0.097 and 0.973 Nsm^{-2} (Dow Corning) and a Brookfield standard of viscosity 59 Nsm^{-2} (Brookfield Engineering Labs. Inc., Stoughton, Mass., USA). To minimize any differences in the history of sample loading into the rheometer, the samples were sheared at 100 s^{-1} for 5 s prior to analysis.

Low shear measurements (1 to 1000 s^{-1}) were carried out in a narrow gap, cup-and-bob rheometer (Rheomat 30, MSO, Contraves AG., Zurich, Switzerland). Each reading of applied torque was taken 10 s after a change in shear rate. To avoid precipitate sedimentation a maximum time of 300 s was allowed for completion of the shear rate cycles.

Rheological characteristics at high shear rates (up to 23 000 s^{-1}) were obtained from pressure drop/flow characteristics in five stanless steel capillaries 1.02 mm i.d., arranged in parallel. Measurements were taken for capillary lengths of 0.375 m and 1.00 m in order to correct for entrance and exit effects. The results obtained are applicable only in laminar flow. Continuous recirculation of test material was employed throughout as described by Devereux *et al.* [7], and a period of 1800 s was allowed to elapse before readings were taken.

All rheological measurements were carried out at 298 K unless otherwise stated.

4. RHEOLOGY OF PROTEIN PRECIPITATE SUSPENSIONS AND SEDIMENTS

4.1 Creep flow characteristics

Representative creep flow curves showing non-linear, viscoelastic rheological behaviour for protein precipitate suspensions from batch stirred tank and continuous tubular reactors are presented in Fig. 1. The creep flow of a precipitate suspension when exposed to an applied shear stress yields information on the interactive forces between and within the aggregate particles. A single shear modulus, G_t is used to characterize the precipitate:

$$G_t = \tau/\gamma_t \tag{1}$$

where τ is the applied shear stress and γ_t is the resulting deformation or strain within time, $t = 300$ s. For the 8.3 Nm^{-2} curve (Fig. 1) $G_t \approx 3.8$ Nm^{-2} while for the 0.1 Nm^{-2} curve, $G_t \approx 0.03$ Nm^{-2}. This shear modulus incorporates three

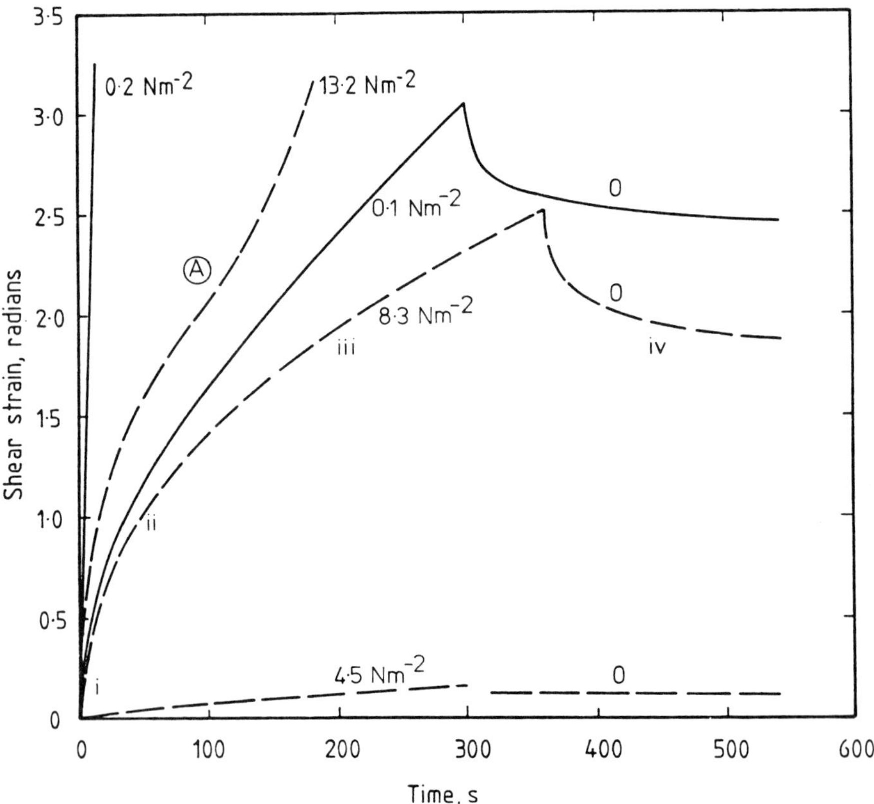

Fig. 1 – The influence of precipitation reactor configuration on the creep rheology of isoelectric soya protein precipitate. Reactor type: ——, batch-stirred tank, 200 l, $\bar{G} = 500$ s^{-1}, $t = 1800$ s [4]; – – – . Continuous tubular ($\bar{G} = 340$ s^{-1}, $\bar{t} = 25$ s). Protein precipitate concentrated by gravity sedimentation to 135 kg m^{-3}. Measurement geometry: cone and plate.

effects. Taking the 8.3 Nm^{-2} curve as an example, region (i) indicates an elastic stretching of the sample with a rapid increase in strain. This is followed in region (ii) by breakage and reformation of the elastic bonds between the structural units and, for a constant applied shear stress, is characterized by a decrease in the rate of shear strain ($\dot{\gamma}_t$) with time. In region (iii) there is viscous flow where $\dot{\gamma}_t$ is constant and where the structural network has been disrupted. Removal of the applied shear stress, region (iv), allows flow reversal due to the elastic behaviour of the structure.

The creep flow curves for the batch-prepared precipitates are consistent with the observed compact and regular-shaped structure of the aggregates (4). The considerable rate of deformation for the application of small shear stresses indicates little inter-aggregate structure although the elastic behaviour, on removal of applied shear stress, does suggest some degree of bonding between the aggregates. For precipitate prepared in the tubular reactor, the increased inter-aggregate contact resulting from their irregular shape is reflected in the larger applied shear stress needed to achieve equivalent deformation compared to batch-prepared material. At point A in the deformation curve for a stress of 13.2 Nm^{-2}, it is probable that the structured network of the suspension has broken down giving a change from viscoelastic- to viscous-dominated flow.

The effect of precipitated protein concentration on the shear modulus is shown in Fig. 2. Sharp increases in shear modulus are observed in the range 200 to 250 kg m^{-3} depending on the level of applied shear stress. The lower values of shear modulus observed at higher applied shear stresses are a result of the greater degree of breakdown of precipitate structure over the test period. The tendency at high protein precipitate concentrations, 300 kg m^{-3}, towards similar values of shear modulus independent of the applied shear stress is indicative of a solid structure with considerable elastic properties as would be expected from a high level of aggregate interaction.

4.2 Viscous flow characteristics

Creep rheology measurements indicate structural characteristics of protein precipitate sediments but low and high shear rheology measurements are also necessary to characterize the properties of flowing suspensions for design purposes. The effect of shear break-up of precipitate aggregates is not significant for exposure to low shear rates ($\dot{\gamma} < 200\ s^{-1}$) for short times (up to 600 s) [3].

Figure 3 shows the variation of apparent viscosity with shear rate for various concentrations of precipitate suspensions and sediments. Consistent trends are observed between apparent viscosities measured in the low shear rate range using a cup-and-bob rheometer and in the high shear rate range using a capillary rheometer. Observed differences are probably due to shear break-up in the capillary rheometer.

Pseudoplastic behaviour is observed for all precipitate concentrations studied with a trend towards Newtonian behaviour at increased shear rates for

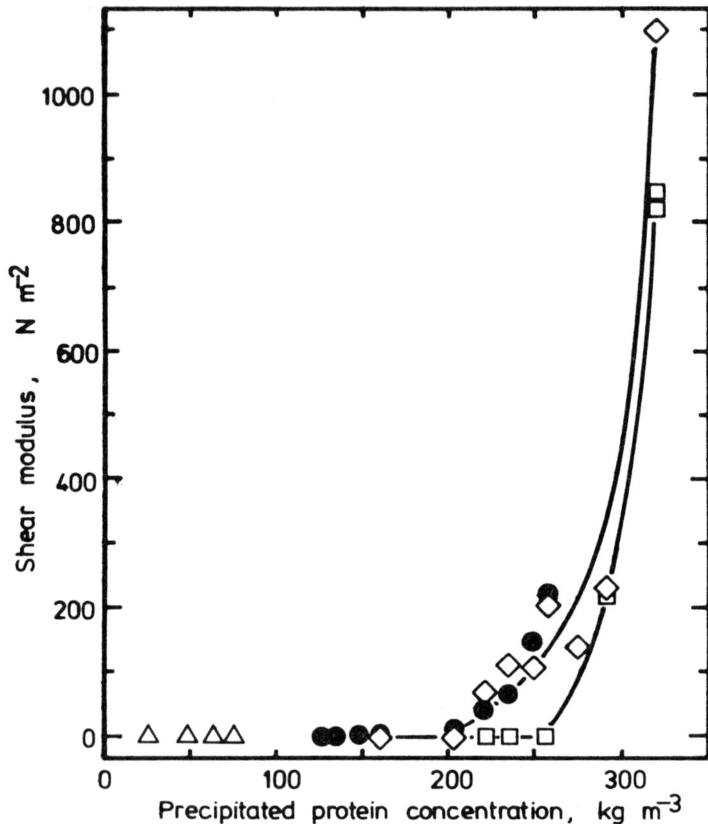

Fig. 2 – The relationship between shear modulus (equation (1)) and isoelectric soya protein precipitate concentration. Precipitate prepared in a batch-turbine-stirred reactor – see Fig. 1 for ageing conditions. Applied shear stress, Nm^{-2}; △, 0.02; ●, 0.2; ◇, 1.0; □, 10.0.

low concentration suspensions. The decrease in apparent viscosity on increasing shear is consistent with the disentanglement of the aggregates. The most concentrated suspensions examined, 202 and 215 kg m^{-3} precipitated protein, are prepared by different methods, centrifugation and ultrafiltration respectively. The ultrafiltration membrane retains the soluble protein, the final retentate containing 40 kg m^{-3} soluble protein compared with 7 kg m^{-3} for the sediment prepared by centrifugation. However, the soluble protein will only play a minor role in determining the retentate or sediment rheology [8] at the temperature studied, 298 K.

An important property of a protein precipitate with respect to recovery is its resistance to movement from rest. The yield stress, τ_0, of the material (Fig. 4) is determined using the Bingham plastic model:

$$\tau = k\dot{\gamma} + \tau_0 \qquad (2)$$

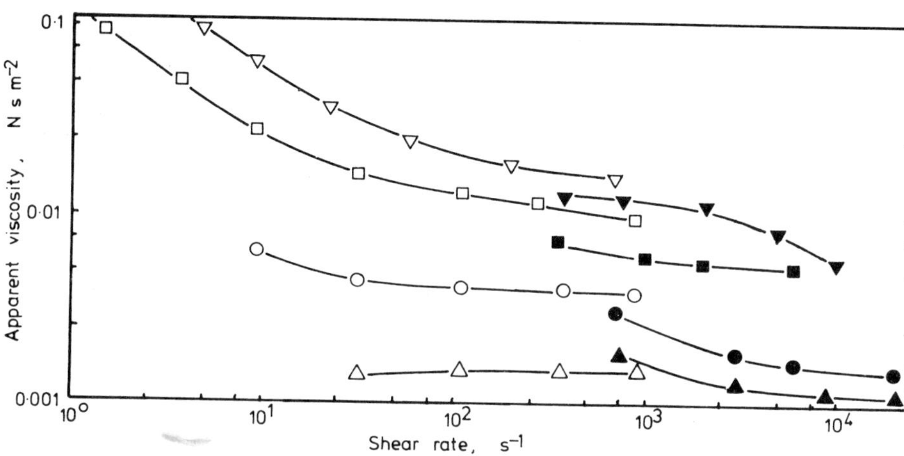

Fig. 3 – The viscosity of precipitate suspensions as a function of shear rate and concentration. Open symbols, cup-and-bob rheometer; closed symbols, capillary rheometer. Precipitated protein concentration, kg m^{-3}; △, 25; ▲, 35; ○, 74; ●, 95; □, 118; ■, 170; ▽, 202; ▼, 215. Precipitate suspensions concentrated by; ○, ●, □, ■, gravity sedimentation; ▽, batch centrifugation; ▼, ultrafiltration, PM 50. Precipitate particle sizes and standard deviations, sd, μm; △, ○, □, ▽, $d_m = 8.5$, $d_{90} = 6.2$, $sd = 3.3$; ▲, $d_m = 4.5$, $d_{90} = 3.2$, $sd = 1.3$, ●, $d_m = 4.0$, $d_{90} = 2.8$, $sd = 1.3$; ■, $d_m = 6.9$, $d_{90} = 3.3$, $sd = 1.9$; ▼, $d_m = 7.6$, $d_{90} = 4.5$, $sd = 1.5$. Average measurements for increasing and decreasing shear rate cycles are shown for the cup-and-bob rheometer. All measurements carried out at 298 K.

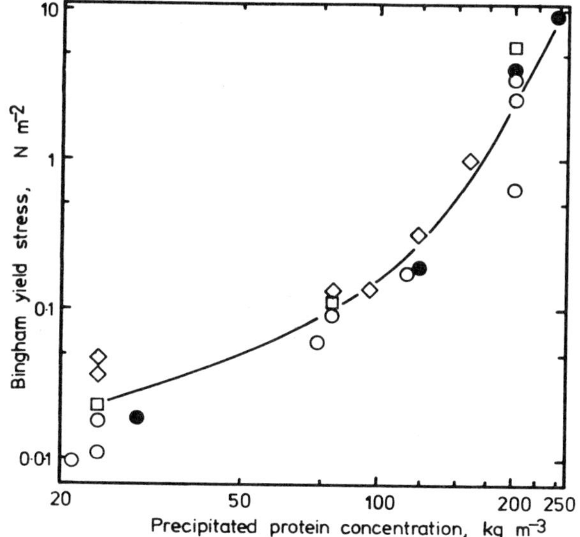

Fig. 4 – The effect of protein precipitate concentration on the suspension Bingham yield stress. Aggregate mean size, d_m; ○, ●, 6–8 μm; □, 13–16 μm; ◇, 22–38 μm. Yield data from creep rheology for $G_t = 0.001$ Nm^{-2}, ●.

fitted to the data obtained at low shear rates ($< 1000\ s^{-1}$). Good agreement is observed between yield stress values and the applied shear stress needed to give a shear modulus of $10^{-3}\ Nm^{-2}$. As with the yield stress, this level of shear modulus marks the onset of viscous flow; that is where the material flows with little resistance arising from its structural make-up.

5. PRECIPITATE RECOVERY AND DEWATERING

5.1 Centrifugation

Industrial-scale centrifugal recovery of protein precipitate generally requires the use of continuous machines. These include nozzle- and intermittent-discharge, disc-stack centrifuges and scroll-discharge, solid-bowl centrifuges. There are few rheological problems in the presentation of the precipitate feed stream to the centrifugal field other than those associated with shear break-up of the aggregates. This may occur in pumps and valves and in the centrifuge feed zone [3, 9]. The main rheological problem is concerned with the movement of the precipitate sediment within the centrifuge bowl and during discharge. For all machines a high apparent viscosity of the flowing suspension will impair sedimentation. A high value of apparent viscosity of the sediment will hinder sediment flow from continuous or intermittent-discharge machines. A substantial yield stress or shear modulus will also hinder flow from rest at the start of the discharge cycle for an intermittent-discharge machine.

For isoelectric soya protein precipitates, the particle size and density differences of ~ 150 to 250 kg m^{-3} [5] are such that the disc-stack centrifuge studied (model SAMR 3036, intermittent-discharge, Westfalia Separator Ltd, max. RCF 9215) gave a very high separation efficiency at feed rates up to $1.7 \times 10^{-4}\ m^3\ s^{-1}$ and for feed concentrations of 35 kg m^{-3} precipitated protein. Indeed the performance at this flow rate leads to either excessive dewatering of the sludge or demands solids discharge at such short intervals as to be an inefficient use of the centrifuge. This is the case both with respect to the relative times of clarification and sludge discharge and with regard to premature wear of the centrifuge motor. Excessive dewatering also can lead to a non-homogeneous sediment which is difficult to handle in subsequent processing.

For a readily sedimented protein precipitate the use of a horizontal bowl scroll-discharge centrifuge (Sharples P600, Pennwalt Ltd., Camberley, Surrey, UK, max. RCF 2130) is more appropriate. Two indices are used to characterize performance; the dewatering or solids dry weight of the sediment and the clarification number or reduced efficiency, E'_T:

$$E'_T = (E_T - R)/(1 - R) \qquad (3)$$

where R is the underflow to throughput volume ratio and E_T is the total efficiency, equal to the mass separated per unit mass of feed. To ensure consistent conditions for comparative studies and a continuous solids discharge the centrifuge was operated with a flooded beach.

The creep rheology data (Fig. 1) show a suspension of tubular reactor prepared precipitate to be more resistant to deformation compared to the batch-prepared precipitate; that is higher shear stresses are necessary to disrupt the intra-aggregate bonds in the precipitate suspension prepared in the tubular reactor. As a consequence the aggregates making up this precipitate are more susceptible to shear-induced break-up as compared with the batch prepared precipitate [4]. Thus the effects of shear stresses on a precipitate suspension or sediment within a centrifuge bowl are characterized by suspension or sediment deformation and aggregate disruption, the former relating to disruption of intra-aggregate bonds, the latter to break-up of inter-aggregate bonds.

The action of the scroll within a scroll discharge centrifuge exposes the sediment to high shear stresses. These disrupt the sediment intra-aggregate structure causing viscous flow through the bowl and allowing drainage. Sediment deformation will occur for both tubular and batch-prepared material, the more irregular shape and greater susceptibility to shear disruption of the former allowing better drainage and a drier solid as compared with the regular and compact structure of the latter (Fig. 5). The small increase in solids dry weight

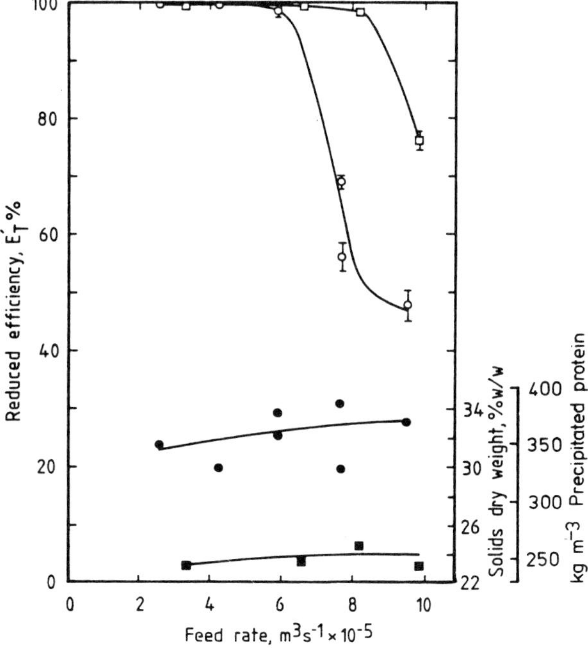

Fig. 5 – Comparison of the clarification and sludge solids content of precipitate prepared in continuous tubular and batch-stirred tank reactors. The centrifuge was a scroll discharge Sharples P 600 (bowl speed 83 rps, differential scroll speed 0.58 rps). Total protein concentration in feed, 35 kg m^{-3}. Reactor used for precipitate preparations; □, ■, 200 l batch-stirred tank; ○, ●, continuous tubular reactor. Open symbols, efficiency; closed symbols, sediment concentration.

with increased centrifuge throughput is consistent with a limited amount of drainage caused by greater compressive forces arising from a deeper sediment layer. A greater resistance to shear break-up in the feed zone is shown by the batch prepared precipitate which has 30% greater throughput (at 98% efficiency) compared with the more fragile precipitate prepared in the tubular reactor (Fig. 5).

The dominance of the shear-induced drainage mechanism of sediment dewatering is further supported by studies of scroll centrifuge operation with feeds of different protein concentrations (Fig. 6). Operating at approximately full reduced efficiency there is a 40% lower output of sediment for a concentrated feed as compared with the more dilute feeds. Thus the maximum feed rate at the

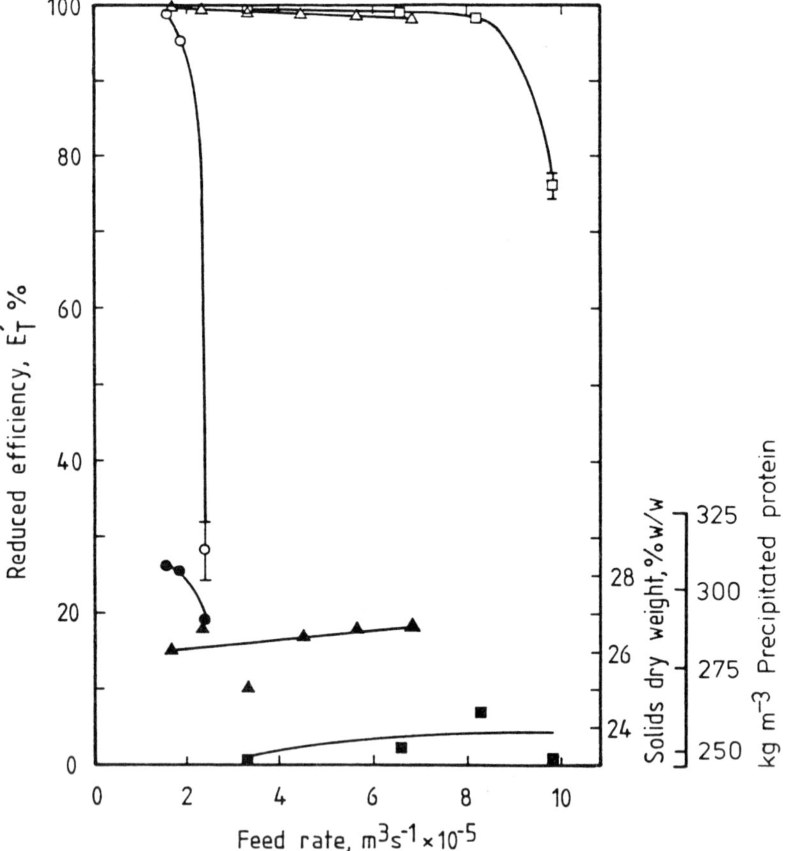

Fig. 6 – The effect of feed concentration on the clarification efficiency and sludge solids content of a scroll discharge centrifuge. The centrifuge was a Sharples P 600 (bowl speed 83 rps, differential scroll speed 0.58 rps). Precipitate prepared in a batch-stirred tank. Precipitated protein concentration; □, ■, 35 kg m^{-3} total protein; △, ▲, 45 kg m^{-3} total protein; ○, ●, 92 kg m^{-3} total protein. Open symbols, efficiency; closed symbols, sediment concentration.

high feed concentration prior to a serious decline in performance is ~ 2×10^{-5} $m^3\ s^{-1}$ as compared with 9×10^{-5} $m^3\ s^{-1}$ at lower concentrations and the corresponding sediment flow rates are 4×10^{-6} $m^3\ s^{-1}$ and 7×10^{-6} $m^3\ s^{-1}$ respectively. Hindered settling of the precipitate aggregates is probably limiting the centrifuge capacity at the highest feed concentration studied. For the same rotational speeds of the centrifuge bowl and scroll the increased feed concentration is accompanied by an increase in the sediment total solids content. This is evidently because the lower throughput limit at higher concentration results in a reduced sediment height in the bowl which allows better drainage.

The pseudoplastic nature of the sediment (Fig. 3) will lead to lower apparent viscosities at higher scroll speeds with the accompanying structural changes which allow release of interstitial supernatant. This results in increased sediment dewatering at higher scroll differential speeds (Fig. 7). Also the lower sediment heights at increased scroll speeds allows easier drainage. The reverse result would be expected if compressive forces were mainly responsible for dewatering the sludge.

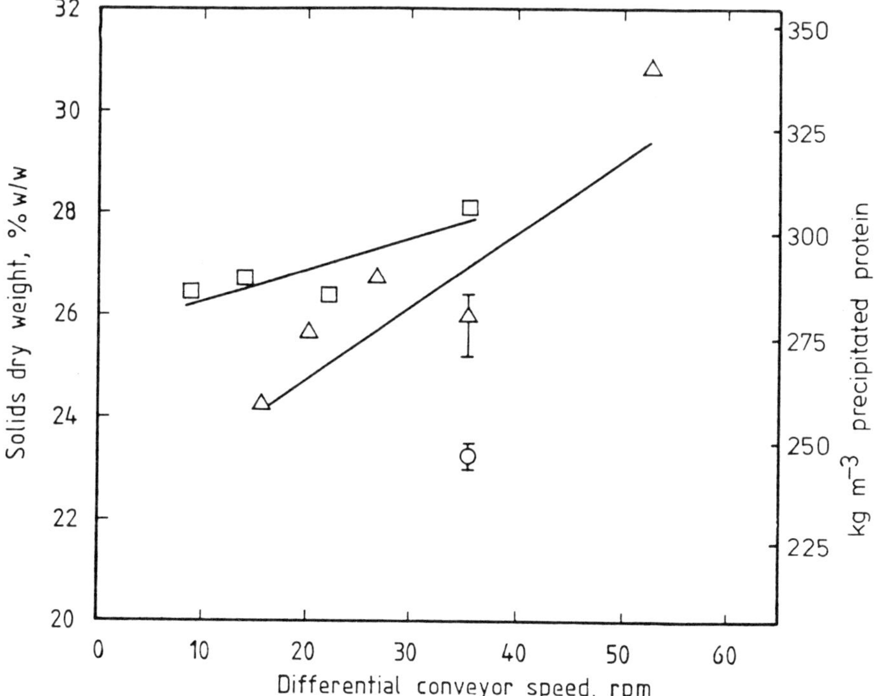

Fig. 7 – The influence of the differential conveyour speed on the sludge solids content of precipitate recovered in a scroll discharge centrifuge. The centrifuge was a Sharples P 600 (bowl speed 83 rps). Precipitate prepared in a batch-stirred reactor. Precipitated protein concentration and feed rate to centrifuge; ○, 35 kg m^{-3} and 3.3×10^{-5} $m^3\ s^{-1}$; △, 45 kg m^{-3} and 3.3×10^{-5} $m^3\ s^{-1}$; □, 92 kg m^{-3} and 1.8×10^{-5} $m^3\ s^{-1}$.

The extent of dewatering achieved in the disc-stack or scroll discharge centrifuges is related to the sediment rheological behaviour at high solids content. Solids dry weights ranging from 23 to 35% w/w (250 to 390 kg m^{-3}) protein precipitate are obtained in the scroll discharge centrifuge. This is in a region of concentration where the sediment exhibits a high shear modulus (Fig. 2) and a significant plastic behaviour (Fig. 4). The rheopectic behaviour at these protein precipitate concentrations [4, 10] indicates the formation of a sediment structure which would prove to be difficult to dewater. This behaviour probably sets the upper limit which may be achieved in terms of sediment dry weight. De (11) has reported sediment concentrations of ~ 210 kg m^{-3} precipitated isoelectric soya protein for an intermittent-discharge disc-stack centrifuge. At this precipitate concentration the sediment exhibits a considerable shear modulus (Fig. 2) or yield stress (Fig. 4) and these properties will control its ability to flow from a state of no shear, that is, just prior to discharge.

5.2 Ultrafiltration

Figure 8 shows the flux versus concentration behaviour of a 43 kg m^{-3} precipitated protein during batch concentration using a Romicon HF ISSS pilot

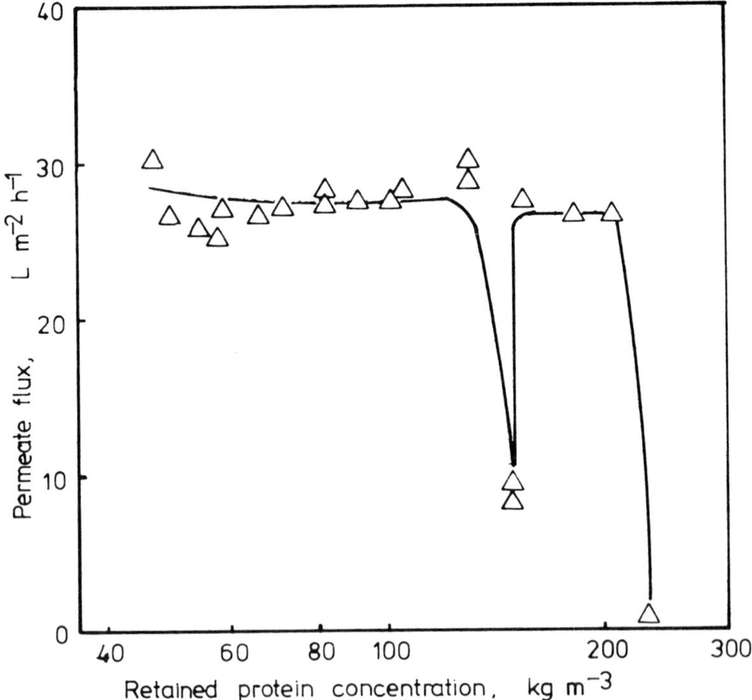

Fig. 8 – Permeate flux during concentration of a 0.17 m^3 batch of 43 kg m^{-3} precipitated protein, pH 4.6, total initial protein concentration 47 kg m^{-3}, average permeate concentration 2.5 kg m^{-3}, average soluble protein rejection 0.6.

rig (Romicon Inc., Woburn, Mass., USA) with PM 50 hollow fibre membrane of effective membrane area 1.4 m^2. There are two main features of this curve; firstly, a flux which is independent of retained protein concentration and secondly a very rapid decline to zero flux over a small range of protein concentration.

Devereux *et al.* [7] have shown that the constant flux versus concentration behaviour is due to a reduction in concentration polarization achieved by isoelectric precipitation of the soluble protein. In the concentration range studied here the precipitated material does not exhibit the flux independent of transmembrane pressure which is associated with concentration polarization [12]. The rig was operated initially with an inlet pressure to the membrane cartridge of 2.1 bar and an outlet pressure of 1.4 bar. At a retained protein concentration of 150 kg m^{-3} the flux declined rapidly.

The inlet pressure was then increased to 2.8 bar and the outlet pressure reduced to 0.3 bar, thus maintaining a similar transmembrane driving force whilst increasing the recirculation driving pressure along the fibre bundle. The effect of the change was to re-establish a permeation rate of 28 l m^{-2} h^{-1}. This rate was maintained until a further rapid decline occurred at 220–230 kg m^{-3} retained protein.

Behaviour of this sort is consistent with a process which only requires that there is some bulk movement of retentate for permeation to occur, and where the rate of permeation is controlled by interaction between the recirculating material and the membrane surface [7]. The maximum concentration obtainable during batch concentration is governed by the rheology of the retentate and, given the relevant rheological properties, a process can be designed to achieve any final concentration, provided the precipitate suspension can be pumped along the fibres without exceeding the maximum inlet pressure rating of the fibre cartridge. Figure 9 illustrates this; as the precipitate concentration increases so the recirculation rate obtainable for a given longitudinal pressure drop decreases. Calculation shows that the pressure drop in associated pipework, even with lower average shear rates, is not significant compared to that within the fibre bundle. Thus, it is the rheology of the suspension within the fibres and header which controls the recirculation rate. The observed rapid decline in permeation rate over a short period of time suggests that below a critical recirculation rate there may be some structure formation within the suspension particularly in local zones of low shear.

The lowest shear rates will be at the entrance to the fibre bundle, where the diameter of the header is equal to the cartridge diameter of 8 cm. At a recirculation rate of 20 l min^{-1} the average shear rate in a pipe of 8 cm diameter is about 3 s^{-1}. At these very low shear rates and at high precipitate concentration considerable structure formation is to be expected with a corresponding increase in apparent viscosity (Fig. 3). If the flow is not distributed evenly across the fibre bundle, local shear rates may drop well below this average value, causing flow to

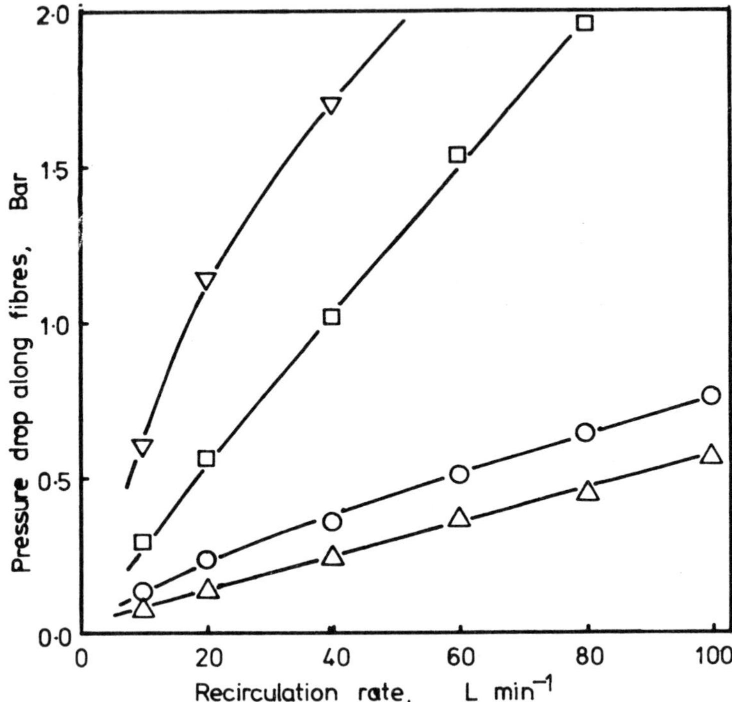

Fig. 9 – Pressure drop as a function of recirculation rate for Romicon HF 1SSS pilot rig with 1.1×10^{-3} m i.d. PM 50 hollow fibres, length 0.7 m. Precipitated protein concentration; △, 35 kg m^{-3}; ○, 95 kg m^{-3}; □, 170 kg m^{-3}; ▽, 215 kg m^{-3}.

stop within some of the fibres. Therefore the design of the header is of great importance. In the present study a non-standard type was used. The expansion from inlet pipe diameter, 3.8 cm, to membrane cartridge diameter was achieved in a distance of 5 mm, compared to a typical value of 2 cm, thus making fibre blockage more likely. This is because of the tendency towards poor flow distribution across the header giving rise to zones of low shear, and hence retentate of high apparent viscosity, in regions remote from the inlet pipe. With a standard header, a higher concentration can be reached before rapid flux decline is observed, owing to the more uniform flow distribution. Once a number of fibres become blocked, the linear velocity in the remaining unblocked fibres will rise, causing an increase in longitudinal pressure drop. If the inlet and outlet pressures are held constant this can only lead to further reduction in recirculation rate, which in turn will cause further fibres to become blocked. If the inlet pressure is increased more force is available to overcome the yield stress in the blocked fibres, and the above trends will be reversed. The initial permeation rate will be re-established, until the blockage mechanism starts again at higher retained protein concentration. Figure 10 demonstrates these effects.

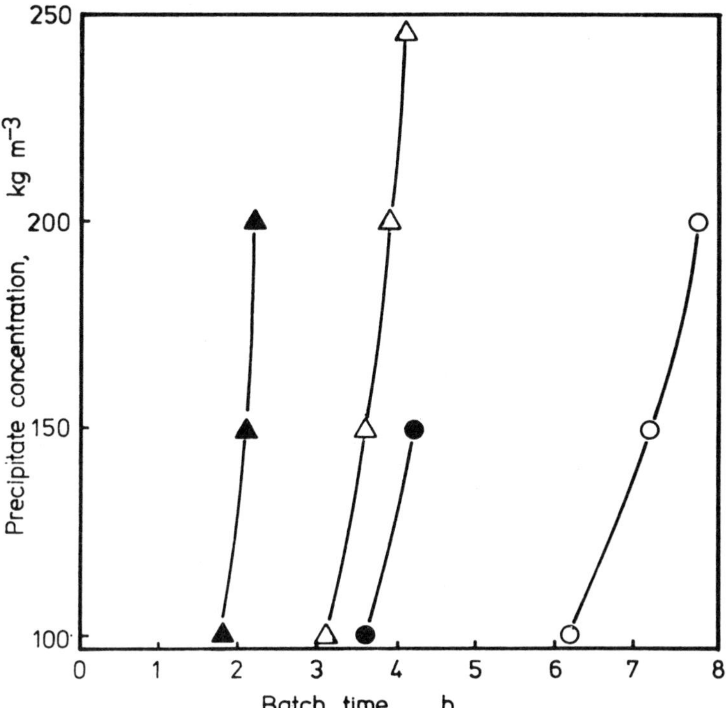

Fig. 10 – Operating constraint diagram for Romicon HF 1SSS pilot ultrafiltration rig, batch volume 0.2 m^3, initial precipitated protein concentration 35 kg m^{-3}, precipitate rejection, $\sigma = 1.0$, PM 50 membrane. Open symbols, 1.5 m^2 membrane area, fibre length 0.7 m; closed symbols, 2.65 m^2 membrane area, fibre length 1.25 m. Average transmembrane pressure; ○, ●, 0.8 bar; △, ▲, 1.7 bar.

6. EQUIPMENT DESIGN FOR RECOVERY OF PRECIPITATES

The rheological studies outlined have been used largely to explain observed behaviour whilst recovering protein as a precipitate in centrifuges and ultra-filters. Of perhaps greater importance are the implications of such studies for equipment design. Where performance is limited primarily by the rheological properties of the precipitate, it is desirable to suggest methods by which these properties may be modified to improve performance, or to design equipment in such a way as to minimize the rheological limitations.

Table 1 summarizes some of the comparisons which can be made between the types of separation equipment studied here. It can be seen that most of the rheological problems occur at high concentration, and consequently any design must also consider the design objective for the separation step. Maximum obtainable protein concentration is not necessarily the only objective, and if further processing is required, such as spray drying, then the desired concentration may be set by precipitate handling characteristics in equipment further downstream.

Table 1

Comparison of separation equipment based on final concentration and capacity

Separation equipment	Rheological problem with feed containing 35 kg m^{-3} precipitate	Maximum concentration limitation	Further handling characteristics	Maximum throughput limitation
Continuous scroll centrifuge	Some precipitate break-up in feed zone	Failure of shear-induced dewatering	Homogeneous sediment	Sedimentation rate
Intermittent-discharge, disc-stack centrifuge	As above	Plastic flow properties of sediment	Non-homogeneous sediment which may be difficult to re-dissolve	Sedimentation rate; discharge frequency
Batch PM 50 ultrafilter	Some precipitate break-up during recirculation	Bulk viscosity in fibres	Homogeneous retentate	Membrane-particle interaction

For scroll discharge centrifugation the final sediment concentration may not be significantly influenced by the throughput (Fig. 5). The technique used for precipitate preparation determines the degree of sediment dewatering. Also changes in the differential scrolling speed (Fig. 7) give an operational flexibility on discharge solids concentration with a range of concentrations from 250 to 350 kg m^{-3}.

For ultrafiltration, similar control over final concentration may be achieved by a combination of operational and rheological manipulation. The maximum concentrations reached during processing, illustrated in Fig. 10, represent a limit set by the bulk rheology of the concentrated material. The use of a membrane cartridge with a large effective area by virtue of greater fibre length allows faster processing times at the expense of lower maximum precipitate concentration. Higher concentrations may be possible with even shorter fibre lengths, although these are unlikely to match the maximum concentration obtainable from the scroll centrifuge.

The detailed choice and design of equipment for particular downstream recovery processes will demand knowledge of the individual protein, but it has been shown that the performance characteristics of both scroll centrifuges and ultrafilters in concentrating precipitated protein can be influenced by manipulation of the rheology of the concentrated precipitate.

ACKNOWLEDGEMENT

The financial support given to N. Devereux by the Science and Engineering Research Council is gratefully acknowledged as is the assistance by M. Twineham in the ultrafiltration experimentation.

NOMENCLATURE

d_m	mean particle size	μm
d_{90}	particle size at 90% by weight cumulative oversize	μm
E'_T	reduced efficiency (equation (3))	(–)
$\bar{G}$	mean velocity gradient	s^{-1}
G_t	shear modulus (equation (1))	Nm^{-2}
k	consistency index (equation (2))	$Ns\ m^{-2}$
sd	standard deviation	μm
t	residence time	s
$\bar{t}$	mean residence time	s
γ	shear strain	(–)
$\dot{\gamma}$	shear rate or rate of shear strain	s^{-1}
τ	applied shear stress	Nm^{-2}

REFERENCES

[1] Bell, D. J., Hoare, M., Dunnill, P., 'Downstream processing' in *Advances in Biochemical Engineering/Biotechnology.* Ed. A. Fiechter, Springer-Verlag, Berlin, vol. 26, 1–72 (1983).
[2] Virkar, P. D., Chan, M. Y. Y., Hoare, M., Dunnill, P., *Biotechnol. Bioeng.,* **24**, 871–887 (1982).
[3] Hoare, M., Narendranathan, T. J., Flint, J. R., Heywood-Waddington, D., Bell, D. J., Dunnill, P., *Ind. Eng. Chem. Fundam.,* **21**, 402–404 (1982).
[4] Bell, D. J., Dunnill, P., *Biotechnol. Bioeng.,* **24**, 2319–2336 (1982).
[5] Bell, D. J., Heywood-Waddington, D., Hoare, M., Dunnill, P., *Biotechnol. Bioeng.,* **24**, 127–141 (1982).
[6] Bell, D. J., Dunnill, P., *Biotechnol. Bioeng.,* **24**, 1271–1285 (1982).
[7] Devereux, N., Hoare, M., Dunnill, P., to be published (1983).
[8] Hermansson, A.-M., *J. Texture Studies,* **5**, 425–439 (1979).
[9] Bell, D. J., Brunner, K.-H., *Filtration and Separation,* **20**, 274–286 (1983).
[10] Hoare, M., Bell, D. J., Dunnill, P., *An. N.Y. Acad. Sciences,* accepted for publication (1983).
[11] De, S. S., 'Technology of production of edible flours and protein products from soy bean', *Agricultural Services Bull.,* No. 11, Food and Agric. Org. United Nations, Rome, pp. 112–125 (1974).

[12] Blatt, W. F., Dravid, A., Michaels, A. S., Nelson, L., in *Membrane Science and Technology*, Ed. J. E. Flynn, Plenum Press, New York, pp. 47–97 (1970).

CHAPTER 12

Particle Counting and Distribution Shape Assessment in Coagulation Processes by Laser Light Scattering

P. G. CUMMINS, A. L. SMITH, E. J. STAPLES and **L. THOMPSON,** Unilever Research, Port Sunlight Laboratory, Quarry Road East, Bebington, Wirral, Merseyside, UK

ABSTRACT

A modified form of a flow ultramicroscope is described which is capable of yielding particle number concentrations and, in addition, size distribution in the range 0.1 μm to 5 μm for particles of spherical and other simple shapes.

Such an instrument can be used to follow quantitatively particle flocculation and deposition process. Data and histograms are reported here for the evolution of aggregate size distributions during the flocculation by electrolyte of latex dispersions.

1. INTRODUCTION

In this work we demonstrate an instrument for measuring particle number concentrations and size distributions in colloidal dispersions. The technique involves measuring the light scattered from individual particles as they pass through a laser-illuminated volume. The dynamic range of the pulses detected is increased using a feedback mechanism within the laser and the experiment may be performed in a coherent (scattering) or incoherent (fluorescence) detection mode. The data are displayed as a frequency distribution of pulse heights.

Light scattering intensity is greatly dependent on the detection angle chosen. The relationship between particle size and scattered light intensity at any angle is accessible for spheres and simple shapes via the Mie theory, though the scattered intensity does not increase monotonically with particle size at all angles.

To retain a monotonic increase in scattering cross section with size, small angles (or small refractive index increments) should ideally be used. In the initial version of this instrument (described elsewhere [1]) a large collection solid angle, centred around 90°, was used. This has the effect of damping out fluctuations which would result from a single defined angle [2, 3]. When analysing flocculating systems, however, these problems, combined with the effects of aggregate morphology and orientation [4–6] make impossible a clear separation between singlet particles and aggregated species.

In the earlier work [1] this problem was avoided by using a fluorescent latex dispersion. This had the advantage of reducing the angle dependence of the scattering process, removing the background scatter and increasing the dynamic range of the instrument. In this way it was possible to clearly separate singlets from higher aggregates and in the particular aggregation process studied, the results showed deviations from the aggregate size distributions predicted by the Smoluchowski theory in that the distribution contained more singlet particles and more very large aggregates than expected.

A critical test of the Smoluchowski theory [7] requires an ability to analyse aggregate size distributions in some considerable detail. Here we discuss the further development of the fluorescence work as well as the construction and application of a low angle modification.

2. EXPERIMENTAL

2.1 Material and Procedures

Polystyrene latex dispersions were obtained from the Dow Chemical Company or were made in this laboratory using an emulsifier-free method [8]. The fluorescent latex was made in a similar but modified manner in which the fluorescer, perylene (1% w/w on the monomer), was added before the polymerization commenced. Particle sizes were measured by photon correlation spectroscopy (PCS).

Distilled water was obtained by a double distillation procedure involving a 'hot spot' to reduce particle carry-over.

The aggregation of the fluorescent latex was brought about by addition of 10^{-2} mol dm^{-3} La $(NO_3)_3$ (analytical reagent grade). Samples taken at appropriate intervals were diluted by a known factor between 10 and 50 depending on degree of aggregation into 10^{-3} mol dm^{-3} La $(NO_3)_3$. This had the effect of slowing the aggregation so that it remained essentially frozen in the time scale of the measurement.

2.2 The Flow Ultramicroscope

A detailed description of the apparatus has been given elsewhere [1]. It is depicted schematically in Fig. 1. It consists of a flow ultramicroscope in which the particles of the dispersion are injected into a stream of flowing water and hydrodynamically focused so as to pass through a laser beam. The scattering is detected by a photomultiplier and electronically processed as a series of pulse heights.

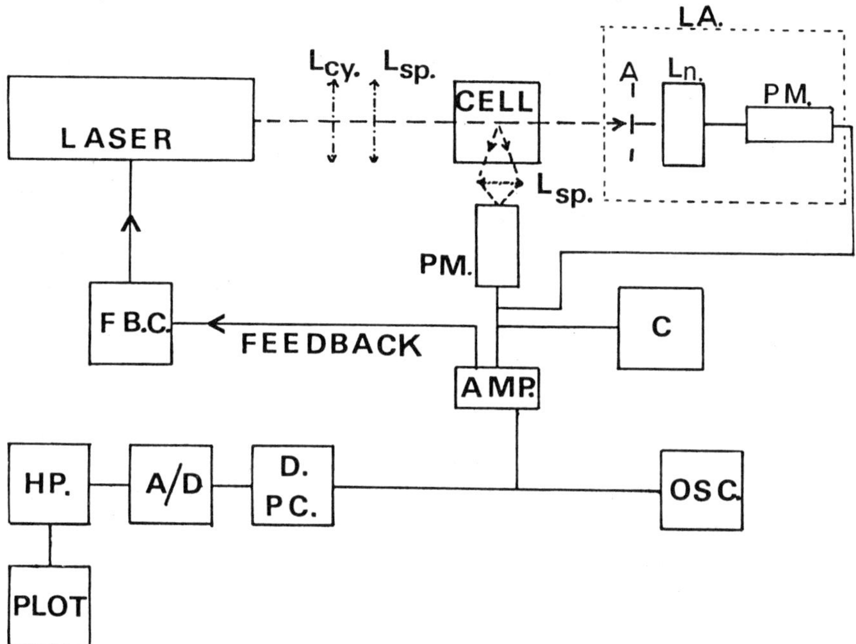

Fig. 1 – Schematic of experimental arrangement. L_{cy}, L_{sp}, cylindrical and spherical lens respectively; PM, photomultiplier; C, pulse counter; FBC, feedback loop control; D/PC, discriminator and peak catcher; A/D, A to D converter; HP, calculator. LA, alternative low angle detection apparatus comprising: A, annulus; Ln, lens assembly; PM, photomultiplier.

Similar instruments have been described by several workers [2, 9–12]. A problem is that the intensity of scattered light is highly dependent on the particle size which means that the dynamic range of the photomultiplier can be exceeded by samples of relatively low polydispersity. Here we have extended the range by a factor of about 700 by using a feedback system from the photomultiplier to the laser.

2.3 Low Angle Modification

The original cell design was preserved and converted to low angle detection by positioning a travelling microscope coaxially with the transmitted laser beam (Ln in diagram).

A long distrance objective, using suitable masking (A in diagram) in the form of an annulus, enabled variation of the scattering angle from 4° through to approx. 45°. The blackened centre of the objective formed the beam stop and also defined the lowest scattering angle available. (Beam divergence was the order of 2°.)

It was possible with this arrangement to form a sharp image of the scattering volume and to use an aperture within the eyepiece to mask (much of) the out-of-focus diffuse scatter arising from the cell walls.

Using non-fluorescent particles it is important to reduce dust contamination in both the sample and in the outer 'hydrodynamic focusing' fluid.

All samples were therefore prepared using Millipore filtered (0.2 μm) solvents and the use of a Waters Associates M6000A chromatography pump permitted the insertion of an in-line high pressure Millipore filter holder (Patent No. XX45 04200) and filter (0.22 μm) whilst maintaining constant flow.

The instrument is interfaced to a Hewlett-Packard 9826 calculator using a BCD interface which permits real time particle counting and histogram production.

3. RESULTS AND DISCUSSION

3.1 Fluorescent Particles

Figure 2 shows the histogram of a fluorescent latex before aggregation. The particle size as measured by PCS was 0.61 μm. Monodisperse particles are essential in such an experiment since the width of the peaks associated with dimers, trimers and n-mers increases monotonically with the value of n, causing the peaks to overlap. The aggregation was carried out in 10^{-2} mol dm^{-3} La $(NO_3)_3$ to

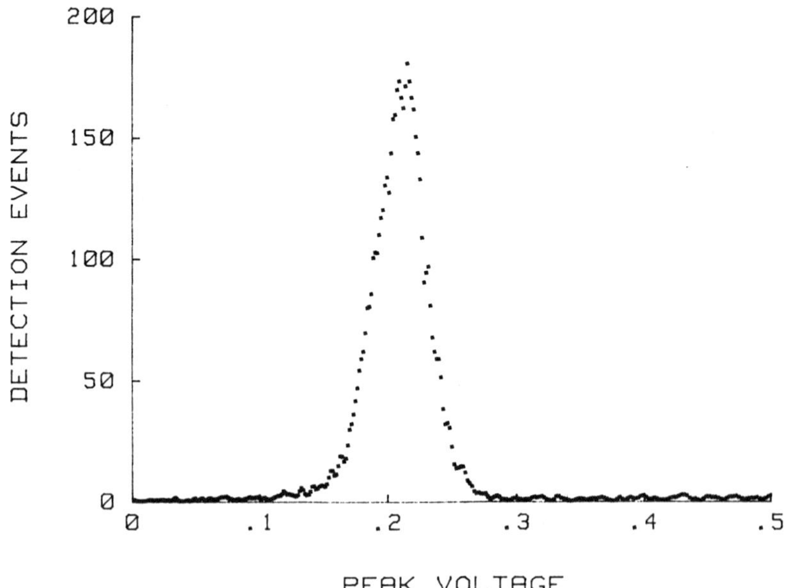

Fig. 2 – Histogram of unaggregated 0.61 μm diameter fluorescent latex.

give a 'rapid' aggregation rate. The time evolution of the aggregate distribution is in the form of histograms and the concentration of all aggregates (including singlets) is determined using the instrument's counting facility. From Fig. 3 the growth of dimers and trimers is clearly seen. The data here are significantly better than in the earlier work, largely because new hardware permits faster data acquisition and many more particles are processed. The improvement allows a direct and unequivocal count of the singlet and doublet concentrations to be made as well as a more uncertain estimation of the triplets.

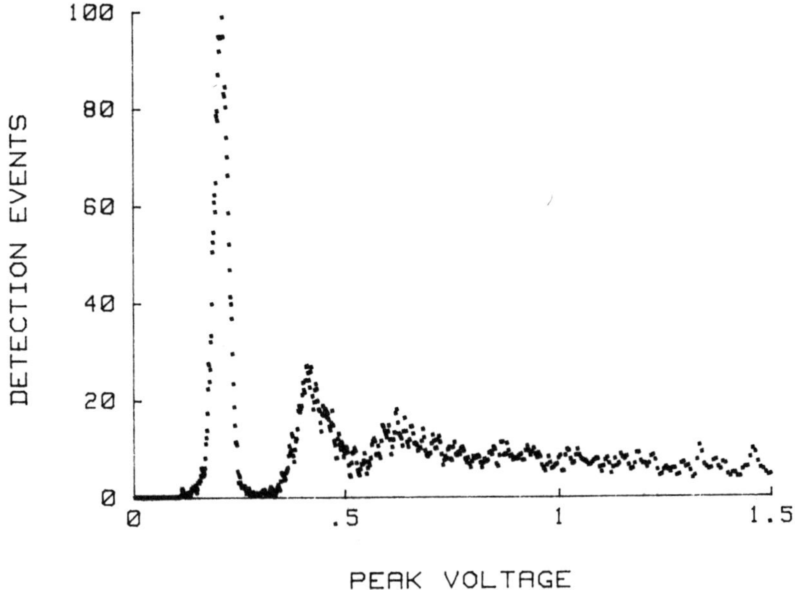

Fig. 3 – Histogram of partially aggregated fluorescent latex showing doublet and triplet peaks.

An analysis of the aggregation data appears in Figs. 4 and 5. Figure 4 shows a 'second order' plot of the reciprocal particle concentration against time. The initial slope gives a rate constant of 2.5×10^{-12} cm^3 s^{-1} which is consistent with the predictions of Smoluchowski modified for viscous interactions [13]. It is a feature of the fluorescent particles we have used, that even in large amounts of electrolyte the depth of the potential energy minimum responsible for aggregation appears restricted, perhaps by surface plasticization of the polymer by the fluorescer. This can result in flocculation equilibria with low levels of aggregation. Here we have used a sufficiently high initial particle concentration that equilibrium favours a high level of aggregation. In effect, aggregation proceeds to 'completion'. In Fig. 5 the time evolution of measured concentrations of singlets,

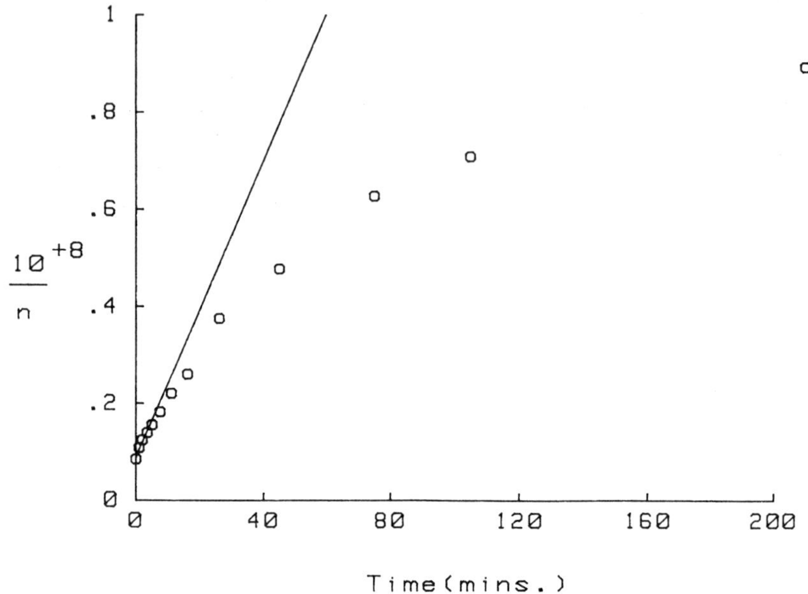

Fig. 4 – Experimental plot showing aggregation of the fluorescent latex. The drawn line is a linear least square fit to the first five points which was used to calculate the initial rate constant.

doublets and triplets are compared at the same overall degrees of aggregation to the predictions of Smoluchowski's treatment. The data are presented in dimensionless units of time t/T where t is the elapsed time and T is the half life of the aggregation process. The deviations from the theoretical treatment are smaller than in the earlier work where a tendency to a condensation type of process, involving large aggregates and singlets, was observed.

The observation of different aggregate distributions from one experiment to another is not entirely surprising. It seems likely that the different distributions which can occur are responsible for the various forms of deviations from second-order kinetics that have been observed [14, 15] for the flocculation process (collisions between different aggregates having different rate constants). Ultimately the distribution itself must be determined by these individual rate constants and the effect that interaction differences between the various aggregates might have on them.

In Fig. 3 the triplet peak overlaps the doublet peak on one side and merges into the higher aggregates on the other. To investigate the prospects of improving this situation the shape of the singlet peak has been fitted to a Gaussian distribution and the theoretical histogram for a 'Smoluchowski' aggregate distribution generated at an appropriate level of aggregation. In Fig. 6 this histogram is compared to the data already presented in Fig. 3. It shows that the theoretical

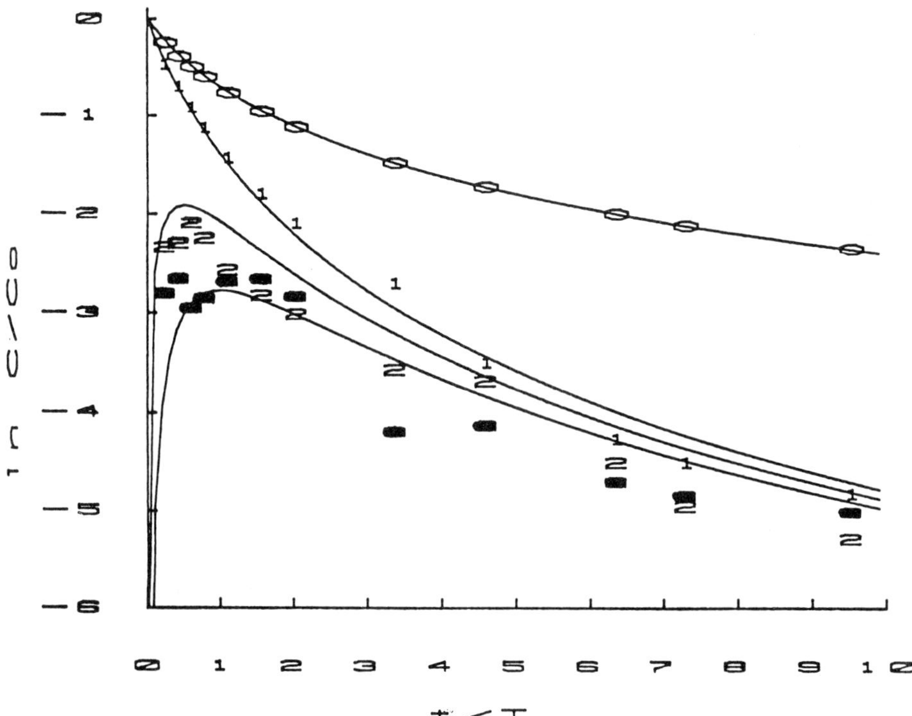

Fig. 5 – Time evolution of singlet (1), doublet (2), and triplet (•) concentrations at the overall levels of aggregation shown by the total aggregate concentration (○). The full lines represent the predictions of von Smoluchowski's treatment.

level of peak overlap is little better than the data. Significant improvement of the peak separation is therefore not possible for this latex because its intrinsic polydispersity is too great.

3.2 Low Angle Scattering

The advantages of low angle scattering have been discussed in the Introduction. The main disadvantage is that the background scatter arising from cell walls and from particles in the carrier fluid is very much greater than with the original 90° orientation. This problem has been largely overcome by changing the carrier liquid's original gravity feed to a pumped system (using a Waters chromatography pump) in which the fluid passes through an appropriate filter (0.22 μm) before reaching the cell.

Figure 7 shows an analysis of a mixture of four monodisperse latices with diameters between 0.2 and 0.6 μm. The separation obtained is rather better than had been achieved with the 90° arrangement.

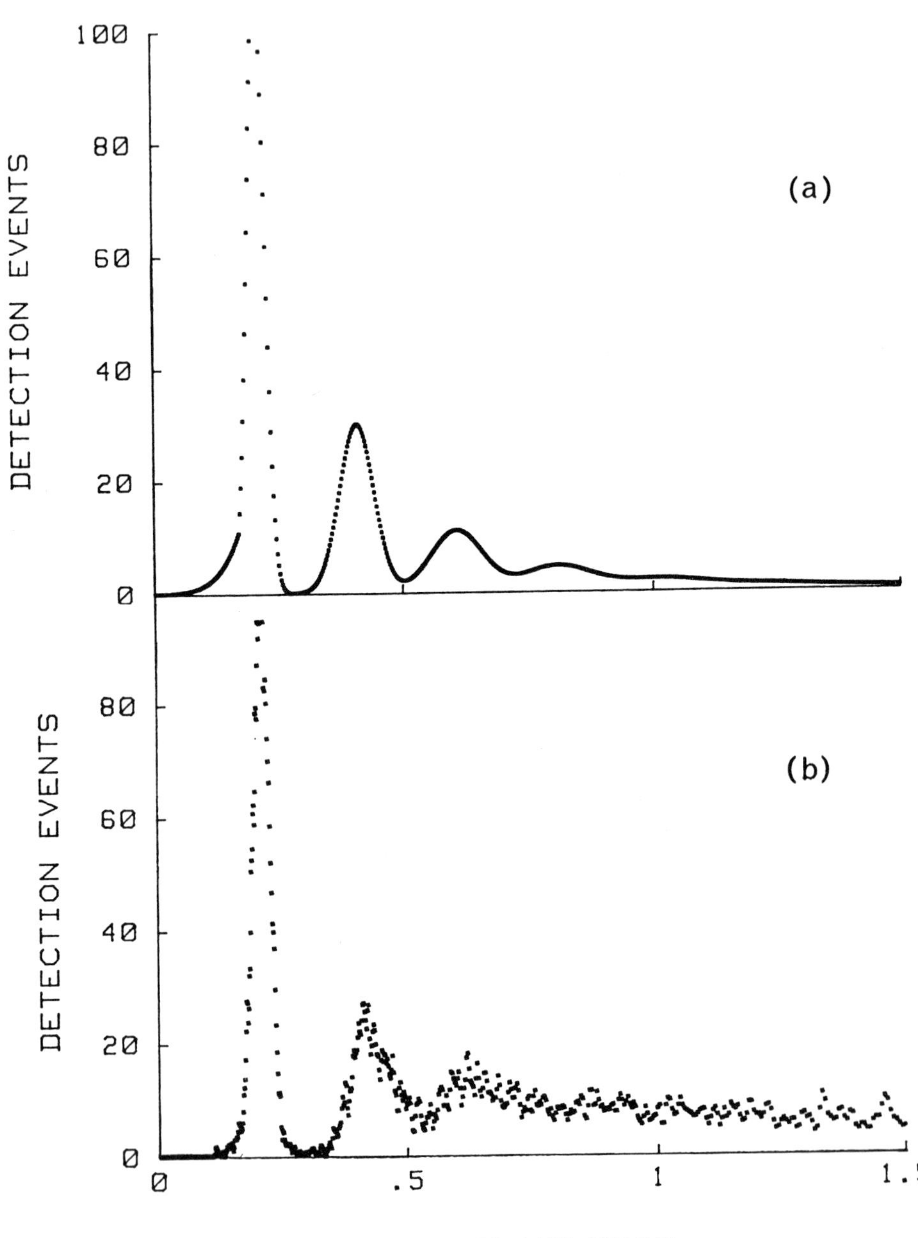

Fig. 6 – Comparison of theoretical (a) and experimental (b) aggregate species separation for fluorescent latex.

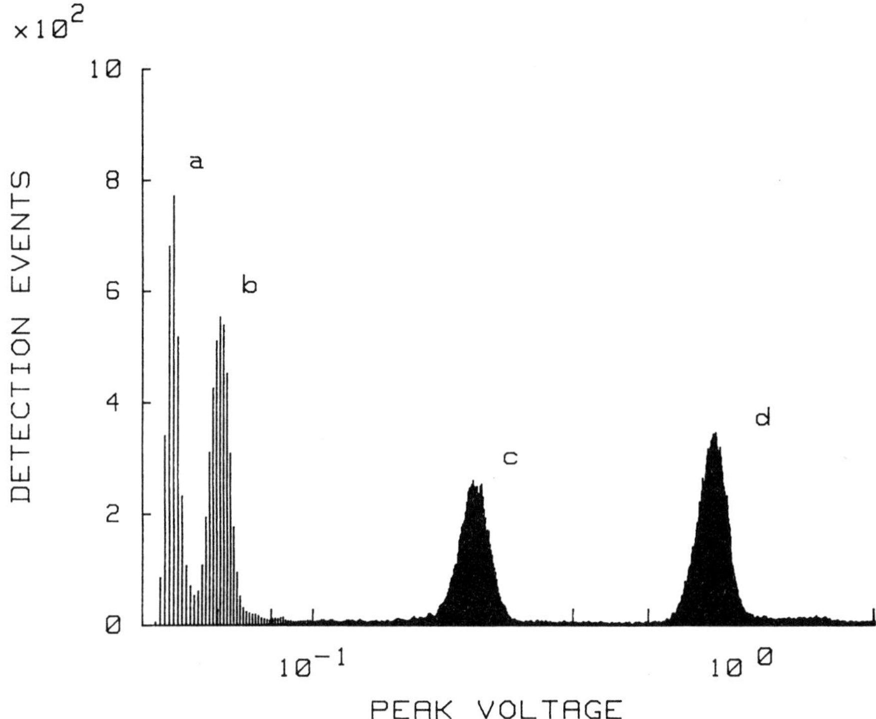

Fig. 7 – Histogram of a mixture of polystyrene latex dispersions of diameter (a) 0.2 μm (b) 0.34 μm (c) 0.5 μm (d) 0.6 μm.

Separation of different aggregate species of a single latex is also greatly improved. Figure 8 shows a partially aggregated polystyrene latex dispersion in which the singlets alone were found by PCS to have a diameter of 0.5 μm. Aggregates containing up to five singlet particles are clearly separable.

The histogram in Fig. 8 is plotted such that the ordinate is linearly related to peak intensity, i.e. the laser feedback correction has been made. This can be used to demonstrate that, at these low angles, the relationship between scattered light intensity and aggregate size is close to that predicted by Rayleigh theory (intensity $\propto$ Mass^2). This relationship is quantitatively expressed in Fig. 9 where a logarithmic plot of relative aggregate mass against intensity is shown to have a least-squares slope close to 2.

In the future it is our intention to conduct a more complete analysis of the aggregation process using the low angle instrument. In addition we intend to employ simultaneous detection at 90° and at low angle. In this way, the particle form factors can be obtained and the pulse height can be more directly related to particle size.

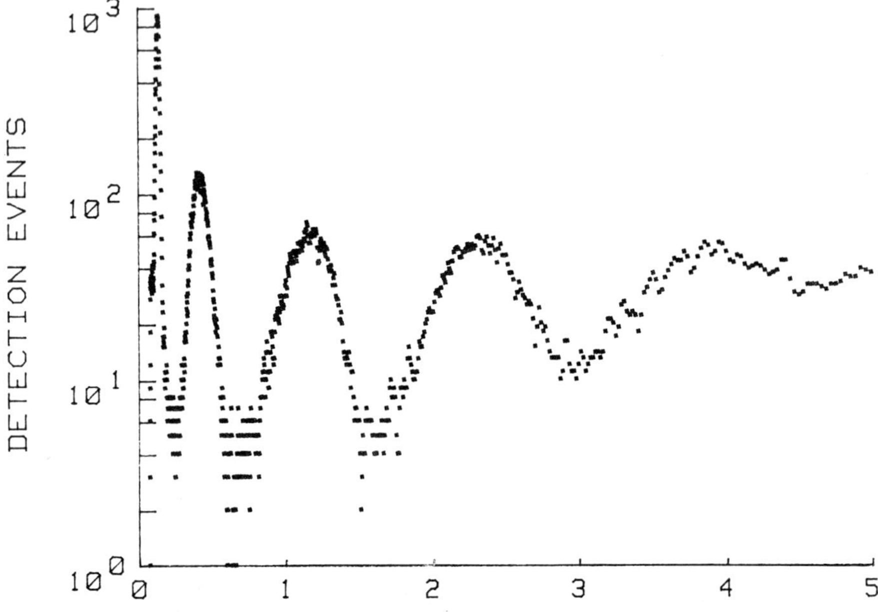

Fig. 8 – Histogram of partially aggregated 0.5 μm polystyrene latex obtained by detection of low angle scatter. Peak voltage corrected for laser feedback.

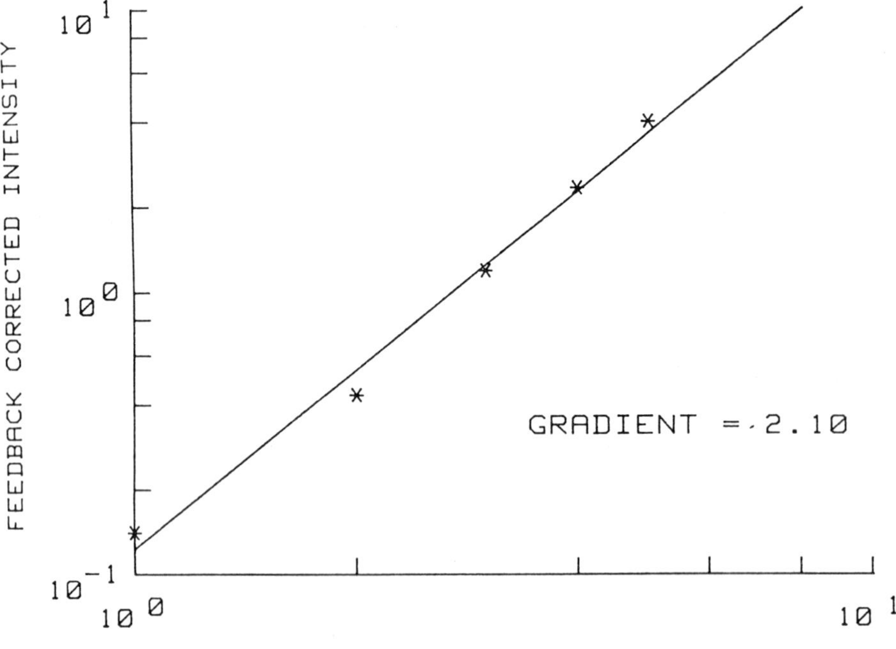

Fig. 9 – Relationship between aggregate size and scattered light intensity.

REFERENCES

[1] Cummins, P. G., Staples, E. J., Thompson, L. G. Smith, A. L. and Pope, L., *J. Colloid Interface Sci.,* **92**, 189 (1983).

[2] Walsh, D. J., Anderson, J., Parker, A., and Dix, M. J., *J. Colloid Polymer Sci.*

[3] Gucker, F. T. and Tuma, J., *J. Colloid Interface Sci.,* **27**, 402 (1968).

[4] Kerker, M., Wang, D. S., and Chew, H. W., *Appl. Optics,* **19**, 2315 (1980).

[5] Chew, H., McNalty, P. J., and Kerker, M., *Phys. Rev.,* **A13**, 396 (1976).

[6] Wang, D. S., Kerker, M., and Chew, H. W., *Cytometry,* **1**, 161 (1980).

[7] Smoluchowski, M. von, *Z. Phys. Chemie,* **92**2 129 (1918).

[8] Goodwin, J. W., Hearn, I., Ho, C. C., and Ottewill, R. H., *Brit. Polymer J.,* **5**, 347 (1973).

[9] Buske, N., Gedan, H., Lichtenfeld, H., Katz, W., and Sonntag, H., *Colloid Polymer Sci.,* **258**, 1303 (1980).

[10] Cooke, D. D., and Kerker, M., *J. Colloid Interface Sci.,* **42**, 150 (1973).

[11] Bartholdi, M., Saltzman, G. C., Hiebert, R. D., and Kerker, M., *Applied Optics,* **19**, 1573 (1980).

[12] Gedan, H., Lichtenfeld, H., and Sonntag, H., *Colloid and Polymer Science,* **260**, 1151 (1982).

[13] Spielman, L. A., *J. Colloid Interface Sci.,* **33**, 562 (1970).

[14] Thompson, L., Thesis, Liverpool Polytechnic (1977).

[15] Duckworth, R. M., Personal Communication.

CHAPTER 13

A New Optical Method for Flocculation Monitoring

J. GREGORY and **D. W. NELSON,** Department of Civil and Municipal Engineering, University College London, Gower Street, London WC1E 6BT, UK

1. INTRODUCTION

Detailed studies of the flocculation process require a method of assessing the state of aggregation of suspensions, preferably one which is rapid and does not cause significant disturbance of the sample. Particle counting techniques such as those based on the Coulter principle [1], and light blockage as used in the Hiac counter [2] or various light-scattering counters [3–5], have been used to follow changes in particle size distribution during flocculation and all can give valuable information. There are problems, however, such as the need for considerable dilution in many cases, to avoid coincidence effects, and the possibility of floc breakage during passage through the Coulter orifice [1] or through the sensing zone of the Hiac counter [6]. In modern counters based on light scattering such as that recently described by Cummins *et al.* [5], fairly high particle concentrations can be measured directly (up to about 10^8 particles per cm^3), but, to avoid coincidence effects, very slow flow rates and hydrodynamic focussing have to be used.

In many practical applications it would be sufficient to have an overall measure of floc formation, rather than a detailed size distribution. Measurements closely related to the desired properties of the flocs, such as settling rate [7], supernatant clarity [8], settled floc volume [7] and filtrability [9] can be made. However, these are usually quite time-consuming procedures which do not lend themselves to rapid, on-line measurements.

During the development of a rapid flocculation test procedure based on laminar tube flow [10], the need arose for a monitoring technique which could detect flocs in the early stages of formation. A method has been found that gives a sensitive indication of floc formation, using a rather simple technique, which is very well suited to on-line applications. An outline of the basic principles of the method will be given here, together with some examples of the results that can be obtained.

2. OPTICAL MONITORING TECHNIQUE

2.1 Basis of the New Technique

In the present apparatus, the suspension under test flows through a glass capillary tube of 1 mm internal diameter. A narrow beam of light passes transversely through the tube and is monitored by a photodetector. The light source is a light-emitting diode which gives a beam about 0.1 mm wide, with a wavelength of 820 nm. The detector is a photodiode with a closely matched spectral response. The presence of particles in the beam causes some of the light to be scattered (and possibly absorbed), giving a reduced intensity at the detector. However, with fairly dilute suspensions (< 100 mg/l solids) and the short optical path length (1 mm), the reduction in intensity is quite small (a few per cent at most). Flocculation of the particles would cause a small change in the average transmitted light intensity but this would not be a reliable monitoring technique for a number of reasons. These include the 'fouling' of the optical surfaces by adhering particles and drift in the electronic components, both of which would cause changes in the apparent transmitted light intensity.

The volume of suspension illuminated is of the order of 0.02 mm^3, so for a typical suspension concentration of 10^8 particles per cm^3 there would be, on average, about 2000 particles in the light beam. Because there are inevitable local fluctuations in particle concentration and the illuminated sample is continually being renewed by flow of suspension through the capillary, the actual number of particles in the beam shows random variation about the mean value. This leads to random fluctuations in transmitted light intensity and the new technique is based on measurement of these fluctuations.

It is worth emphasizing that this method has very little in common with other recent techniques based on optical monitoring of flowing dispersions, such as the light scattering counters mentioned previously and the transmitted light method of Ditter *et al.* [11]. In these methods, a very much smaller sample volume is illuminated, using a laser beam. Essentially, single particle events are monitored and processed by some form of pulse height analysis. The novelty of the present technique lies in the fact that large numbers of particles can be simultaneously present in the sensing zone and the method of processing the fluctuating signal.

2.2 Nature of the Fluctuations

The signal from the photodiode consists of a large d.c. component, corresponding to the average transmitted light intensity, together with a small a.c. component or 'ripple', due to the fluctuations mentioned above (see Fig. 1). The d.c. component can be eliminated by passing the signal through a suitable capacitor and the a.c. component can then be amplified and analysed. The simplest measure of a fluctuating signal is its root mean square (r.m.s.) value, which is related to the total power in the signal. In the present method, the signal is passed to an r.m.s.-to-d.c. converter, which gives a d.c. output proportional to the r.m.s. value of the input waveform. The output from this device can be displayed on a digital voltmeter or chart recorder, or digitized and processed by an on-line microcomputer.

The fluctuating signal has a characteristic frequency which is related to the transit time of the particles through the light beam. This time depends on the flow rate through the tube which, in the current work, is of the order of 5–10 cm^3/min, giving a transit time of about 1–2 ms and a frequency of a few hundred hertz. The measured r.m.s. value should not depend on the frequency and so should be independent of flow rate.

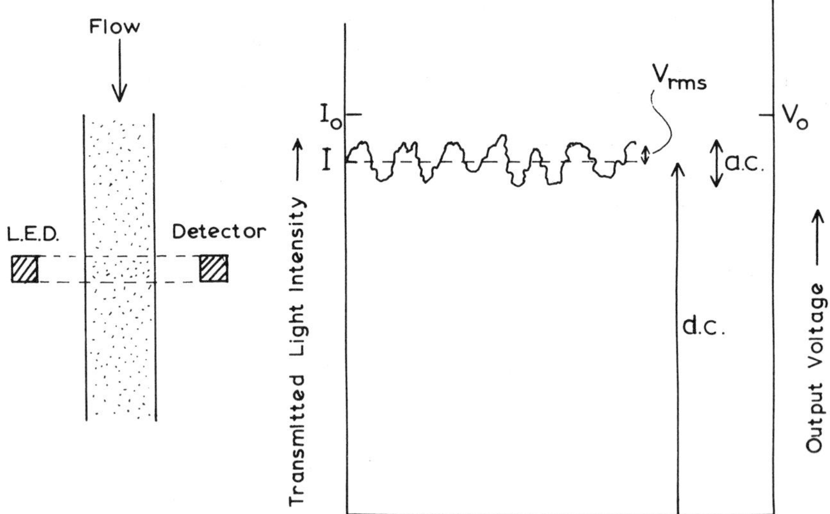

Fig. 1 – Principle of the monitoring technique. The intensity of light transmitted through a flowing dispersion shows random fluctuations about the mean. The r.m.s. value of the fluctuations is monitored continuously.

It is well known [12] that local variations in composition of a suspension follow the Poisson distribution, so that the standard deviation about the mean is equal to the square root of the average value. Thus, in the example mentioned previously, the number of particles in the beam would mostly be in the range 2000 ± 45. The standard deviation is equivalent to the root mean square variation about the mean and can be related to the r.m.s. intensity measurement, as shown below.

The intensity, I, of light transmitted through a suspension of particles is given by:

$$I = I_0 \exp(-NKl) \tag{1}$$

where I_0 is the intensity of the incident beam (i.e. that transmitted through particle-free water), N is the number of particles per unit volume, K is the average scattering cross-section of the particles and l is the optical path length.

For our purposes, it is more convenient to express the result in terms of the number of particles in the beam, n, rather than the number per unit volume, N. If the beam has an effective cross-sectional area A, then the volume of suspension illuminated is lA and the number of particles in this volume is simply given by $n = NlA$.

Equation (1) can then be re-written:

$$I = I_0 \exp(-nK/A) \tag{2}$$

We shall be mostly concerned with quite dilute suspensions, where only a small fraction of the light is scattered, so that $I/I_0 \approx 1$ and $nK/A \ll 1$. The exponential term can then be replaced by its linear approximation, giving:

$$I \approx I_0 (1 - nK/A) \tag{3}$$

This step considerably simplifies the treatment, but is not essential to the analysis.

In the method outlined previously, the r.m.s. value of the intensity fluctuations is measured, which arises from random fluctuations in the number of particles in the beam. Since the r.m.s. value of the number fluctuations (or the standard deviation about the mean) is $\sqrt{n}$ (assuming a Poisson distribution) it follows that:

$$V_{rms} \approx V_0 n^{1/2} K/A \ . \tag{4}$$

Since the intensities are measured as voltages, equation (4) has been written in terms of the r.m.s. voltage for the flowing suspension, V_{rms}, and the d.c. voltage corresponding to the incident light intensity, V_0 (see Fig. 1).

For given experimental conditions, where the effective area and incident intensity of the light beam are constant, the measured r.m.s. value should depend on the square root of the average number of particles in the beam and on their scattering cross-section K. When only the concentration of particles is changed, the r.m.s. value should depend on the square root of the concentration.

Figure 2 shows measured r.m.s. values for a series of kaolin suspensions ranging in concentration from 3–140 mg/l. The r.m.s. value gives a very good straight line when plotted against the square root of the clay concentration, confirming the validity of equation (4), and the underlying assumption of Poisson statistics. Since the kaolin particles have a range of sizes up to about 2 μm, it is clear that the foregoing analysis is not restricted to uniform particles. The line in Fig. 2 does not pass through the origin and there is a small r.m.s. reading at zero clay concentration. This is due to the presence of particles in the laboratory deionized water which was used to make up the suspensions and can be greatly reduced by filtration through a 0.22-μm membrane filter.

The linear behaviour in Fig. 2 is not observed at much higher kaolin concentrations; in fact, the r.m.s. value passes through a maximum. This is a consequence of the exponential form of equation (2) and it can be shown that the maximum occurs when the total scattering cross-section of the particles is half the cross-sectional area of the beam (i.e. $nK/A = 0.5$). This aspect will be dealt with more fully in a subsequent publication. Provided that the reduction in light transmission by the particles is quite small and that the approximation in equation (3) is valid, then the linear behaviour shown in Fig. 2 should be observed. In practice, this covers most suspensions of interest up to concentrations of at least 100 mg/l in a 1-mm cell.

The simplicity of the r.m.s. measurement and its sensitivity to quite low concentrations makes the technique attractive as a means of monitoring the level of suspended solids in water. One very significant advantage over traditional flow-through turbidity measurements is the virtual absence of problems due to

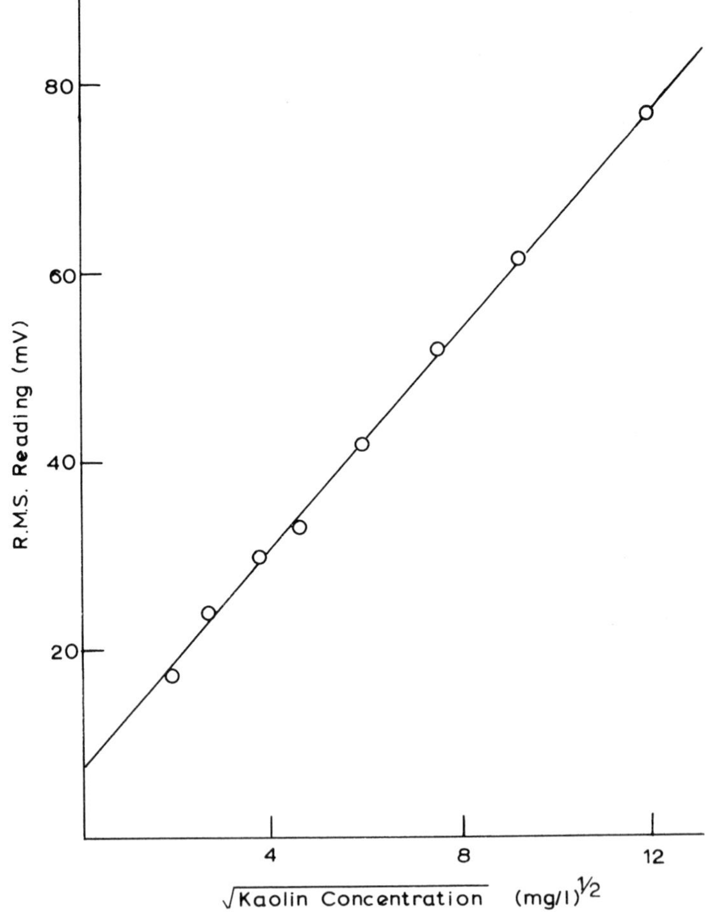

Fig. 2 – Measured r.m.s. value plotted against the square root of kaolin concentration.

fouling of the tube walls and electronic drift. Both of these predominantly affect the d.c. component of the transmitted light and have very little influence on the r.m.s. value of the fluctuations.

Because there are pronounced effects on the r.m.s. value caused by changes in the state of aggregation of particles, it may not be possible to interpret the results unambiguously in terms of suspended solids concentration. However, this sensitivity to aggregation makes the r.m.s. measurement very useful in monitoring flocculation, as will be shown in the next section.

2.3 Effect of Particle Aggregation

Consider, for simplicity, a suspension of monodisperse, spherical particles of radius a, at a concentration giving, on average, n particles in the light beam. Then, for dilute suspensions, the r.m.s. value is given by equation (4). In principle, the scattering cross-section K could be calculated from light scattering theory, but this is difficult except in certain limiting cases. If the particles are very small (so that the radius is very much less than the light wavelength), then Rayleigh theory can be applied. However, for the particles we are concerned with, especially aggregates, this would not be a realistic approach. For much larger particles a simple diffraction approach becomes acceptable (see, for example, Friedlander [13]) and the scattering cross-section is then just twice the geometric cross-sectional area of a particle. Intermediate particle sizes require application of the much more elaborate Mie theory.

For illustrative purposes, we shall assume that the particles are large enough (a few μm) for the diffraction or 'light blockage' approach to be valid, in which case:

$$K = 2\pi a^2 \tag{5}$$

When particles aggregate, the r.m.s. value should change for two reasons:

(a) The number concentration decreases.
(b) The scattering cross-section increases.

These have opposite effects on the r.m.s. value, but it can be shown that, in virtually all cases of practical interest, the net effect is a substantial increase in fluctuations and hence in the measured r.m.s. value.

Light scattering by aggregates is very difficult to predict, not the least of the problems being the need to define the form of the aggregate. The simplest assumption is that the particles comprising the aggregate coalesce to form a sphere of the same total volume, in which case the calculation is quite straightforward. However, this is a very unrealistic assumption and 'real' aggregates must be considered. For fairly small spherical particles, it can be shown [14] that 'real' doublets scatter less light than the equivalent coalesced sphere, but the opposite effect is expected for larger particles [15]. On a simple light-blockage picture, it seems intuitively obvious that a 'real' aggregate would block more light than a sphere of the same total volume, since the latter presents less cross-sectional area to the beam. In the present device, where aggregates flow through a capillary tube, it is probable that chain-like aggregates would tend to align with the flow and hence block more light than randomly-oriented aggregates. These concepts are illustrated schematically in Fig. 3.

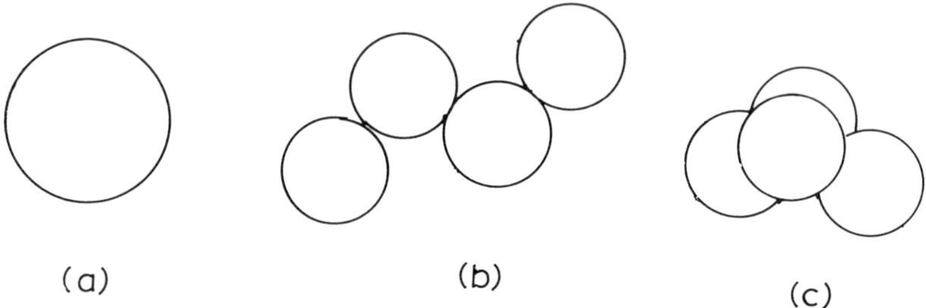

Fig. 3 – Forms of aggregates composed of four equal spheres. (a) Coalesced spheres, (b) Extended aggregate (e.g. aligned by flow), (c) Randomly oriented aggregate.

No great care has been taken in the current work to ensure a narrow light acceptance angle at the photodiode, so that some forward-scattered light must be detected. This considerably complicates the interpretation of transmitted light measurements [15] and tends to reduce the effect of aggregation on measured extinctions. It seems reasonable to treat K as an 'effective' scattering cross-section, which includes both extinction and forward scattering.

The effect of aggregation on the r.m.s. value will be estimated for two extreme cases, (a) coalescence and (b) extended aggregates, in which all of the cross-sectional area of the individual particles is presented to the light beam. The simple diffraction result, equation (5) will be used.

Suppose that the particles form j-fold aggregates (i.e. on average, j single particles combine to form one aggregate). In that case, the average number of particles in the beam will be reduced to (n/j) and the scattering cross-section will be, for the two cases above:

(a) $K = 2j^{2/3}\pi a^2$ (coalescence)

(b) $K = 2j\pi a^2$ (Extended aggregate)

The r.m.s. values, from equation (4), can then be calculated for the two cases, by inserting the appropriate value of K and replacing $n^{1/2}$ by $(n/j)^{1/2}$. The results are:

(a) $$V_{\text{rms}} = [2\pi a V_0 n^{1/2}/A]\, j^{1/6} \tag{6}$$

(b) $$V_{\text{rms}} = [2\pi a V_0 n^{1/2}/A]\, j^{1/2} \tag{7}$$

The terms in square brackets are identical and depend on experimental conditions and the properties of the initial suspension. During flocculation these terms remain constant and the measured r.m.s. value should vary as either the sixth root or the square root of the aggregation number, j, depending on whether coalesced spheres or extended aggregates are assumed. In practice, behaviour intermediate between these two extremes would be expected, although probably nearer to equation (7). For 60-fold aggregates the r.m.s. value should increase by a factor of between about 2 and 8 depending on the nature of the aggregates.

2.4 Experimental Results

A detailed investigation of the flocculation of uniform, spherical particles and its effect on measured r.m.s. values has been initiated and should enable the technique to be thoroughly evaluated. Results of these studies will be published separately. In the meantime, some typical observations of changes in r.m.s. value during flocculation of kaolin suspensions are presented in Fig. 4.

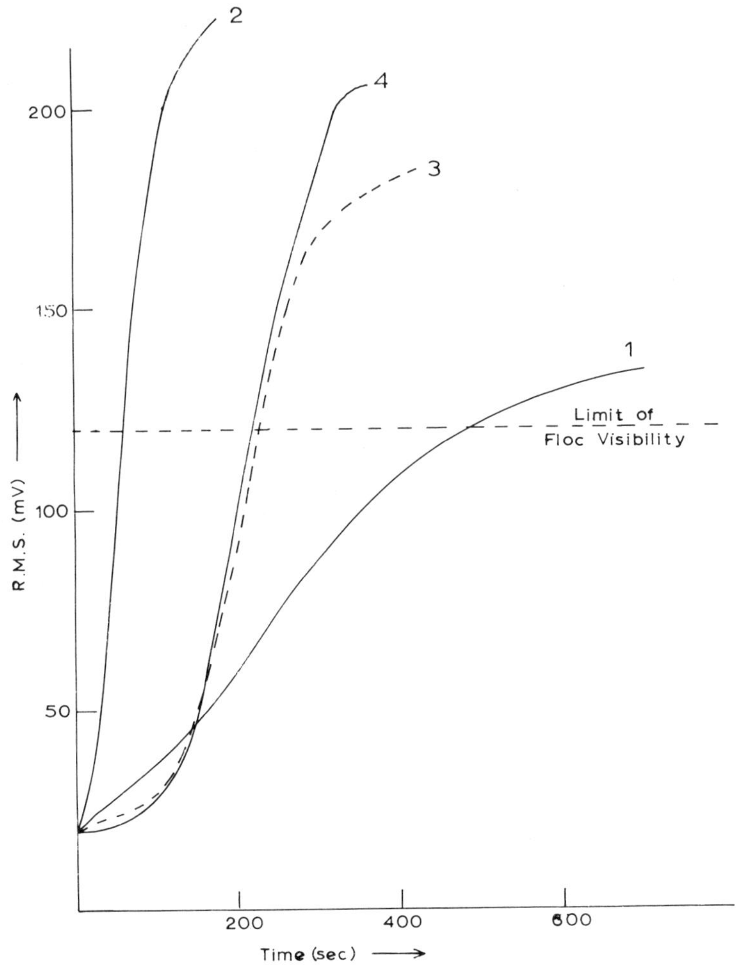

Fig. 4 – Effect on r.m.s. reading of kaolin flocculation by: (1) Calcium nitrate, (2) aluminium sulphate, (3) low molecular weight cationic polymer, (4) high molecular weight cationic polymer (see text).

In these tests, the kaolin concentration was fixed at 35 mg/l and flocculation was carried out under conditions of low-speed magnetic stirring. The suspension flowed by gravity from the flocculation vessel through a short length of tubing to the capillary flow cell, where the fluctuations in transmitted light intensity were monitored continuously. The flow rate was about 5 cm^3/min and the lag time between the flocculation vessel and the capillary cell was small (2–3 sec), so that any aggregation of particles was detected almost immediately as a change in r.m.s. value. Flocculants were rapidly pipetted into the stirred kaolin suspensions and the r.m.s. value was displayed on a chart recorder.

Four different flocculants were used, each at one concentration (chosen to be within the optimum range for the particular additive):

(1) Calcium nitrate at a concentration of 10^{-2} M.
(2) Aluminium sulphate at a concentration of 4×10^{-5} M (about 2.2 mg/l as Al).
(3) A low molecular weight cationic polymer, poly(dimethylaminoethylmethacrylate), with a molecular weight of about 5000 and fully quaternized. The concentration was 0.1 mg/l.
(4) A high molecular weight cationic polymer, poly(1-ethyl 2-methyl 5-vinyl pyridinium bromide) molecular weight about 1 million, at a concentration of 0.1 mg/l.

In the case of the cationic polymers the kaolin suspension was made up in 10^{-3} M NaCl, to control the ionic strength, which is known to influence flocculation by cationic polymers [16]. For aluminium sulphate, the kaolin was suspended in London tap water diluted with an equal volume of deionized water, to give sufficient alkalinity and optimum pH for hydroxide precipitation.

It is not our intention here to discuss these results in detail, but simply to illustrate that the new monitoring technique provides a very sensitive means of following the flocculation process and clearly shows up the different behaviour of various flocculants. All of the results in Fig. 4 were obtained under similar conditions, so that useful comparisons can be made. The stirring speed was constant (about 50 r.p.m.) and the instrument settings were such that the r.m.s. reading for the unflocculated kaolin was 20 mV and could rise to about 200 mV for flocculated samples. As indicated on the figure, flocs just visible to the unaided eye gave r.m.s. values of about 120 mV, i.e. a six fold rise from the unflocculated suspension.

With calcium nitrate, the negatively charged clay particles are destabilized by a reduction in double layer repulsion. The r.m.s. value rises steadily after addition of the salt and reaches a limiting value of about 140 mV, indicating that a limiting floc size has been attained in the stirred suspension. It is well known that flocs grow to a limiting size in sheared suspensions, as a result of a competition between floc growth and break-up. With simple salts, such as calcium salts, floc strength is not very great.

Cationic polymers are effective with negative particles either because of a bridging mechanism or as a result of charge neutralization. With the kaolin suspension, they clearly give very different behaviour to that observed with a calcium salt. The r.m.s. value rises much more rapidly and reaches higher limiting values, indicating larger floc sizes and hence stronger flocs. The fact that the two polymers give rather similar behaviour, despite a large difference in molecular weight, indicates that polymer bridging is not of major significance in this case. The slight differences between the two polymers might be explained in terms of 'electrostatic patch' effects [16]. There is some evidence, especially with the higher molecular weight polymer, of a short time lag (50–100 seconds) between addition of the flocculant and the onset of rapid flocculation, probably due to the relatively slow adsorption of polymers in a stirred suspension [17].

The behaviour with aluminium sulphate is quite unlike the other cases. The r.m.s. value rises at an extremely rapid rate immediately after dosing and reaches a limiting value after only about 2 minutes. Precipitation of aluminium hydroxide occurs under these experimental conditions (and in water treatment practice), giving an increase in the amount of suspended material as well as flocculation of the clay particles, both of which should increase the r.m.s. reading.

3. CONCLUSIONS

By measuring fluctuations in the intensity of a light beam transmitted through a flowing suspension, information can be obtained on the concentration of suspended particles and their state of aggregation. The root mean square value of the fluctuating signal can be easily derived and provides a very sensitive indication of floc formation, long before the flocs grow sufficiently large to be visible to the unaided eye.

This technique has great promise as an on-line detector in laboratory flocculation tests and in the monitoring of suspensions in a number of industrial processes.

ACKNOWLEDGEMENTS

This work was supported by a grant from the Science and Engineering Research Council. The monitoring technique described here and its use in an automated flocculation test method form the subject of a patent application.

REFERENCES

[1] Matthews, B. A. and Rhodes, B., *J. Colloid Interface Sci.* **32**, 339 (1970).

[2] Jeckel, M. R., *GWF Wasser/Abwasser,* **123**, 555 (1982).

[3] Walsh, D. J., Anderson, J., Parker, A., and Dix, M. J., *Colloid and Polymer Sci.,* **259**, 1003 (1981).

[4] Buske, N., Gedan, H., Lichtenfeld, H., Katz, W., and Sonntag, H., *Colloid and Polymer Sci.,* **258**, 1303 (1980).

[5] Cummins, P. G., Staples, E. J., Thompson, L. G., Smith, A. L., and Pope, L., *J. Colloid Interface Sci.*, **92**, 189 (1983).
[6] Gibbs, R. J., *Env. Sci. Tech.*, **16**, 298 (1982).
[7] Le Bell, J. C., *Colloids Surf.*, **5**, 285 (1982).
[8] Letterman, R. D. and Vanderbrook, S. G., *Water Res.*, **17**, 195 (1983).
[9] de Moor, A. E. L. and Gregory, J., An Automated Filtrability Test. *Proc. Water Filtration Symposium,* Antwerp (1982).
[10] Gregory, J., *Chem. Eng. Sci.*, **36**, 1789 (1981).
[11] Ditter, W., Eisenlauer, J., and Horn, D., Laser optical method for dynamic flocculation testing in flowing dispersions, in *The Effect of Polymers on Dispersion Properties* (T. F. Tadros, Ed.) p. 323, Academic Press, London (1982).
[12] Freundlich, H., *Colloid and Capillary Chemistry,* p. 352, Methuen, London (1926).
[13] Friedlander, S. K., *Smoke, Dust and Haze,* p. 127, Wiley–Interscience, New York (1977).
[14] Lips, A., Smart, C., and Willis, E., *Trans. Faraday Soc.*, **67**, 2979 (1971).
[15] Latimer, P. and Wamble, F., *Applied Optics,* **21** 2447 (1982).
[16] Gregory, J., *J. Colloid Interface Sci.*, **42**, 448 (1973).
[17] Gregory, J., Polymer flocculation in flowing dispersions, in *The Effect of Polymers on Dispersion Properties,* (T. F. Tadros, Ed.) p. 301, Academic Press, London (1982).

CHAPTER 14

Influence of Shear and Salt on Flocculation in Laminar Tube Flow

J. EISENLAUER and **D. HORN,** BASF Aktiengesellschaft, Abt. ZHV/D, D-6700 Ludwigshafen/Rhein, FRG

ABSTRACT

A computer-controlled, laser optical flocculation test in laminar tube flow enables a rapid *in-situ* measurement of the degree of flocculation, thus providing a sufficient characterization of the mode of action of flocculation aids in terms of: *O*ptimum (*C*ritical) *F*locculation *C*oncentration, OFC (CFC) under defined shear conditions (shear rate, G, contact opportunity, $G{\cdot}t$). Furthermore, variations in the mode of dosing (rapid mixing) or the reflocculation behaviour after excessive shearing can be studied.

Flocculation testing is thus reduced to the most important parameters relevant to practice and to modern orthokinetic flocculation theories.

The following typical results are presented:

- There is a close correlation between the degree of flocculation and the corresponding dewatering behaviour in a typical paper stock suspension.
- Dosing of a polycationic flocculant into an anionic latex dispersion produces, in combination with indifferent salts and shear (G, $G{\cdot}t$=const.), odd flocculation phenomena. The interpretation is placed within the frame of the charge-patch flocculation mechanism and orthokinetic flocculation theory.
- Combined dosing of hydrolyzable metal salt coagulants and polyelectrolyte flocculants can be optimized in terms of dosage (metal salt, polyelectrolyte) and kinetics (variation of $G{\cdot}t$ at constant G).

1. INTRODUCTION

The synthesis of tailor-made flocculation aids [1] and their optimum dosage in solid–liquid separation technology requires the identification of the operating flocculation and dewatering mechanisms [2]. To these ends flocculation testing methods are necessary which enable an *objective* evaluation of the flocculant efficiency relevant to practical problems.

Classical coagulation and flocculation tests [3], i.e. sedimentation behaviour in the jar (velocity, sediment volume, turbidity in the supernatant) or drainage time of the flocculated system, can be characterized as follows:

- Very extensive in respect to material and manpower. For the evaluation of the optimum dose, as a rule 10 jars at least are required.
- The results are usually dependent on the testing procedure and are very often not identical with the optimum dosing conditions in practice [4, 5].

From a macroscopic point of view the latter difficulties arise from an inappropriate scale down. A microscopic analysis based on principles of colloid and interface science reveals the following details which can help to overcome the problems concerned with the optimum handling of flocculants:

- Effective flocculants are, as a rule, polymeric (organic or inorganic) substances [1]; they start their mode of action by practically irreversible, high affinity adsorption at the substrate interface [2]. This means that rapid mixing is required for a homogeneous distribution of the flocculant, thus promoting efficient destabilization [6].
- Whereas first destabilization (formation of microflocs) is a perikinetic process (only controlled by diffusion), the morphology and size of the macroflocs is dependent on the hydrodynamics (shearing) of the system (orthokinetics) [6, 7]. This means that results from flocculation tests have only practical relevance if measured under identical shear conditions.

An extensive characterization of auxiliary efficiency, therefore, requires the evaluation of the *O*ptimum *F*locculation *C*oncentration (OFC) at different modes of dosing (rapid mixing) and shearing (shear stress, G; contact opportunity, Camp Number, power input, $G \cdot t$).

For this objective a computer-controlled laser optical technique in a tube flow system has been developed which enables a rapid *in-situ* measurement of the degree of flocculation as a function of the auxiliary dose [8]. This dynamic flocculation test reduces the characterization of flocculation processes to the most significant parameters, thus improving the transparency of test results and their scaling-up to practical problems.

Details of the method and examples of the broad application spectrum are described elsewhere [8, 9]. In this paper further typical results are presented, especially for the complicated flocculation behaviour observed when metal salt coagulants, either indifferent or charge determining, are dosed in combination with polyelectrolytes. Although this combined dosing is frequently used in water and wastewater treatment the mechanistic details are not yet understood to such an extent that an optimization is feasible. Furthermore, the possibility of a systematic variation of shear in terms of G and $G \cdot t$ enables experimental tests of modern orthokinetic flocculation theories [10, 11].

2. EXPERIMENTAL

The principle and the experimental details of the flocculation test are illustrated in Fig. 1.

The dispersion (Q, c_s) can be destabilized by turbulent mixing with a coagulant solution (dosage I, q_1, c_{Fo}, $q_1 < Q/100$). The resulting microfloc formation takes place under defined shear conditions (G, $G{\cdot}t_1$) in a laminar tube flow section (Q, d_i, L_1). Macrofloc formations can be initiated by a programmed polymer dosage II (q_2, c_{Po}, $q_2 < Q/100$).

The *in-situ* detection of the degree of flocculation, F, is done after a second laminar tube flow section (L_2, d_i, Q) by a modified pluse height analysis (SMA) of the light blockage signals produced by particles and/or flocs passing a laser focus [8]. The measurement and calculation of F by a desk computer (C) takes less than ten seconds and is correlated (C, MP, V_2) with the stepwise polymer dosage II. The development of the flocculation experiment can be followed on the screen. After typical measuring times of 10 minutes the optimum conditions, OFC or CFC, are determined from the flocculation diagram, F over c_P, as shown in Fig. 1.

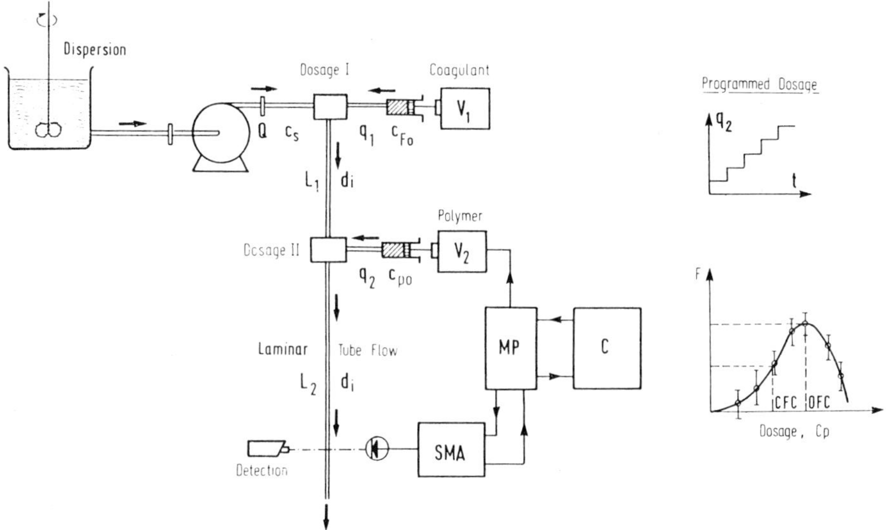

Fig. 1 – Laser optical, dynamic flocculation testing in tube flow for optimizing the dosing of coagulants/flocculants in water and waste-water treatment.

This experimental set up allows an optimization of combined coagulation–flocculation processes in respect to the mode of dosage (q_1, c_{Fo}, q_2, c_{Po}, geometry of the rapid mixing devices) and the shear conditions, G, $G{\cdot}t \propto L$ (kinetics). The coagulation/flocculation behaviour of either of the two single components can be established as well.

3. SHEARING IN LAMINAR TUBE FLOW

For laminar tube flow the shear conditions during floc formation can be easily approximated in terms of the parameters G and $G{\cdot}t$ when averaged over the tube cross section; especially $G{\cdot}t$ is, independent of the volume stream, Q, only a function of goemetry (L, d_i) [12–14].

Laminar tube flow, therefore, can be used for systematic variations of the shearing during flocculation experiments as follows:

Shear stress G can be varied by Q at constant $G{\cdot}t$ (L), provided that rapid mixing is independent of Q.

On the other hand flocculation kinetics (with underlying adsorption and mixing kinetics) can be studied at constant G (Q) by taking measurements at different positions, L ($\propto G{\cdot}t$).

The laser beam in the optical set up for flocculation detection according to the light blockage principle in Fig. 1 is focussed into the tube axis [8]. Due to the depth of focus and mixing effects during the tube flow the average values of G and $G{\cdot}t$, respectively, should be a good approximation for the actual shearing conditions in the tube flow section detected [12].

4. RESULTS AND DISCUSSION

4.1 Correlation between flocculation and dewatering behaviour

There are differing opinions in the literature concerning the correlation between flocculation and the dewatering behaviour of flocculated systems [2, 15].

In Fig. 2 the degree of flocculation F in a typical paper stock suspension of newspaper quality is compared with the drainage time, t_{70}, from a standardized dewatering test (Schopper Riegler Test), while the auxiliary dosing, q_2 ($q_1 = 0$) is stepwise increased. At each single dosing step one litre of the flocculated suspension was filled into the dewatering tester; the time, t_{70}, was measured when the filtrate volume reached 70%.

There is a close correlation between F and t_{70}; higher polymer doses than 0.6 mg/g produce asymptotic values for both parameters. In additional experiments using polyethylenimine (PEI) auxiliaries of lower molar mass (Polymin® P, G 500, G 20 [16]) the same asymptotic behaviour is established; flocculation and dewatering efficiency are, however, drastically reduced with decreasing molar mass. These features are in agreement with the experience at paper machines and support a flocculation model along the lines of polymer bridging [2, 15].

4.2 Influence of indifferent salts and shearing on the flocculation of anionic latex dispersions with polycationic flocculants

Changing the substrate from cellulose fibre, see Fig. 2, to a well defined, anionic latex† in Fig. 3 changes the mode of flocculation with polyethylenimine type aids.

† 50% polystyrene, 50% butylacrylate, $\phi_h = 0.86\ \mu m$, SD = 30%, $\sigma_0 = 5.7\ \mu C/cm^2$.

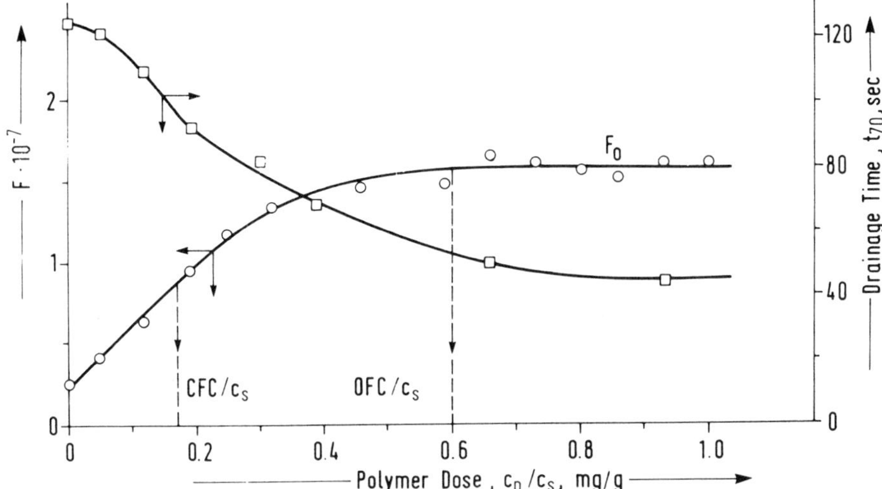

Fig. 2 – Correlation between degree of flocculation, F (laser optical, dynamic flocculation test) and drainage time, t_{70}, of the flocculated system (Schopper Riegler Test). Typical paper stock suspension for newspaper quality with polycationic auxiliary (programmed dosage I). $c_S = 2.5$ g/dm^3; pH 6.5; $Q = 100$ cm^3/min; $G = 419$ s^{-1}; $G \cdot t = 18916$; Polymin® G 500 primary flocculant

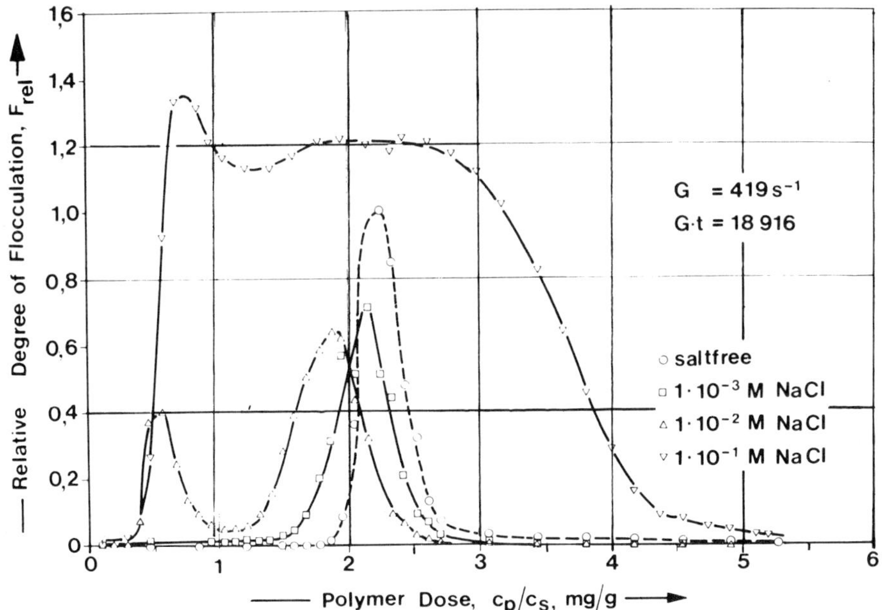

Fig. 3 – Influence of NaCl on the flocculation of an anionic latex dispersion with a polycationic flocculant (programmed dosage I). $c_S = 0.25$ g/dm^3, pH 4.5, $Q = 100$ cm^3/min; $G = 419$ s^{-1}; $G \cdot t = 18916$; Polymin® G 500 primary flocculant.

The latex dispersion can be totally restabilized by overdosing of the flocculant whereas in the paper stock suspension an asymptotic value, F_0, is reached. Restabilization in the overdose region was also established in the salt free latex system up to pH 9 and substrate concentrations of the same range as in the paper stock suspension.

Additional experiments show that flocculation efficiency is (nearly independent of molar mass) only a function of the pH-dependent charge density of the polyethylenimine molecules. Together with former results on influence of pH-value, shear resistance of the flocs, reflocculation kinetics and adsorption isotherm [8, 16] a highly charge controlled flocculation mechanism in terms of a charge-patch mechanism can be postulated, at least for pH $\leqslant$ 7. The adsorption of the PEI-molecules at the latex surface is obviously of a high affinity type with rapid adsorption kinetics, producing a charge-pattern of flatly adsorbed molecules before first particle contacts take place [17].

Additional dosing of NaCl leads to a broadening of the flocculation peaks which is not surprising for a charge-patch mechanism. The distinct double peak at 1×10^{-2} M NaCl is, however, unexpected and needs further interpretation.

If the same system is flocculated under mild shearing conditions of the classical jar test (mixing the latex dispersion with a diluted PEI solution using magnetic stirring), as shown in Fig. 4, only a broad flocculation region is indicated

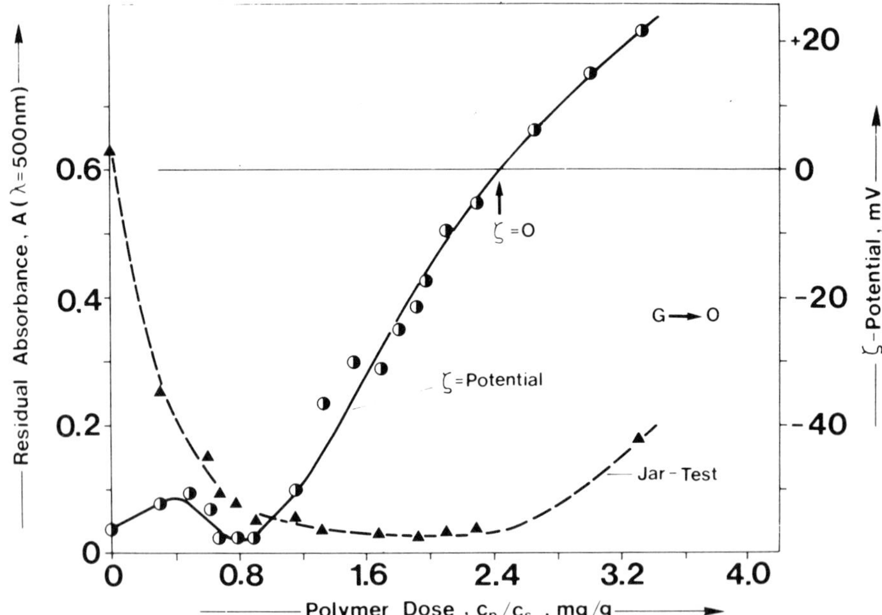

Fig. 4 – Flocculation behaviour of the system in Fig. 3 at 1×10^{-2} M NaCl under mild shearing conditions of the jar test compared to electrophoretic mobility; ζ-potential is calculated using the Smoluchowski Helmholtz Equation; A is measured after 24 h of settling.

by the residual absorbance in the supernatant. As already established for the salt free system [8, 16] recharging of the system (ζ-potential) together with restabilization is observed in the overdose region. There is, however, only a close correlation between OFC and the isoelectric point ($\zeta = 0$) in the salt free system whereas at 1×10^{-2} M $NaCl_2$ flocculation is already observed at ζ-potentials of approx. −50 mV; this phenomenon is in accordance with the charge-patch flocculation mechanism [9].

The double peak is observed when the systems for the jar test are flocculated under the shear conditions of laminar tube flow in Fig. 1. In Fig. 5 the good correlation between the *in-situ* test and the corresponding jar test is demonstrated. By additional variation of G ($200 \leqslant G \leqslant 500$ s^{-1}) at constant $G \cdot t$ the shape of the flocculation peak can be changed from the broad single peak of Fig. 4 to the double peak of Fig. 5.

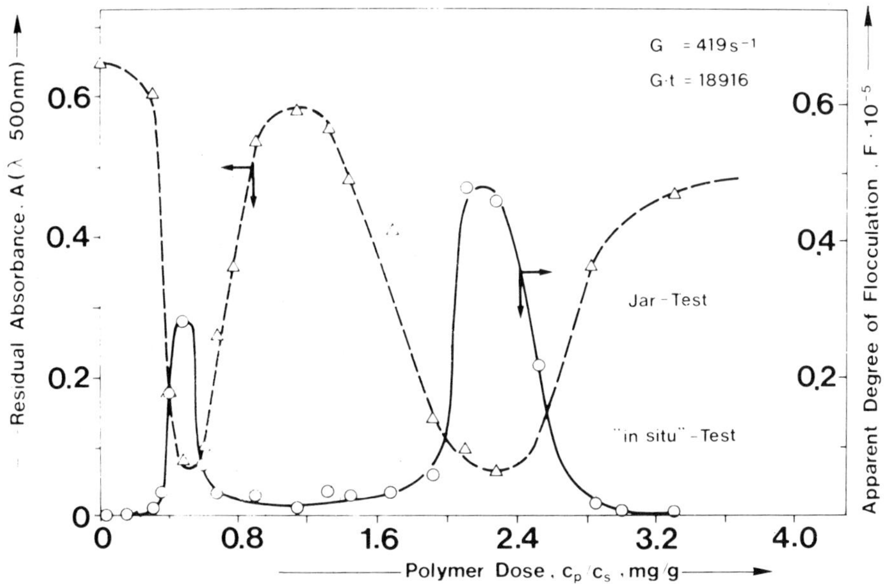

Fig. 5 – Flocculation behaviour of the system in Fig. 4 under the shearing of laminar tube flow corresponding to $G = 419$ s^{-1}; $G \cdot t = 18916$. Comparison between *in-situ*- and jar-test.

Experiments at pH 7 show that dosing of NaCl in the same range induces only a broadening of the flocculation peaks; flocculation efficiency in terms of flocculant consumption is, however, reduced by a factor of three due to the lower charge density. Further results with other salts (Na_2SO_4, $CaCl_2$) have already been discussed elesewhere [9].

The drastic influence of shear can be qualitatively interpreted in terms of modern orthokinetic flocculation theories [10, 11]. In these extensions of the classical Smoluchowski theory a complicated dependence of the collision efficiency factor, α, on shear stress, ζ-potential and ionic strength is proposed for coagulation with indifferent salts. There are very specific combinations of these parameters postulated where α tends towards zero although ζ-potential and ionic strength would indicate totally destabilized systems.

Similar conditions are obviously valid for the highly charge controlled system under discussion. At constant shearing (G, $G{\cdot}t$) and ionic strength, increasing polymer dosing leads to variations of the ζ-potential, thus inducing variations of α as indicated in the double peak behaviour.

Only preliminary formulations have been published for a corresponding theory dealing with polymeric flocculants (polymer bridging) [18, 19].

4.3 Flocculation of an anionic latex dispersion with $FeCl_3$

The dosage of hydrolysable metal salt coagulants is, especially in combination with polyelectrolyte flocculants, of outstanding importance for water and wastewater treatment. The mode of action is dominated, contrary to indifferent salts, by interfacial adsorption and is closely correlated with the hydrolysis kinetics. During dosage different steps of hydrolysis starting from the hexa-aquo complex through polycationic hydroxo-complexes up to precipitated metal hydroxides can be reached; these processes are mainly controlled by temperature, salt concentration and pH value [20, 21].

Figure 6 shows a complete flocculation picture in terms of a three dimensional diagram, F, c_F/c_s, $G{\cdot}t$, when a diluted $FeCl_3$-solution is dosed as primary coagulant (q_1, c_{Fo}) into a diluted anionic latex dispersion of pH 7.

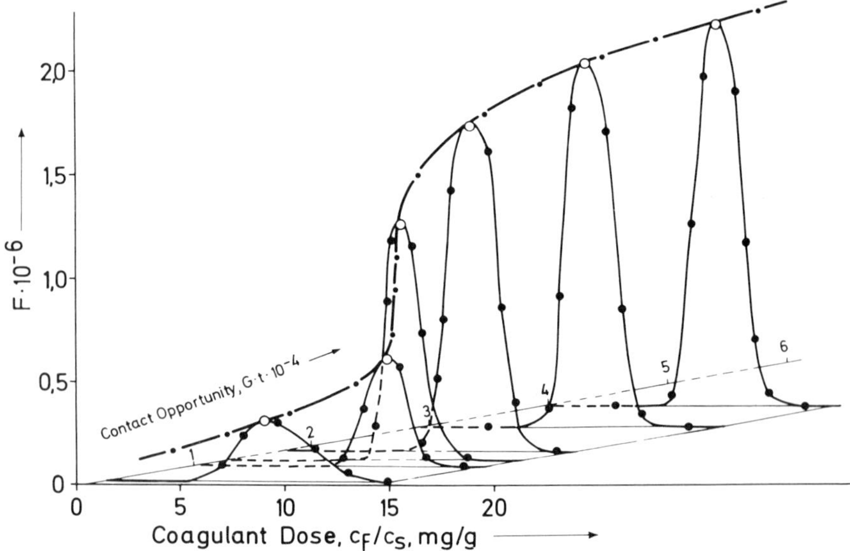

Fig. 6 – Coagulation kinetics in an anionic latex dispersion using $FeCl_3$ as primary coagulant (programmed dosage I). c_S = 0.25 g/dm^3; Q = 100 cm^3/min; pH 7; c_{Fo} = 0.73 g/dm^3; $q_1 < Q/100$; G = 419 s^{-1}.

The variation of $G \cdot t$ at constant G (Q) corresponds to experiments with identical dosing ranges but different positions, L, in the tube. The diluted $FeCl_3$-solution (pH 2) was freshly prepared from a three molar standardized stock solution. Because of $q_1 < Q/100$ the pH value decreases in the whole dosing range only by 0.4 pH units.

First microflocs can already be detected about $L < 1$ m ($G \cdot t < 2 \times 10^3$). For longer tubes F sharply increases. At $G \cdot t$-values of about 2×10^4 (11 m) asymptotic microfloc sizes are formed dependent on the specific dosage. Destablilization starts irrespective of the tube position at doses of about 4 mg/g. An optimum is reached at about 10 mg/g; total restabilization is detected for higher doses than 15 mg/g.

The mode of action of the $FeCl_3$-coagulant is in many respects similar to that of polyethylenimine. Highly cationic, polymeric hydroxo-complexes should, therefore, be the flocculation active species with this special set of parameters.

4.4 Flocculation of anionic latex dispersions by combined dosing of $FeCl_3$ and polycationic flocculants

Microflocs with asymptotic sizes corresponding to a constant dosage $I(q_1, c_{FeCl_3})$ and $L_1 = 10$ m ($G \cdot t_1 = 17.800$) can be further flocculated into macroflocs by the programmed dosage II of a polycationic flocculant, e.g. Polymin® G 500 as shown in Fig. 7. Flocculation detection is done after a second laminar tube flow section of $L_2 = 6$ m ($G \cdot t_2 = 10.700$).

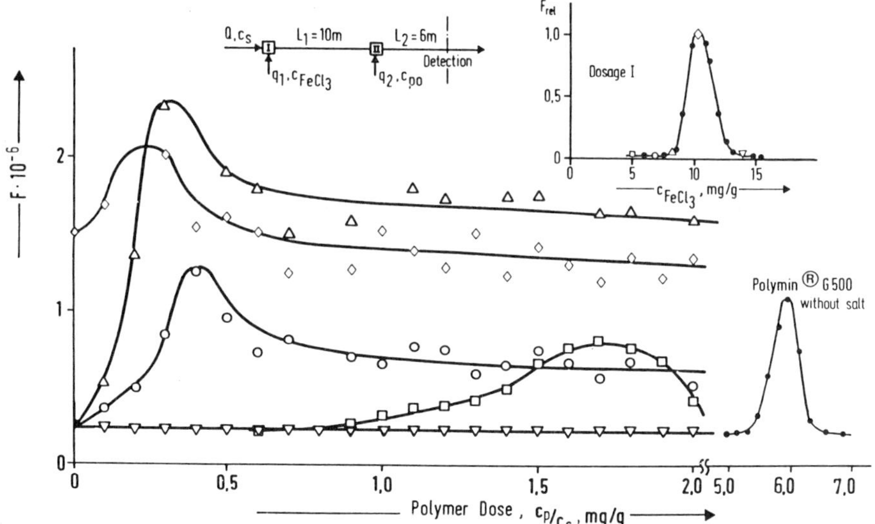

Fig. 7 – Flocculation of an anionic latex dispersion by combined dosing of $FeCl_3$ as a metal salt coagulant (constant dosage I) and a polycationic flocculant (programmed dosage II). $c_S = 0.25$ g/dm³; $Q = 100$ cm³/min; pH 7; $c_{Fo} = 0.73$ g/dm³; Polymin® G 500.

With a single programmed dosage I of PEI as a primary flocculant a sharp flocculation peak at OFC/c_s = 6 mg/g is detected at a tube length of $L_1 + L_2$ = 16 m ($G \cdot t$ = 28400) (see also section 4.2). Changing to a single $FeCl_3$-coagulant at identical experimental parameters produces the flocculation diagram inserted into Fig. 7 (see also section 4.3).

The combined dosing of a $FeCl_3$-coagulant and a PEI flocculant corresponding to the experimental parameters in Fig. 7 increases the flocculation efficiency (F-values) very drastically at a highly reduced level of flocculant consumption.

$FeCl_3$-dosages, too small for inducing single primary coagulation (□, ○, △) produce sensitized latex particles which can be effectively flocculated into large macroflocs when a PEI flocculant is added; with increasing $FeCl_3$-dosage the F-values increase while the values of OFC/c_s decrease. An optimum of the combined dosing can be established at $FeCl_3$-doses close to the optimum of the single dosing experiment (◇).

In the $FeCl_3$-overdose region (▽) the PEI flocculant is inefficient because recharging of the latex surface hinders anchoring of the cationic PEI molecules.

The tube lengths used, L_1 = 10 m and L_2 = 6 m, correspond from the kinetic point of view to asymptotic floc sizes for microfloc as well as for macrofloc formation. From an additional variation of L_1 information about the adsorption kinetics of the hydroxo-complexes should be obtainable.

4.5 Flocculation of anionic latex dispersions by combined dosing of $FeCl_3$ and polyanionic flocculants

Replacing the PEI-flocculant in the experiment of Fig. 7 by a modified, anionic polyacrylamide† produces flocculation results as shown in Fig. 8.

Coagulant doses of up to 7 mg/g (□) are not yet sufficient to provide cationic adsorption sites on the anionic latex surface for anchoring the polyanionic flocculant; this is a prerequisite for polymer flocculation. Higher coagulant doses at a level sufficient for microfloc formation (△, ◇), give highly efficient macrofloc formation when additional polymeric flocculant is dosed.

For $FeCl_3$-dosage in the restabilization range (▽) flocculation efficiency in terms of F and OFC/c_s is however reduced. Again a coagulant dose corresponding to the optimum coagulation range is obviously the optimum for the combined dosing experiment.

5. ACKNOWLEDGEMENT

Part of this work has been supported by grant No. 02–WT 037 of the Bundesministerium für Forschung und Technologie (BMFT).

† Sedipur® AF 400, high molecular copolymer of acrylamide and salt of acrylic acid; degree of modification 40%.

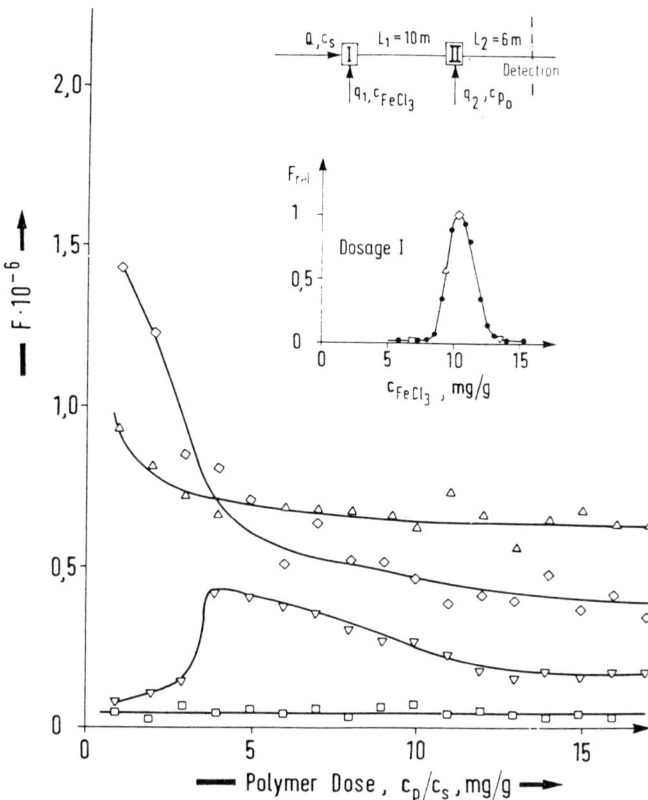

Fig. 8 – As Fig. 7 but using a polyanionic flocculant. $c_{Po} = 0.5$ g/dm³; Sedipur® AF 400.

SYMBOLS, TERMS AND ABBREVIATIONS

Symbol	Term	Unit
c_{Fo}	concentration of coagulant stock solution	g/dm³
c_{Po}	concentration of flocculant stock solution	g/dm³
c_F	actual coagulant concentration in the tube	g/dm³
c_P	actual flocculant concentration in the tube	g/dm³
c_s	substrate concentration	g/dm³
F	degree of flocculation [8]	–
F_{rel}	relative degree of flocculation	–
G	shear stress	1/s
$G \cdot t$	contact opportunity (power input, Camp Number)	–
L	tube length	m
d_i	inner tube diameter	mm
Q	volume flow rate	cm³/min
q	volume flow rate of coagulant/flocculant	cm³/min

t	residence time in the tube	s
ϕ_h	hydrodynamic particle diameter from Photon Correlation Spectroscopy	μm
α	collision efficiency factor	—
σ_0	surface charge density from conductometric titration	$\mu C/cm^2$
ζ	zeta-potential	mV
CFC	critical flocculation concentration	g/dm^3
OFC	optimum flocculation concentration	g/dm^3
C	deskcomputer	
MP	interfacing	
SMA	modified pulse height analysis [8]	
V	programmable dosing pump	

REFERENCES

[1] Halverson, F. and Panzer, H. P., *Encyclopedia of Chemical Technology,* 3rd. Ed. Vol. 10; pp. 489–523, (Kirk-Othmer, ed.), Wiley, New York (1980).

[2] Vincent, B., The effect of adsorbed polymers on dispersion stability, *Adv. Colloid Interface Sci.,* **4**, 193–277 (1974).

[3] Bratby, J., *Coagulation and Flocculation,* pp. 263–287, Uplands Press, Croydon/England (1982).

[4] Akers, R. J., *The Scientific Basis of Flocculation,* (Ives, K. J., ed.), Series E. Applied Science No. 27, pp. 131–163, Sijthoff and Noordhoff, The Netherlands (1978).

[5] Ives, K. J., as ref [4], pp. 165–217.

[6] Bratby, J., as ref. [3], pp. 172–259.

[7] Ives, K. J., *Solid Liquid Separation* (2nd ed.), (Svarovsky, L., Ed.), pp. 86–119, Butterworths, London (1981).

[8] Ditter, W., Eisenlauer, J., and Horn D., *The Effect of Polymers on Dispersion Properties,* (Tadros, Th.F., ed.), pp. 323–342, Academic Press, London (1982).

[9] Eisenlauer, J. and Horn, D. Dynamische Flockungsmessung in strömenden Systemen zur Optimierung der Anwendung von Flockungshilfsmitteln, *Z. Wasser/Abwasser Forsch.,* **16**, 9–15 (1983).

[10] Mason, S. G. and van de Ven, T. G. M., The microrheology of colloidal dispersion. VII. Orthokinetic doublet formation of spheres, *Colloid Polym. Sci.,* **255**, 468–479 (1977).

[11] Van de Ven, T. G. M., Interaction between colloidal particles in simple shear flow,*Adv. Colloid Interface Sci.,* **17**, 105–127 (1982).

[12] Gregory, J., Flocculation in laminar tube flow, *Chem. Eng. Sci.,* **36**, 1789–1794 (1981).

[13] Gregory, J., *The Effect of Polymers on Dispersion Properties,* (Tadros, Th.F., ed.), pp. 301–321, Academic Press, London (1982).

[14] Gregory, J., Particle Interaction in Flowing Dispersions, *Adv. Colloid Interface Sci.*, **17**, 149–160 (1982).
[15] Smellie, R. H. and La Mer, V. K., Flocculation, subsidence and filtration of phosphate slimes, *J. Colloid Sci.*, **11**, 720–731 (1956).
[16] Horn, D. *Polymeric Amines and Ammonium Salts,* (Goethals, E. J., ed.), pp. 333–355, Pergamon Press, Oxford and New York (1980).
[17] Bratby, J., *Coagulation and Flocculation,* pp. 136–172, Uplands Press, Croydon/England (1982).
[18] Van de Ven, T. G. M. and Mason, S. G., Comparison of hydrodynamic and colloidal forces in paper machine headboxes, *TAPPI,* **64**, 171–175 (1981).
[19] Takamura, K., Goldsmith, H. L., and Mason, S. G., The microrheology of colloidal dispersion. XIII Trajectories of orthokinetic pair collisions of latex spheres in a cationic polyelectrolyte, *J. Colloid Interface Sci.*, **82**, 190–202 (1981).
[20] Gregory, J., *The Scientific Basis of Flocculation,* (Ives, K. J., ed.), pp. 89–99, Series E. Applied Science No. 27, Sijthoff Noordhoff, The Netherlands (1978).
[21] O'Melia, C. R. as ref. [20], pp. 219–269.

CHAPTER 15

Experiments in Orthokinetic Flocculation

K. J. IVES, Professor of Public Health Engineering, University College London, Gower Street, London WC1E 6BT, UK

1. INTRODUCTION

Orthokinetic flocculation is the aggregation of particles in suspension by collisions due to fluid motion. It differs from perikinetic flocculation where particle collisions are caused by Brownian motion, which in water is confined to particles less than a few microns in diameter.

The fluid motion may be laminar or turbulent, but in either case is characterized by the velocity gradient which only approaches a uniform value in Couette flow. Otherwise a mean value is required, averaged through the fluid volume. In turbulent, transitional or non-uniform laminar flow the mean velocity gradient in a fluid volume is given by

$$\bar{G} = (P/V\mu)^{1/2} \tag{1}$$

where P/V is the power dissipated by fluid motion per unit volume, and μ is the fluid dynamic viscosity.

The Smoluchowski equation describes the rate of reduction of particles by collision in a velocity gradient field of G. In its simplest form for uni-sized particles of diameter d, number concentration N, this is:

$$\frac{-\mathrm{d}N}{\mathrm{d}t} = \frac{2G}{3} d^3 N^2 \ . \tag{2}$$

More complex forms of this equation take account of the change of particle size, and the change of number in each size fraction, together with a collision efficiency (the fraction of collisions which result in permanent contact) and a limit set by the action of fluid shear restricting the maximum size of aggregates (flocs) [1].

This paper describes a series of laboratory experiments carried out at University College London. One set employed a near-monodisperse polystyrene latex suspension in water, destabilized with $Ca(NO_3)_2$, with velocity gradients provided by Couette laminar flow, laminar flow through a fixed bed, laminar flow through a fluidized bed, and non-laminar flow caused by a rising bubble swarm. The other set used kaolin suspensions in water, flocculated with hydrolysed aluminium sulphate ($Al_2(SO_4)_3 16H_2O$), in beakers stirred mechanically with paddles of various shapes. This is the laboratory 'jar test', widely used in practice in water treatment control.

2. COUETTE FLOCCULATION

The apparatus, which has been described in detail by Elson [2], consists of coaxial cylinders, with an annular gap of 3 mm, and an inner rotating cylinder 27 mm diameter (Fig. 1). The annular gap is 150 mm high, and the lower fixed end effect is confined to the bottom 10 mm. The velocity gradient is almost linear, and mean values less than 10 s^{-1} are in uniform laminar shear.

Suspensions of polystyrene latex microspheres† (peak diameter 1.2 μm by Coulter Counter analysis, 30 μm orifice) were in triple membrane filtered (0.45 μm) distilled water containing 0.1-M $Ca(NO_3)_2$ for destabilization.

The total numbers of particles remaining in the suspension, as a function of time of operation of Couette flocculation, is shown in Fig. 2. It can be seen that a significant number reduction takes place with no velocity gradient; this is attributed to perikinetic flocculation. Analysis of the particle size distributions showed that aggregates up to 1600-fold were being formed, particularly at the higher velocity gradients [3]. Measurements of total volume of particles present were constant for about 50 minutes, after this the volume decreased slowly. This was attributed to the settling out of the larger aggregates (flocs) which were being formed.

A recent re-analysis of these data [4] combining equation (2) with a perikinetic term, has used the equation:

$$\frac{-\mathrm{d}N}{\mathrm{d}t_{(\text{Couette})}} = \frac{2}{3} f_o G d^3 N^2 + \frac{4}{3} f_p \frac{kTN^2}{\mu} \tag{3}$$

where f_o and f_p are the orthokinetic and perikinetic collision efficiencies respectively, k is Boltzmann's constant and T the absolute temperature. This analysis has yielded the values f_o = 0.228 (standard deviation 0.01) and f_p = 0.35 (standard deviation 0.02). These values will be compared with some other flocculation data later.

3. FIXED BED (FILTER) FLOCCULATION

A laboratory scale granular (deep-bed) filter was constructed of 21-mm i.d. perspex tube, containing 20 mm depth of 0.55 mm glass spheres (ballotini)

† Dow Chemical Corp.

Fig. 1 – Couette flocculator.

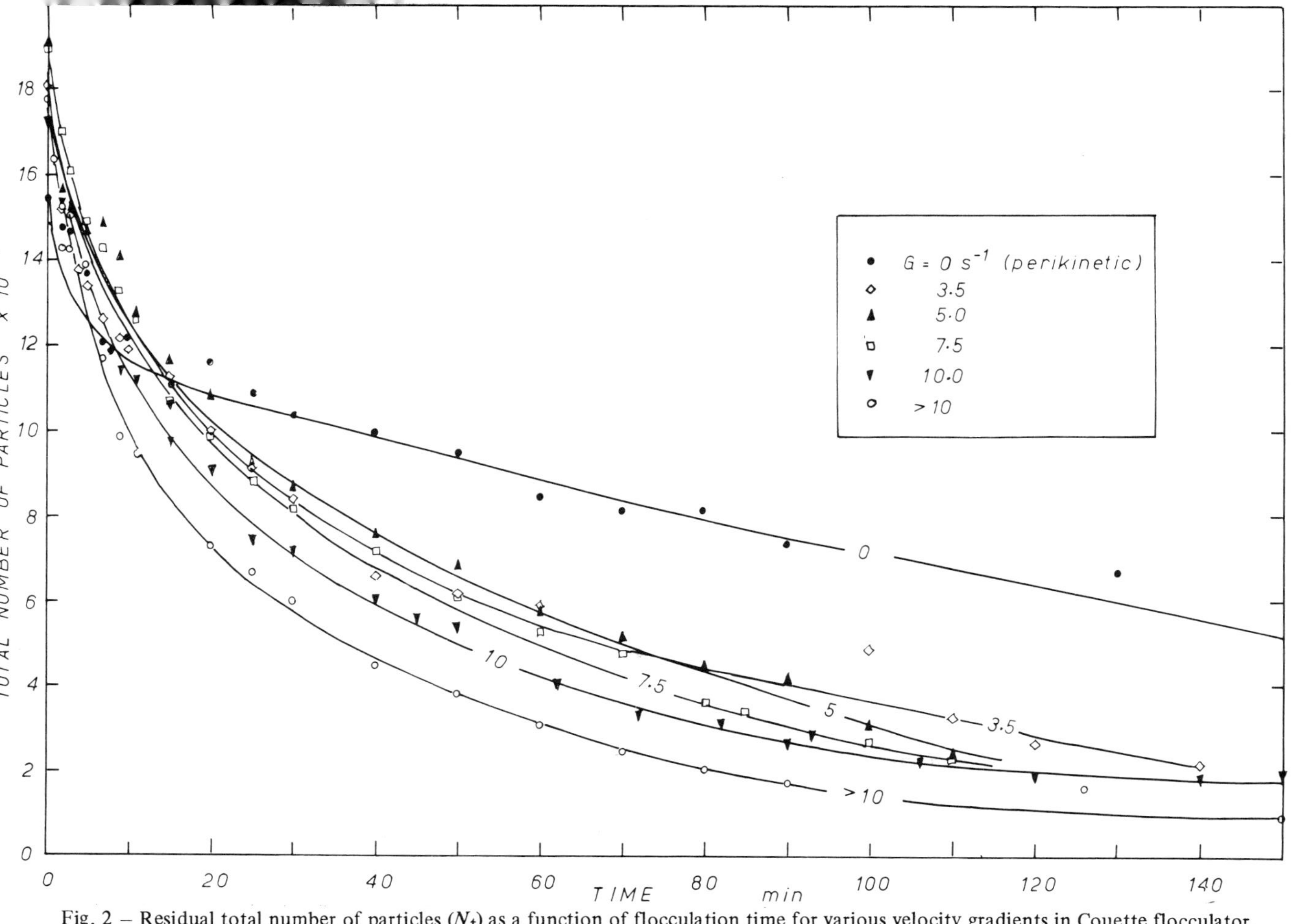

Fig. 2 – Residual total number of particles (N_t) as a function of flocculation time for various velocity gradients in Couette flocculator.

randomly packed. Equal flows of polystyrene latex (1.2 μm) suspension in distilled water and $Ca(NO_3)_2$ were mixed just prior to their flow into the fixed bed.

The mean velocity gradient through the bed is given by equation (1), with the power being dissipated as head loss (H) through the bed thickness (L) at an approach velocity (v_a).

$$\bar{G} = \left(\frac{P}{V\mu}\right)^{1/2} = \left(\frac{\rho g H v_a}{\epsilon L \mu}\right)^{1/2} \tag{4}$$

where ϵ is the porosity of the bed.

As the flow is laminar, but not uniform, the Kozeny equation for hydraulic gradient applies:

$$\frac{H}{L} = \frac{5\mu v_a(1-\epsilon)^2}{\rho g \epsilon^3}\left(\frac{6}{D}\right)^2 \tag{5}$$

Therefore

$$\bar{G} = 13.4\,\frac{v_a(1-\epsilon)}{\epsilon^2 D}\ . \tag{6}$$

Using $v_a = 0.1$ mm s^{-1}, and $\epsilon = 0.4$, giving a residence time of 80 s, $\bar{G}$ was 10 s^{-1} which is comparable to the Couette velocity gradients.

A problem which occurs in any system which has collectors for the particles, is that total particle number reduction by flocculation is simultaneously accompanied by a reduction due to collection. The collection effect was particularly strong in the fixed bed, due to filtration, and almost masked the flocculation effect. In less efficient particle removal systems, as described later with fluidized beds and rising bubble swarms, it is possible to differentiate the collection and flocculation effects, by using variable particle concentrations.

In a fixed bed filter, the single pass collection produces insufficient deposit for any scour of aggregates to occur from the grain surfaces. Therefore any coarsening of the particle size distribution as the suspension passes through the bed, must be due to orthokinetic flocculation within the filter pores. This is shown in Fig. 3 to be a very slight, but positive, effect. The slightness of the effect may be attributed to the low product of $\bar{G}t$ (800) and the diminishing particle concentration due to filtration. In similar flowthrough flocculation of low concentrations, values of $\bar{G}t$ in the range 10^4–10^5 have been required for adequate practical flocculation. It was noted that in the Couette flocculator, where no particle collection took place, very marked flocculation occurred at Gt values of 3×10^4, for $G = 10$ s^{-1} [3].

Recently, a technique has been developed at Imperial College, London [5], where surface chemical treatment of the filter grains has reduced the filtration effect enabling the flocculation effect to be observed better, with a suspension of methylated silica particles flocculated with a non-ionic high molecular weight polyacrylamide.

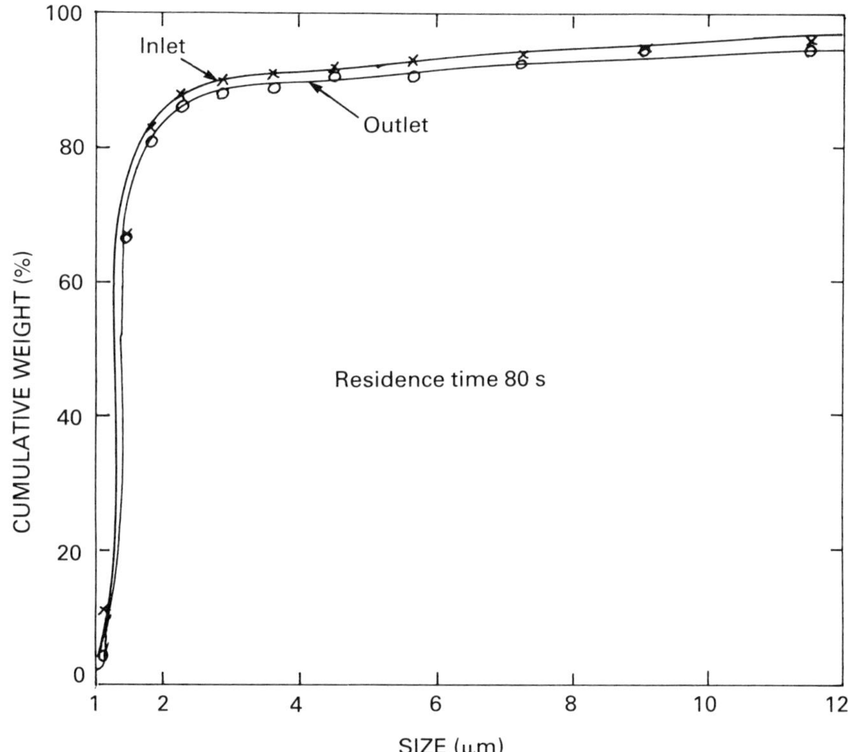

Fig. 3 – Cumulative weight-size distribution before and after filtration with residence time 80 s.

4. FLUIDIZED BED FLOCCULATION

A small scale fluidized bed apparatus, comprised a 25-mm i.d. perspex cylinder, 150 mm deep, mounted on a conical inlet, with a feed of latex suspension and $Ca(NO_3)_2$ solution being mixed at the inlet to the cone (Figs. 4 and 5). Verticality of the fluidized bed vessel was critical, to maintain uniformly distributed upflow. The fluidized bed consisted of PVC/PVA copolymer spheres 125–150 μm diameter, density 1350 kg m^{-3}.

The power dissipated in flow through a fluidized bed is due to the drag of the fluid on the suspended spheres, maintaining them in a fluidized state against settling.

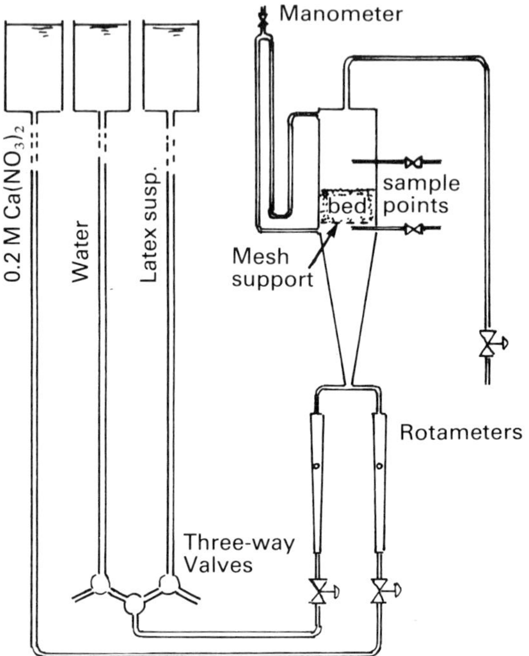

Fig. 4 – Diagram of fluidized-bed flocculator.

Consequently, equation (1) becomes [1]:

$$\bar{G} = \left(\frac{P}{V\mu}\right)^{\frac{1}{2}} = \left(\frac{\nu_a(1-\epsilon)(\rho_s-\rho)g}{\epsilon\mu}\right)^{\frac{1}{2}} \tag{7}$$

With an expanded bed porosity of 0.6 (which for a given bed particle size and density, also defines ν_a at a given water temperature) $\bar{G}$ was nearly 22 s^{-1}. This is in non-uniform laminar flow, and was higher than the 10 s^{-1} maximum velocity gradient in the Couette flocculator. The residence time, t, could be varied by using different bed lengths;.the maximum for a single pass was 300 s, giving $\bar{G}t = 6.6 \times 10^3$. By recycling, multiple passes gave residence times up to 1500 s, with $\bar{G}t = 3.3 \times 10^4$. The effects of the recycling tubing and peristaltic pump were not evaluated separately, but the enhanced flocculation with increasing residence times is very marked, as shown on Fig. 6. This shows the presence of up to 230-fold aggregates [3].

The masking of flocculation reduction of total particle number, by a collection reduction (noted as filtration effect in fixed bed flocculation), is also true in fluidized beds. This collection is desirable in practical systems such as floc blanket clarification of water, where a fluidized bed of (usually metal hydroxide)

Fig. 5 – Fluidized-bed flocculator.

flocs retains particles from the upflowing suspension. However, to determine if flocculation processes are significant, separation of the collection and flocculation effects is required. As collection effects are first-order with respect to suspension particle numbers [6], and flocculation effects are second-order (equation (2)), they can be separated by analysis of the particle concentrations N [7].

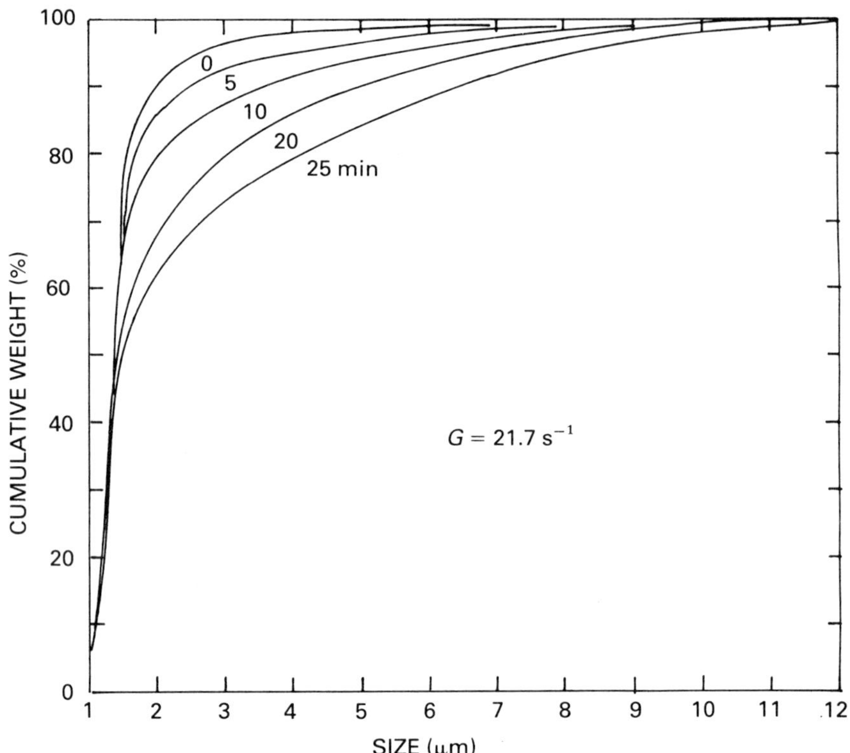

Fig. 6 – Cumulative weight–size distributions after various residence times in a fluidized bed flocculator, multiple pass.

The removal of primary particles by collection, as in a deep bed filter, and presumably in a fluidized bed, is given by [6]:

$$-\delta N_{\text{collect}} = \beta_c N_o \tag{8}$$

Where $\beta_c = [1 - \exp(-\lambda L)]$, and λ is the 'filter' coefficient.

Similarly, the removal by flocculation is

$$-\delta N_{\text{floc}} = \beta_f N_o^{\,2} \tag{9}$$

where $\beta_f = \dfrac{4\bar{G}d^3}{3}\dfrac{\epsilon L}{v_a}$, and $\dfrac{\epsilon L}{v_a}$ is the residence time.

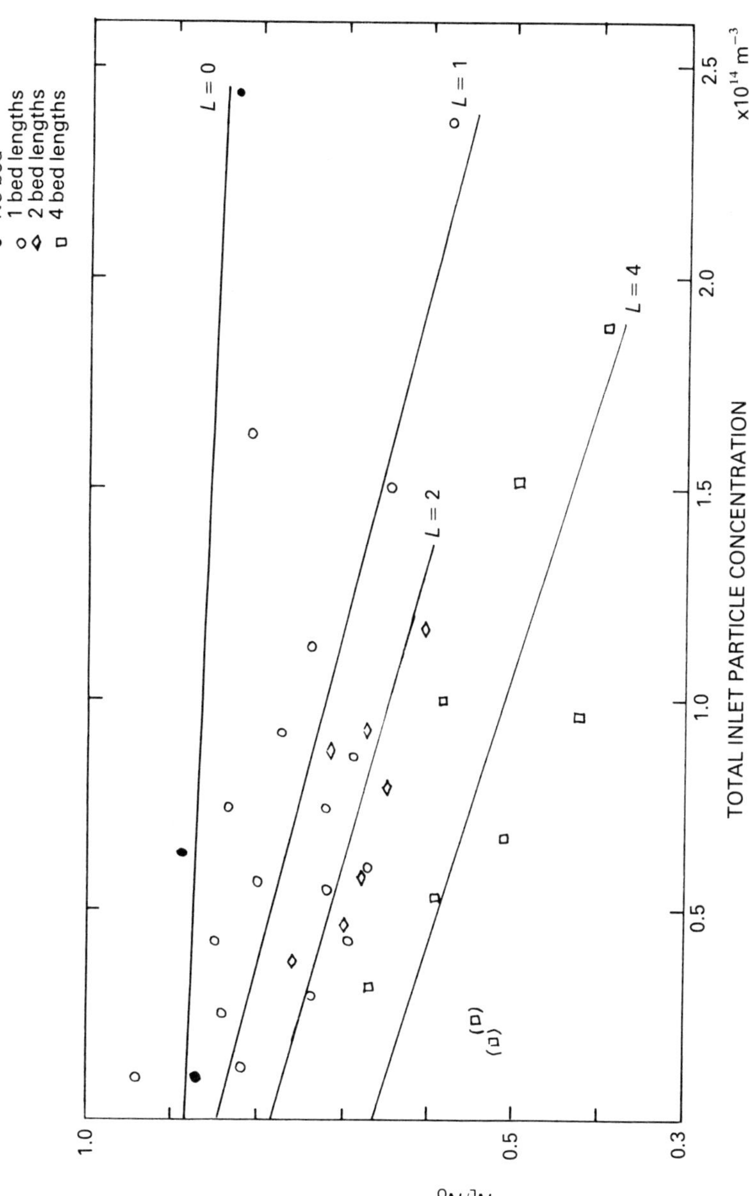

Fig. 7 – Ratio of outlet to inlet singlet particle concentration (N_L/N_O) against total inlet particle concentration.

If L is small, the removal of particles by these two processes is approximately additive.

$$N_L = N_o - \delta N_{collect} - \delta N_{floc} = N_o - \beta_c N_o - \beta_f N_o^2 \ .$$

It follows that:

$$\frac{N_L}{N_o} = 1 - \beta_c - \beta_f N_o \tag{10}$$

which is a linear relationship, with slope of $-\beta_f$ and intercept $(1 - \beta_c)$. Using different inlet concentrations of latex particles N_o, in the range 0.1×10^{14} to 2.5×10^{14} particles m^{-3}, and with different bed lengths L, the values shown in Fig. 7 were obtained, with least squares straight lines fitted through the data. The slopes $(-\beta_f)$ are plotted against length (L) in Fig. 8 indicating the validity of the expression for β_f in equation (9). Also the intercepts $(1 - \beta_c)$ are given by $(1 - \beta_c) = \exp(-\lambda L)$, and, from filtration theory:

$$\frac{N_L}{N_o} = \exp(-\lambda L) \tag{11}$$

or

$$-\ln\left(\frac{N_L}{N_o}\right) = \lambda L \ . \tag{12}$$

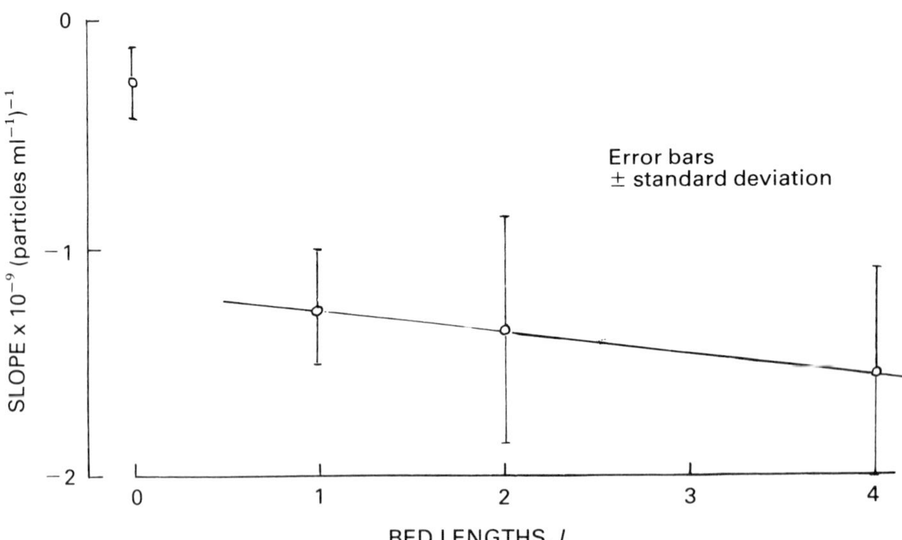

Fig. 8 – Flocculation effect (singlet microspheres).

The values have been plotted in Fig. 9 indicating the validity of equation (12) and, therefore, equation (8).

It can be concluded from Fig. 7–9 that the flocculation and filtration effects were both present and separately identified, with their relationships with bed length confirmed.

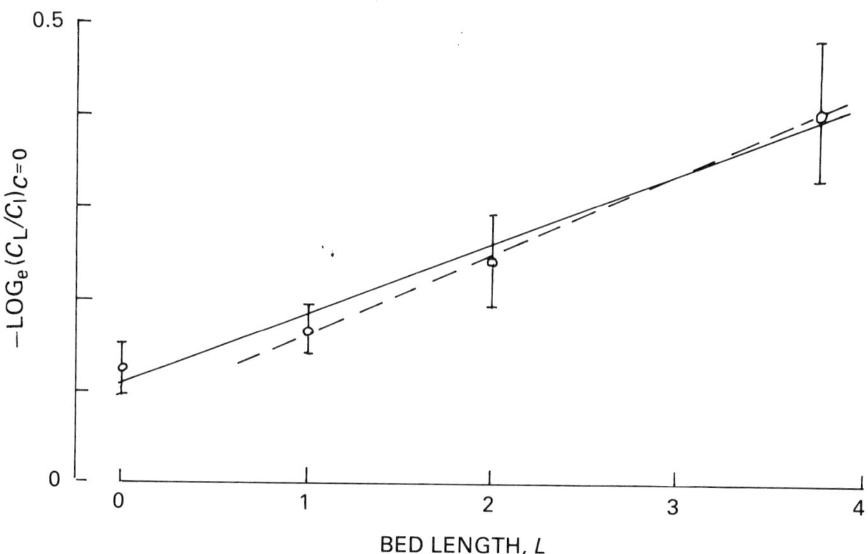

Fig. 9 – Filtration effect (singlet microspheres). Semi-logarithmic plot.

5. BUBBLE FLOCCULATION

The possibility of using bubble swarms as flocculators was discussed in 1955 by Camp [8], but there is no evidence that this was ever investigated experimentally.

A laboratory bubble flocculator using dispersed air from a glass sinter, was 30 mm i.d. and 150 mm high (Fig. 10). The polystyrene latex suspension, destabilized with 0.1 M $Ca(NO_3)_2$, was the same as used in the other experiments but with a 1:50 addition of ethanol to reduce bubble size. Two manometers attached to the bubble flocculation vessel enabled the air bubble fraction to be measured in a depth of 100 mm.

The power dissipated per unit liquid volume by drag as the bubbles rise through the liquid is given by

$$\frac{P}{V} = (\rho - \rho_g)gj_g \tag{13}$$

where ρ_g is the gas (air) density, which is negligible compared with ρ, and j_g is the gas (air) flux $= Q_g/A$, being the volumetric gas flow rate per unit plan

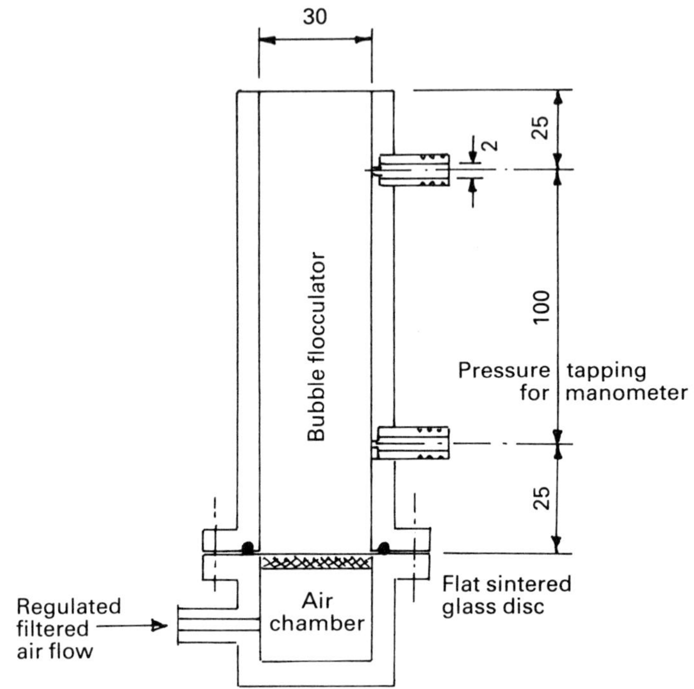

Fig. 10 – Vertical section through cylindrical bubble flocculator (all dimensions in millimetres).

area of the vessel. Consequently, equation (1) for the mean velocity gradient, in this transitional flow system, becomes:

$$\bar{G} = \left(\frac{P}{V\mu}\right)^{1/2} = \left(\frac{\rho g j_g}{\mu}\right)^{1/2} . \tag{14}$$

For the air flow range used in these experiments, 0–400 ml min^{-1}, the gas flux range was 0–10 mm s^{-1}, giving a corresponding non-linear range of mean velocity gradient, according to equation (14), of 0–300 s^{-1}.

Mean bubble size, measured photographically was 624 μm, with an approximately normal size distribution (standard deviation 82 μm), and was independent of the gas flux. The bubble volume fraction (α) increased with gas flux j_g, with the data conforming to equation (15) which was proposed by Wallis [9].

$$j_g = v_\infty \alpha (1 - \alpha)^{0.75} \tag{15}$$

where v_∞ is the rise velocity of a single isolated bubble, which from the empirical data was 80.3 mm s^{-1}. This correlates with the observed bubble diameter, in the appropriate transitional (non-Stokesian) flow regime (*Re* approx. 50).

The bubble flocculation experiments were batch, as contrasted with the flowthrough flocculation in fixed and fluidized bed. Air-free samples were withdrawn by automatic pipette, for Coulter Counter analysis, at successive time intervals after commencing gas flow, normally up to 1000 s.

A preliminary experiment, with observations up to $t = 2700$ s, showed a successive coarsening from the initially nearly monodisperse (mean diameter 1.22 μm) latex suspension, to a size distribution containing up to approximately 100-fold aggregates during the first 120 seconds. After that time the size distribution appeared to revert towards the initial size during the remaining 2500 seconds. However, during this overall period the total number of particles fell from 4.3×10^{14} m^{-3} to 0.093×10^{14} m^{-3}, a reduction of 98% (Fig. 11). The whole process may be explained as: initially small particles were flocculated by the velocity gradients produced by the bubbles, later the resulting larger flocs were collected and removed by the bubbles, leaving a small residual of original particles (2% in this experiment) unflocculated and uncollected. This is the essence of the bubble flocculation–flotation process.

Subsequent experiments were conducted at lower initial particle concentrations (3.3×10^{12} m^{-3}) to slow up the flocculation and collection, for more detailed analysis during the first 300 seconds. These experiments used different gas flow rates, from 50 to 400 ml min^{-1} (Fig. 12). A further set of experiments, at a constant flow rate of 50 ml min^{-1} used different intial concentrations of particles, from 3.3×10^{12} to 200×10^{12} m^{-3} (Fig. 13). As to be expected, both showed declining particle concentrations with time; the rate of decline increasing both with gas flow rate and initial particle concentration.

A mathematical model of the process, consisting of three terms accounting for a reduction of total particle concentration, contains bubble collection, orthokinetic flocculation and perikinetic flocculation.

$$\frac{\mathrm{d}N}{\mathrm{d}t} = -aj_gN - \frac{2f_o'}{3}\left(\frac{\rho g j_g}{\mu}\right)^{1/2} \bar{d}^3 N^2 - \frac{4f_p'}{3}\frac{kTN^2}{\mu} . \tag{16}$$

The linear dependence of the bubble collection (first) term on gas flux j_g has been confirmed by analysis of the experimental data. Also the square-root dependence on gas flux in the orthokinetic flocculation (second) term has similarly been confirmed. An integrated solution of equation (16) has been fitted to the data, shown as the solid lines on Figs 12 and 13. From this analysis values of the bubble collection coefficient, a, and the orthokinetic and perikinetic collision efficiencies, f_o' and f_p', have been estimated [10].

$a = 1.34\,(0.05)$ m^{-1}; $f_o' = 0.192\,(0.02)$; $f_p' = 0.35\,(0.05)$.

(standard deviations are in brackets).

These flocculation collision efficiencies for bubble flocculation can be compared with those for the Couette flocculator, where $f_o = 0.228\,(0.01)$ and

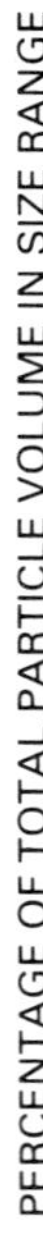

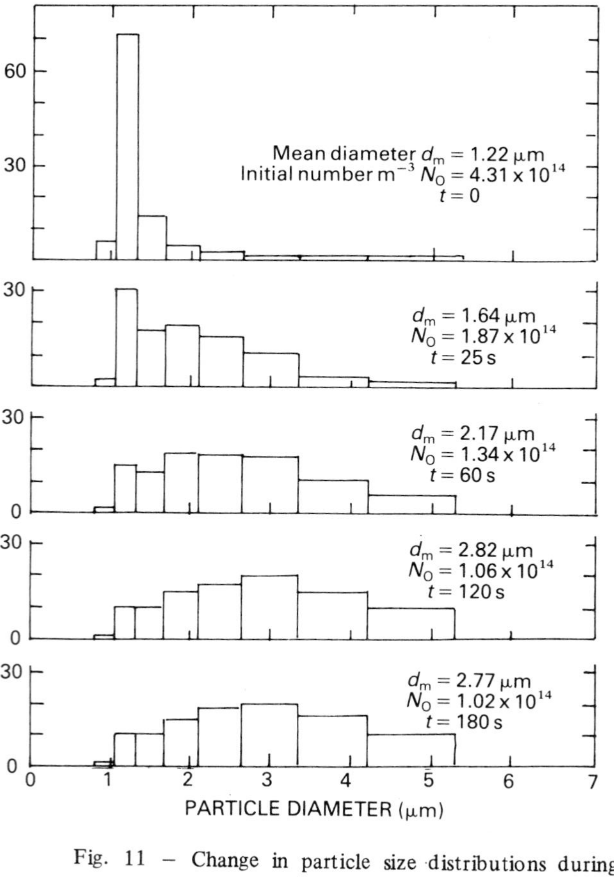

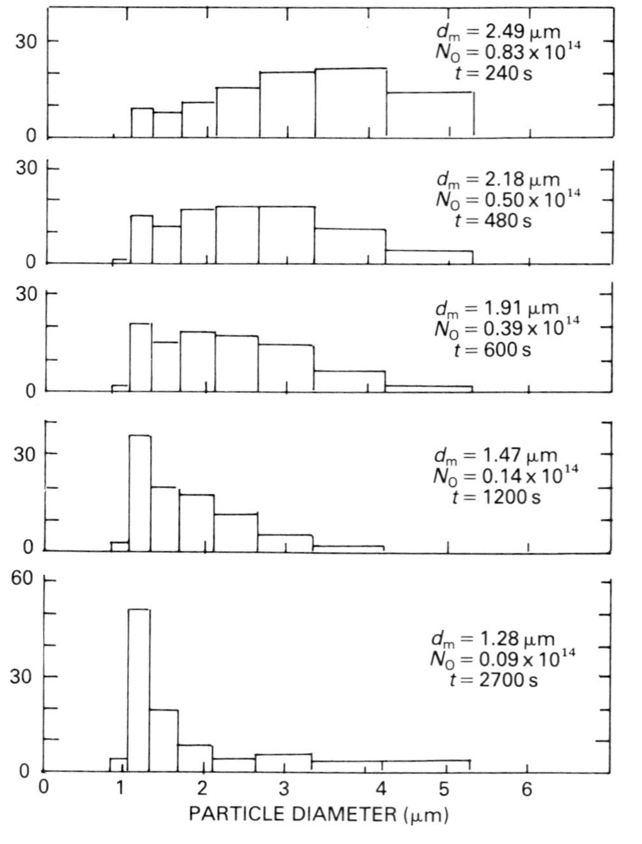

Fig. 11 – Change in particle size distributions during a bubble flocculation experiment Q_g = 500 ml min^{-1}.

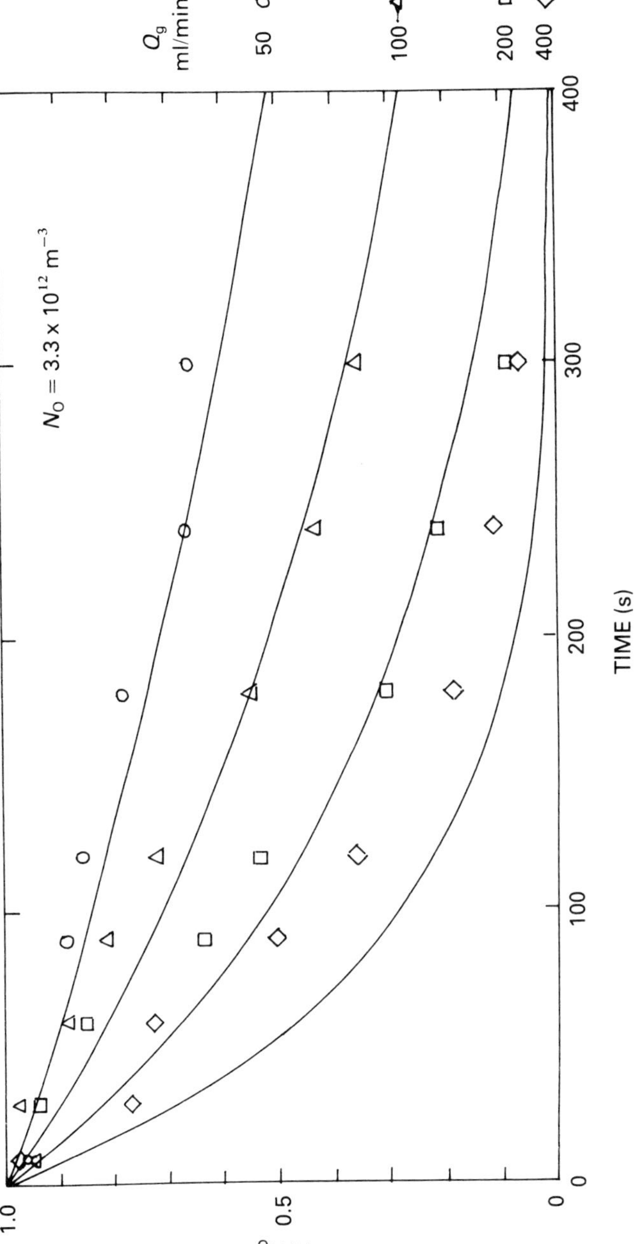

Fig. 12 – Effect of gas flow rate Q_g at constant low initial particle concentration in the bubble flocculator.

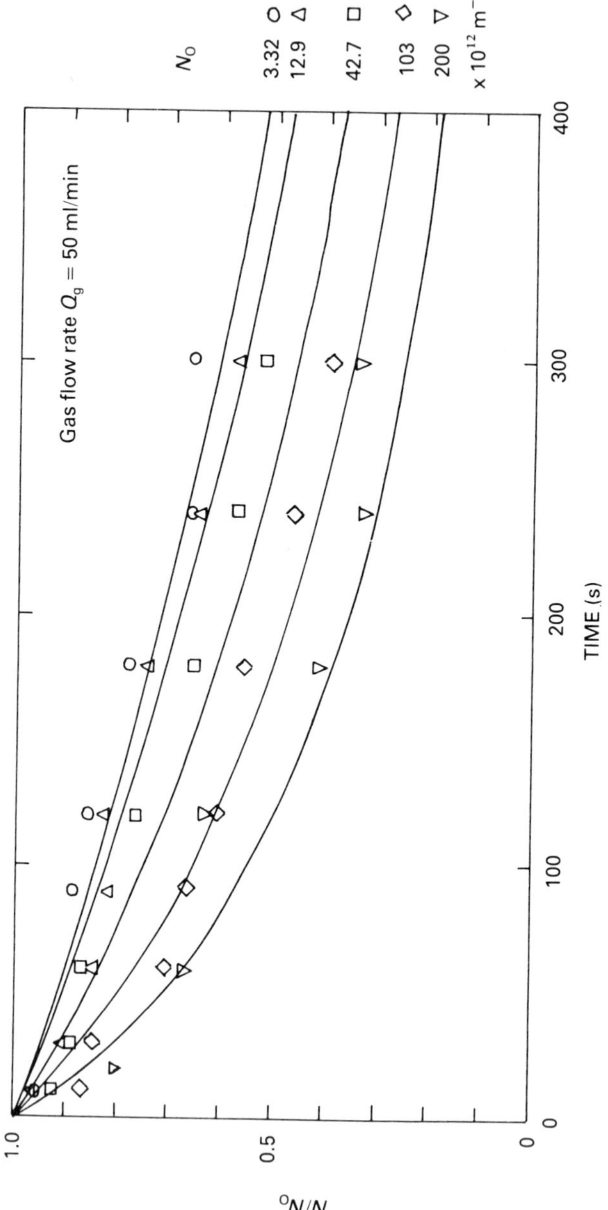

Fig. 13 – Effect of initial particle concentration at constant gas flow rate in the bubble flocculator.

f_p = 0.35 (0.02). Although both were batch experiments, the differences in orthokinetic collision efficiencies (about 17%) may be attributed to the very different flow regimes and velocity gradients, and the removal of larger particles from the system in the case of the bubble flocculator. The agreement of the perikinetic collision efficiencies indicates that the ethanol in the bubble flocculator had no effect on flocculation.

6. PADDLE STIRRER FLOCCULATION (JAR TEST)

The waterworks laboratory test of paddle stirring in beakers, known as the jar test, is widely used for assessing the efficiency of flocculants to produce rapidly settling flocs. A set of beakers (typically six) are stirred simultaneously after the addition of flocculant doses with a rapid mixing period. The set of beakers enables comparative assessments to be made of various flocculant doses, pH values and paddle designs on samples of the same suspension. Some details of the jar test are given in Refs [1] and [11].

The most common design of paddle blade in the UK is flat and rectangular. However, other designs are used elsewhere and a survey of literature from Czechoslovakia, Netherlands, USA, and West Germany revealed the following alternative geometries: propeller, mesh, prongs (6, 8 or 10 mm), fork (picket fence with stators), perforated plate, square wire, small H, large H with stator. These are shown in Fig. 14, where the original geometries, all of which are essentially in one plane, have been scaled to fit in 1-litre tallform beakers as used in the UK conventional jar test. These various designs (although there is no evidence in the literature of any shape being 'designed' on any rational or hydrodynamic grounds) have been compared in the laboratory using a standard kaolin suspension, flocculated with aluminium sulphate solution, which simulates the waterworks flocculation of turbid natural waters.

The suspension was prepared from purified kaolin clay† blended in de-ionized water, and settled for 4 hours. The remaining unsettled suspension was diluted into London tapwater, to 100 mg l^{-1}. Laboratory reagent† $Al_2(SO_4)_3.16H_2O$ (alum) was prepared as a solution in de-ionized water at a stock concentration of 10 g l^{-1}.

Initial tests compared the various paddle geometries for the mean velocity gradients ($\bar{G}$) which they imparted to the water in the 1-litre beakers. By measuring the torque on the paddle shaft (torquemeter‡) the value of $\bar{G}$ could be calculated as a function of paddle rotational speed, using equation (1)

$$\bar{G} = \left(\frac{P}{V\mu}\right)^{1/2} = \left(\frac{2\pi\omega\tau}{V\mu}\right)^{1/2} \tag{17}$$

† Hopkin & Williams Ltd., UK.
‡ Power Instrument Inc., USA, Model 781-B-2.

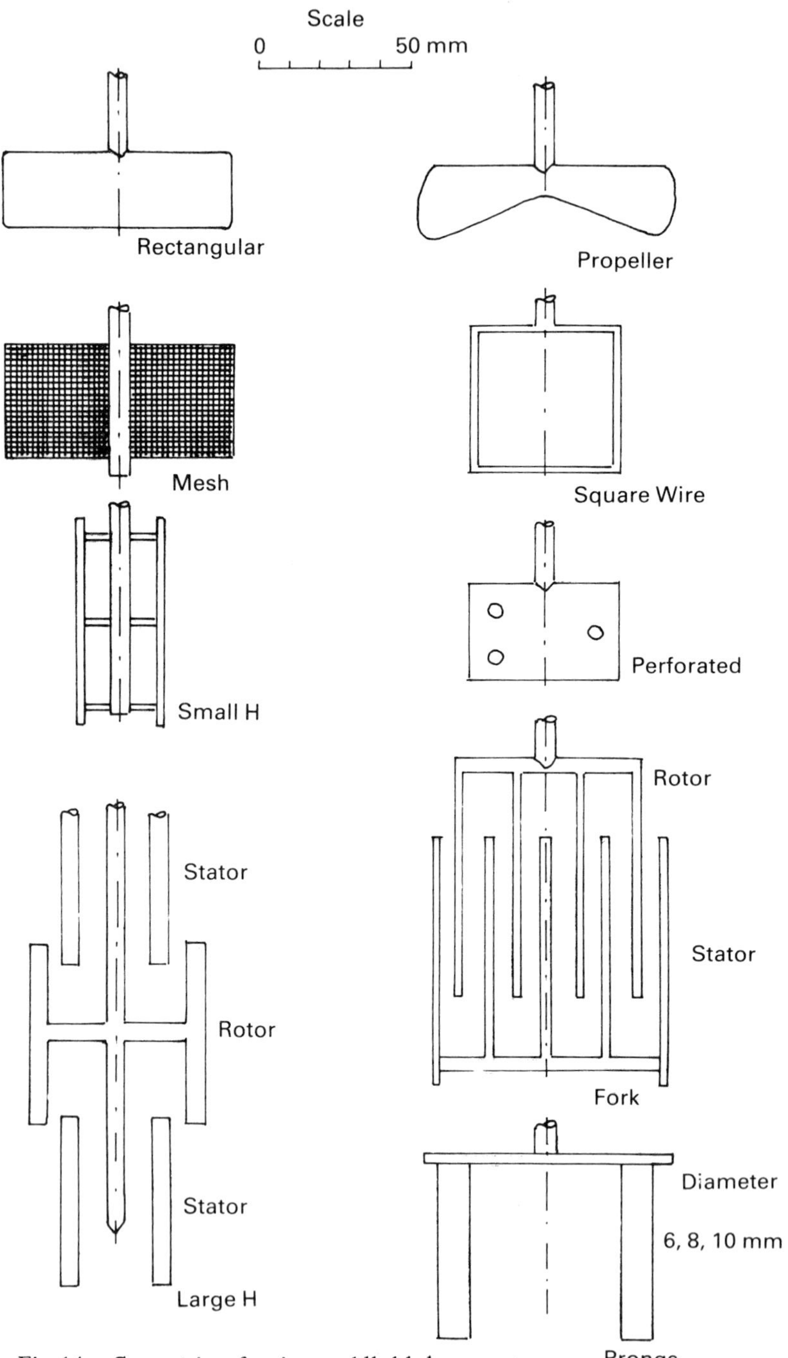

Fig. 14 – Geometries of various paddle blades.

where ω is the rotational speed (revolutions s^{-1}) and τ is the measured torque. The results of these measurements are shown on Fig. 15. If mean velocity gradient were the only criterion of blade design, the propeller would appear to be the best and the small H the worst. Excluding the two blades with stators (fork, large H) the mean velocity gradient correlates linearly with overall horizontal dimension ('diameter' when rotating), which may be due, because of fixed beaker diameter, to the clearance between the rotating edge and the beaker wall.

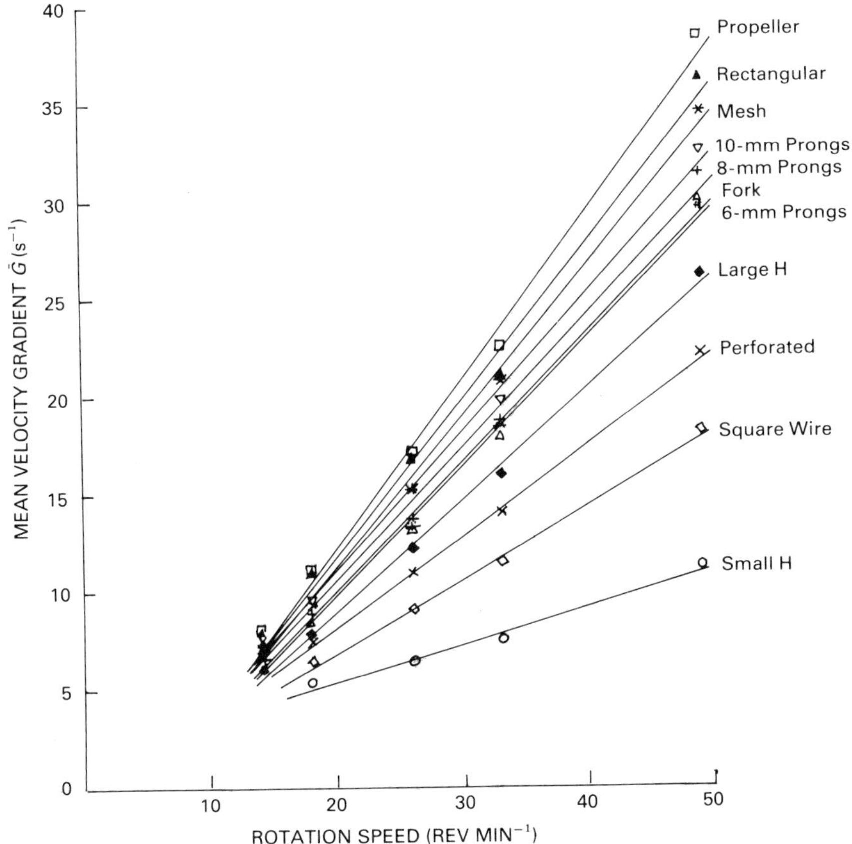

Fig. 15 – Mean velocity gradient vs. rotation speed for various paddle geometries.

A further assessment of these various blades, was made by observing the effects of flocculation on the suspension. Due to the initially unknown size distribution of the clay (much of which was colloidal) and the fluffy, precipitated flocs of hydrous aluminium oxide, particle size analysis of the flocculating suspension was not feasible as with the latex microsphere suspensions. In keeping with the practical nature of the jar test, and of the clay suspension and alum flocculant, practical assessments of settling and filterability were used. These measured the clarification of the flocculated suspension, using nephelometry (Hach turbidimeter†), after sedimentation or filtration respectively. The nephelometer was calibrated with kaolin suspensions of known concentration giving an almost linear relationship with Nephelometric Turbidity Units (NTU). Using the rectangular blades only, the optimum alum doses were determined for settling and filterability, after 15 min stirring at $\bar{G} = 25\ s^{-1}$. Then with these optimum doses, the various paddle blade designs were assessed by the same settling and filterability criteria.

Settling was tested by withdrawing a pipette sample from 30 mm below the water surface in the beaker, taken 10 minutes after the paddle rotation had ceased. Filterability utilized the whole suspension, immediately after the paddle rotation had ceased, filtered through 40 mm depth of 0.5–0.6 mm sand, in the Armfield Filterability Apparatus††. In settling, the residual turbidity is a sufficient measure. In filterability, the head loss H, approach velocity v_a, time t, and inlet and filtrate turbidities C_o and C_F, all enter into the Filterability Number F [12].

$$F = \frac{HC_F}{v_a C_o t} \,. \tag{18}$$

From Fig. 16 it can be seen that for settling, the small H blade gave the worst turbidity which may be related to the fact that it also gave the lowest $\bar{G}$ values. There is little to choose between the rest, with the standard rectangular blade not being outstanding in any way.

The Filterability Numbers are shown on Fig. 17, where the square wire, small H and mesh paddles appear to be the best. There is little to choose among the others.

There appears to be no rational way to choose a paddle blade design on the basis of the mean velocity gradient produced, the resulting settling of flocs, or filterability of the flocculated suspension. Other criteria may be more significant, such as with three-dimensional blades (e.g. a cross-shape in plan), or differently shaped beakers, which were not tested.

7. CONCLUSIONS

There is no doubt that velocity gradients cause orthokinetic flocculation, and that these can be produced in uniform laminar flow (Couette), non-uniform laminar flow (fixed and fluidized beds), and transitional and turbulent flow (bubble swarms, and paddle stirrers). For highly characterized, monodisperse suspensions, such as latex microspheres destabilized with $Ca(NO_3)_2$, some of these systems can be compared and the effects of flocculation and collection

† Hach Chemical Co., USA, Model 2100A.

† Armfield Technical Education Ltd., UK Model W4.

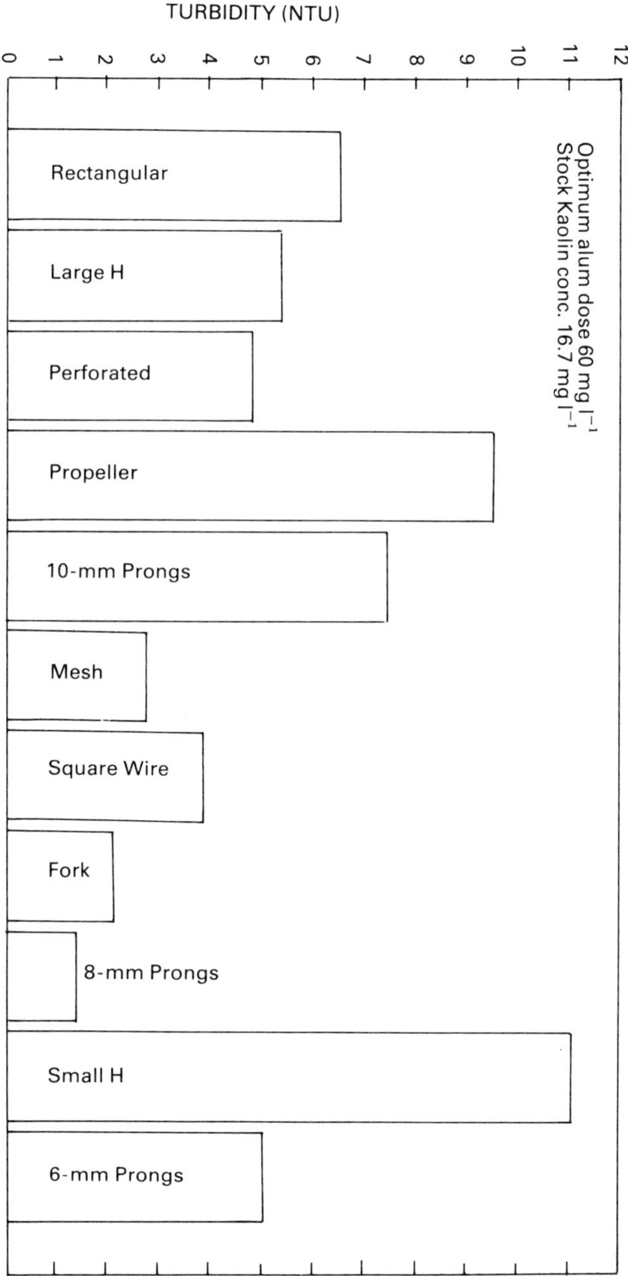

Fig. 16 – Turbidity after sedimentation for various paddle geometries.

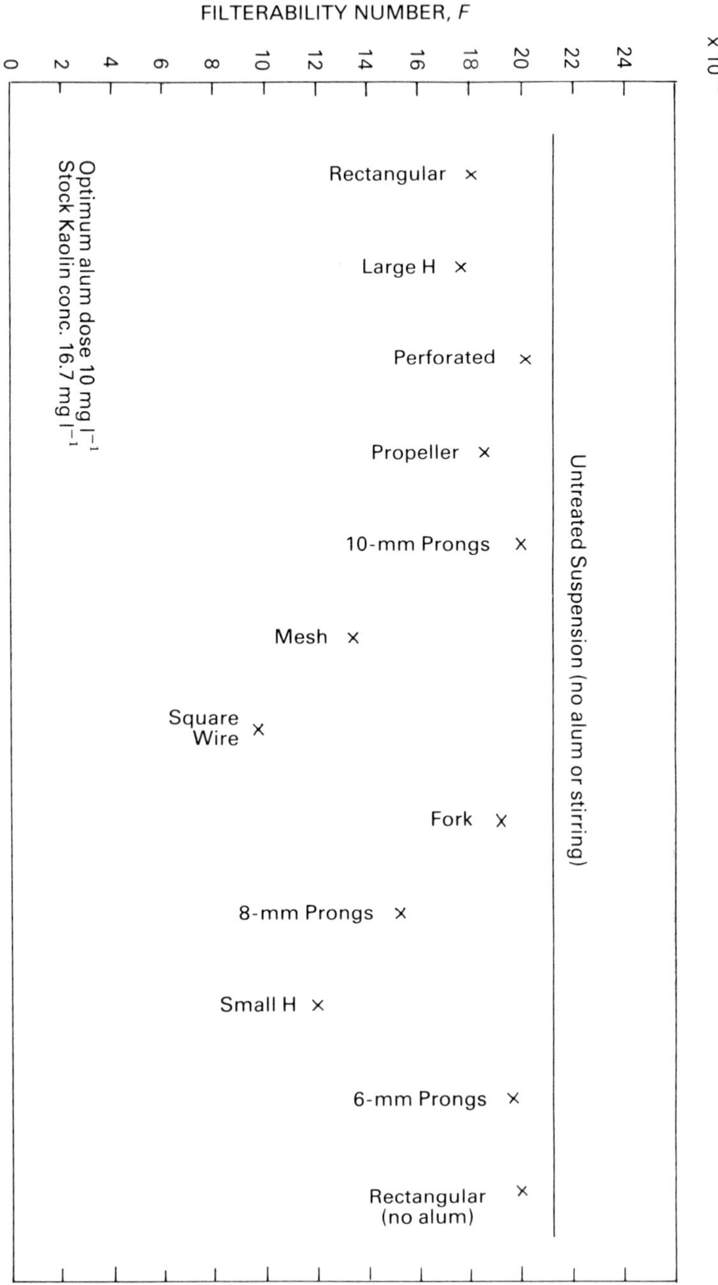

Fig. 17 – Filterability numbers for various paddle geometries.

separated. In certain cases, collision efficiencies have been estimated and compared; identical values have been obtained with perikinetic flocculation, and similar values in orthokinetic flocculation.

In more practical systems using clay suspensions in tapwater, flocculated with alum, in jar tests, the velocity gradient does not appear to be a sufficient criterion to choose among several alternative paddle blade designs.

ACKNOWLEDGEMENTS

The experiments described in sections 2–5 were supported by SERC grants, and I am particularly indebted to my colleagues, Dr T. P. Elson, Dr M. Al-Dibouni and Mr M. J. Kneen, for their collaboration. The experiments in section 6 were made by undergraduates in civil engineering: A. F. Abu Hassan Ashari, K. Rahim, M. Tsui and K. M. Yick.

SYMBOLS

a	collection coefficient in bubble flocculation	m^{-1}
A	plan area of flocculation vessel	m^{-2}
C_o	initial concentration (Filterability test)	NTU
C_F	filtrate concentration	NTU
d	diameter of particles	m
D	diameter of grains	m
f_o	orthokinetic collision efficiency	–
f_p	perikinetic collision efficiency	–
F	Filterability number	–
g	gravitation acceleration (9.81)	$m\ s^{-2}$
G	velocity gradient	s^{-1}
$\bar{G}$	mean velocity gradient	s^{-1}
H	head loss in filtration	m
j_g	gas flux	$mm\ s^{-1}$
k	Boltzmann's constant	$J\ K^{-1}$
L	thickness of filter media	m
N	particle number concentration	m^{-3}
N_o	initial particle number concentration	m^{-3}
N_L	particle number concentration at depth L	m^{-3}
P	power	W
Q_g	volumetric gas (air) flow rate	$m^3\ s^{-1}$
Re	Reynolds number	–
t	time	s
T	absolute temperature	K
v_a	approach velocity	$m\ s^{-1}$
v	rise velocity of isolated bubble	$m\ s^{-1}$
V	volume of suspension	m^3

α	gas bubble volume fraction	–
β_c	collision parameter	–
β_f	flocculation parameter	m^3
ϵ	filter porosity	–
λ	filter coefficient	m^{-1}
μ	dynamic viscosity	Pa s
ρ	liquid density	kg m^{-3}
ρ_g	gas density	kg m^{-3}
ρ_s	solids density	kg m^{-3}
τ	torque	N m
ω	rotational speed	rev s^{-1}

REFERENCES

[1] Ives, K. J. (Ed.), *The Scientific Basis of Flocculation,* NATO ASI Series, Sijthoff-Noordhoff, Alphen a/d Rijn, NL (1978).

[2] Elson, T. P., Velocity profiles of concentric flow between coaxial rotating cylinders with a stationary lower boundary. *Chem. Eng. Sci.,* **34**, 373–377 (1979).

[3] Ives, K. J., and Al-Dibouni M., Orthokinetic flocculation of latex microspheres. *Chem. Eng. Sci.,* **34**, 983–991 (1979).

[4] Elson, T. P., Personal communication. Dept. Chemical Engineering University College London, 1983.

[5] Graham, N. J. *Significance of Filter Pore Particle Flocculation in Direct Filtration.* Ph.D. thesis, University of London (1982).

[6] Ives, K. J. (Ed.), *The Scientific Basis of Filtration,* NATO ASI Series, Noordhoff, Leyden (1975).

[7] Elson, T. P., Flocculation in fluidised beds., *Inst. Chem. Engrs.,* 4th Ann. Res. Meeting, University College London, April 1979.

[8] Camp, T. R., Flocculation and flocculation basins. *Trans. Am. Soc. Civ. Engrs.,* **120**, 1–16 (1955).

[9] Wallis, G. B., *One-Dimensional Two-Phase Flow.* McGraw-Hill, New York (1969).

[10] Ives, K. J., Elson, T. P., and Kneen, M. J., *Bubble Flocculation in Water Treatment.* SERC Final Report GR/B03634, December 1982.

[11] Svarovsky, L. (Ed.), *Solid-Liquid Separation.* Butterworths, London, 2nd Edition (1981).

[12] Ives, K. J., A new concept of filterability. *Prog. Wat. Tech.,* **10** (5/6), 123–137 (1978).

CHAPTER 16

Modelling the Effects of Adsorbed Hydrolysed Aluminium on the Electrical Double-Layer Properties of Aqueous Colloids

RAYMOND D. LETTERMAN and **DHARMARAJAN R. IYER,** Department of Civil Engineering, 150 Hinds Hall, Syracuse University, Syracuse, New York 13210, USA

1. INTRODUCTION

The destabilization of aqueous suspensions by the addition of a hydrolysable metal salt, particularly aluminium sulphate, is an important process in a number of engineered systems. The coagulation process used in many municipal and industrial water clarification systems is a good example.

At the present time the practical application of aluminium salts as coagulants is based almost entirely on empirical relationships. In a few cases, operational data have been used to derive site-specific empirical equations which relate coagulant concentration and solution and suspension characteristics to solid–liquid separation efficiency. However, in most cases, engineers and operators use a rule-of-thumb approach supplemented by time-consuming batch flocculation tests and measurements such as particle electrophoretic mobility.

Computer programs for the interpretation of chemical processes in aqueous systems, particularly in the calculation of multi-component, multi-phase equilibria, are becoming increasingly available. Models exist which can be used to solve simultaneously equilibria involving ionization and complexation reactions at colloid surfaces as well as reactions (ionization, precipitation, complexation) in the bulk solution. One of the newer and more advanced programs with this capability, MINEQL, employs a detailed model of the electrical double layer to account for the distribution of charge and potential at the colloid–solution interface. These advances, combined with a growing knowledge of the effect of hydrolysed metal ions and hydroxide precipitates on colloid stability, strongly suggest that computerized chemical equilibrium models such as MINEQL should be studied as tools for predicting the effect of hydrolysable metal salts on the electrical double-layer properties of aqueous colloids and on the kinetics of flocculation.

2. ADSORPTION OF HYDROLYSED ALUMINIUM

It has been well documented that hydrolysed aluminium adsorbs on a variety of hydrophobic colloidal surfaces [1–5]. A comparison of stability domain diagrams obtained using mineral particles with those obtained using hydrophilic particles such as bacteria [6] and humic sols [7] suggests that adsorption of hydrolysed aluminium is an important mechanism in these systems also.

Evidence exists which suggests that all types of aluminium hydrolysis products are adsorbed. Some investigators [1, 8, 9] have proposed that the predominant adsorbable species are polynuclear compounds such as $Al_8(OH)_{20}^{4+}$ or $Al_6(OH)_{12}^{6+}$. There is evidence [1] that simple monomeric species such as $Al(OH)^{2+}$ adsorb on mineral surfaces by a cation exchange process. Brown and Hem [1] and McAtee and Wells [2] have observed that preformed microcrystalline aluminium hydroxide (gibbsite) adsorbs on mineral surfaces. In any case, the uptake of aluminium increases as hydrolysis of the aluminium increases.

Stumm and Morgan [10] propose that hydrolysis increases the tendency to adsorb by rendering the complex more hydrophobic by reducing the effective charge of the central ion or ions and hence decreasing the interaction between the central Al atom and the remaining aquo groups. This, they propose, might then enhance the formation of covalent bonds between the Al ion and specific sites on the solid surface by reducing the energy necessary to displace water molecules from the coordination sheath. A similar argument was used by James and Healy [11] in developing a thermodynamic model for hydrolysed metal ion adsorption.

James and Healy [11] and Matijevic [4] have presented evidence which indicates that the adsorption of hydrolysis products leads to the formation of a layer or a partial layer (depending on the amounts of aluminium, hydrogen ion, Al-complexing ligands, etc., in the system) of amorphous hydroxide precipitate on the particle surface. James and Healy referred to this as 'surface nucleation or precipitation' and concluded that it results from a lowered stability of the precipitate at the surface due to the interfacial electric field. White [12], using electrophoretic mobility measurements of aluminium-treated particles at various pH values, observed that the desorption of adsorbed aluminium hydrolysis products at low pH is very slow and he concluded that the adsorbed aluminium formed a coating of amorphous aluminium hydroxide precipitate. James [13] has stated recently that one of the challenging problems that remains to be answered is whether the coating forms by surface condensation of simple ionic species, surface condensation of hydrolysed species, including polymers, or whether it is due to adhesion of colloidal hydroxides formed in the solution (i.e. heterocoagulation).

Brown and Hem [1] observed that when mineral surfaces are introduced into a system containing aqueous polynuclear aluminium hydroxide species the polynuclear fraction is rapidly adsorbed. However, the difference in pH between the blank solutions and those containing surfaces is small compared with the difference in the polynuclear aluminium hydroxide species content. Since polymerization upon adsorption would tend to produce H^+ and therefore lower the pH, Brown and Hem concluded that precipitation upon adsorption was minimal.

Figure 1 is an isotherm-type plot for the adsorption of hydrolysed aluminium onto a suspension of silica particles at pH 5. Corresponding values of particle electrophoretic mobility are also shown. The abscissa, C_e, is the molar concentration of total aluminium in the filtrate after the particles of silica with adsorbed aluminium have been removed by a 0.4-μm nucleopore filter. Similar isotherms

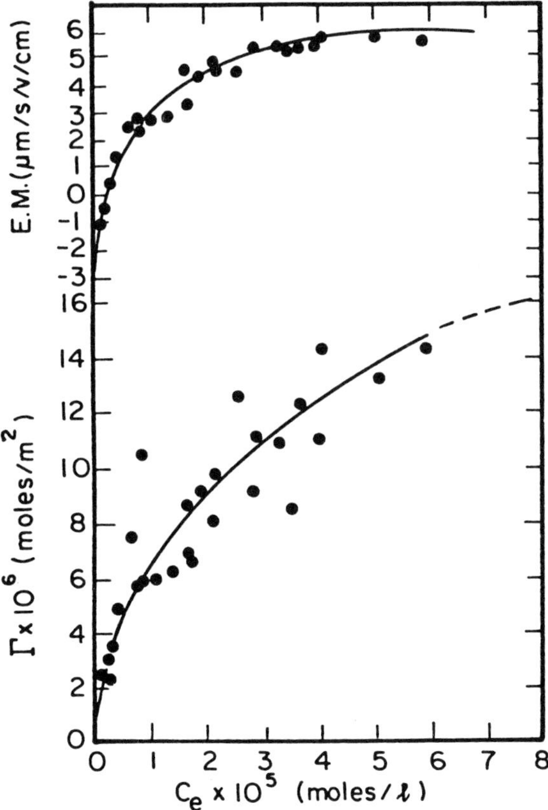

Fig. 1 – Moles of Al adsorbed per m² of silica surface and particle electrophoretic mobility versus molar concentration of aluminium remaining unadsorbed (Min-U-Sil 5 particles and pH = 5).

have been presented by Hahn and Stumm [5] for hydrolysed aluminium(III) and O'Melia and Stumm [14] for iron(III).

According to Fig. 1, under these experimental conditions an electrophoretic mobility of zero is obtained at a surface aluminium concentration of approximately 3×10^{-6} mole/m^2 and a corresponding unadsorbed Al concentration of roughly 3×10^{-6} M. At a surface concentration of approximately 9×10^{-6} mole/m^2 the electrophoretic mobility curve tends to approach an asymptote suggesting that the adsorbed aluminium hydroxide is controlling the surface properties of the particles.

Hahn and Stumm [5] have shown that the pH of the solution has a significant effect on the hydrolysed aluminium(III) adsorption isotherm. In general, as the pH is increased from, say, 4 to 6 the isotherm plots become progressively steeper. In other words, as the pH increases the solution equilibrium concentration corresponding to a given surface aluminium concentration decreases. Above pH 6 data like those plotted in Fig. 1 are very difficult to obtain because a significant fraction of the unadsorbed aluminium hydroxide tends to be removed from suspension by membrane filtration.

At the present time there are no models which can be used to predict quantitatively the adsorption of hydrolysed aluminium(III). Existing models, e.g. James and Healy [11] and the site-binding model used in MINEQL, must assume a certain limiting surface coverage or total site density and this limit is often much less than the surface coverages observed when adsorption experiments are conducted, e.g. Fig. 1 and Hahn and Stumm [5].

3. CHEMICAL EQUILIBRIUM MODEL

The computer program MINEQL, written by Westall, Zachary and Morel [15] was designed to solve for the composition of complex chemical systems which may involve solution species as well as precipitating or dissolving solids. It also is able to deal with purely chemical interactions at aqueous surfaces, i.e. in its basic, unamended form it does not include the effect of the electrical double layer. Davis *et al.* [16] expanded the MINEQL program to include the effect of the electrical double layer on ionization and electrolyte binding (complexation) reactions for surface functional groups.

The surface ionization and complexation model described by Davis *et al.* [16] consists of a set of equations that can be solved numerically by the MINEQL program. Written for a hypothetical oxide/hydroxide on which the ionizable sites are assumed to be amphoteric, $\underline{S}OH$, groups, the ionization and electrolyte binding reactions are, in a NaCl electrolyte:

$$\underline{S}OH \rightleftharpoons \underline{S}O^- + H^+ \quad (1)$$

$$H^+ + \underline{S}OH \rightleftharpoons \underline{S}OH_2^+ \quad (2)$$

$$Na^+ + \underline{S}O^- \rightleftharpoons \underline{S}O^-Na^+ \tag{3}$$

$$Cl^- + \underline{S}OH_2^+ \rightleftharpoons \underline{S}OH_2^+Cl^- \tag{4}$$

The concentrations of surface sites are related to solution species concentrations through intrinsic equilibrium constants. It is assumed that the activities of protons and adsorbing ions are modified by the electrical work necessary to bring them from the bulk solution to the plane in which they react. The concentration at the locus of adsorption, x, is related to the bulk solution concentration by the Boltzmann distribution, i.e.

$$[\text{ion}]_x = [\text{ion}]_{\text{bulk}} \exp\left(-\frac{z\,e\,\psi_x}{kT}\right) \tag{5}$$

where ψ_x is the potential at the locus of adsorption of the ion compared to a bulk solution potential of zero and z is the difference in charge between the complexed and uncomplexed surface site. For the chloride ion (equation (4)) the intrinsic equilibrium constant becomes:

$$K_{\text{Cl}}^{\text{int}} = \frac{[\underline{S}OH_2^+Cl^-]}{[\underline{S}OH_2^+]\,[Cl^-]\exp\left(\frac{e\,\psi_x}{kT}\right)} \tag{6}$$

where $[Cl^-]$ is the bulk solution concentration of chloride ion.

To facilitate computation, the complexation reactions, e.g. equations (3) and (4), are usually written in terms of the uncharged, $\underline{S}OH$, site. For the chloride ion

$$\underline{S}OH + Cl^- + H^+ \rightleftharpoons \underline{S}OH_2^+Cl^- \;. \tag{7}$$

The intrinsic complexation constant, in this case, is given by:

$$*K_{\text{Cl}}^{\text{int}} = K_{\text{Cl}}^{\text{int}} \big/ K_{\text{al}}^{\text{int}} \tag{8}$$

where $K_{\text{Cl}}^{\text{int}}$ is defined by equation (6) and $K_{\text{al}}^{\text{int}}$ is the intrinsic ionization constant for equation (1).

The total molecular concentration of surface sites, S_T, is related to the mass concentration, M, specific surface area, A_s, and surface site density, N_s, of the particles. For the hypothetical case of equations (1) to (4),

$$S_T = M A_s N_s/N_a = [\underline{S}O^-] + [\underline{S}OH] + [\underline{S}OH_2^+] + [\underline{S}OH_2^+Cl^-] + [\underline{S}O^-Na^+] \tag{9}$$

where N_a is Avogadro's number.

An electrical double layer model is needed to solve the surface speciation problem. The major difference among the models which have been used is in the geometric description of the adsorption process, particularly its effect on the expression for coulombic interactions. The simplest case is the one in which all ions (H^+, Na^+, Cl^-) are considered to react with the surface in the same surface plane and, therefore, the same potential, ψ_0, is assumed to apply to all of them. The surface charge is simply the sum of the charges of the surface species. An example of this approach is given by Hohl, Sigg and Stumm [17].

Davis, James and Leckie [16], in formulating their more detailed approach to the surface speciation calculations, applied what is effectively the Gouy–Chapman–Stern–Grahame (GCSG) model of the electrical double layer. In this scheme H^+ and OH^- are considered to react in an inner adsorption layer (effectively in the solid) while specifically sorbed ions, including major electrolytes, are in a layer adjacent to the surface. Two different capacitances are assumed for the inner and outer regions of the compact (sorbed) layer. (Electrical double layer models exist for systems in which the surface of the particle is porous and ions from the bulk solution penetrate the surface [18-20]. Since in the adsorbed hydrolysed aluminium system it is thought that the adsorbed aluminium(III) forms a gel coating on the particles, future study will involve testing the appropriateness of the GCSG model and, if necessary, evaluating a porous surface model).

In the GCSG model as applied by Davis *et al.* [16], the inner layer or surface charge, σ_0, is determined by the surface proton balance, i.e. the net number of protons released or consumed by all surface reactions. For the hypothetical surface,

$$\sigma_0 = \frac{F}{MA_s} \{[\underline{S}OH_2^+] + [\underline{S}OH_2^+ Cl^-] - [\underline{S}O^-] - [\underline{S}O^- Na^+]\} \tag{10}$$

where F is the Faraday constant. The corresponding charge in the mean plane of specifically adsorbed ions, σ_β, is given by:

$$\sigma_\beta = \frac{F}{MA_s} \{[\underline{S}O^- Na^+] - [\underline{S}OH_2^+ Cl^-]\} \tag{11}$$

Electroneutrality requires that

$$\sigma_0 + \sigma_\beta + \sigma_d = 0 \ , \tag{12}$$

where σ_d is the diffuse layer charge. From Gouy–Chapman diffuse layer theory, σ_d is related to the diffuse layer potential, ψ_d, by:

$$\sigma_d = -(8c\epsilon kT)^{1/2} \sinh\left(\frac{ze\,\psi_d}{2kT}\right) \tag{13}$$

where c is the electrolyte concentration and z is the valence of the electrolyte ions.

Davis *et al.* [16] used Grahame's [21] assumption that potential decays linearly between planes of charge in the compact layer. Assuming an analogy with parallel plate capacitors in series they obtained:

$$\psi_0 - \psi_\beta = \sigma_0 / C_1 \tag{14}$$

and

$$\psi_\beta - \psi_d = \frac{\sigma_0 + \sigma_\beta}{C_2} = -\sigma_d / C_2 \tag{15}$$

where C_1 and C_2 are the integral capacitances for the inner and outer parts of the compact layer.

With the application of the GCGS model the coulombic term for H^+ and OH^- becomes $\exp\left(-\frac{ze\psi_0}{kT}\right)$ and $\exp\left(-\frac{ze\psi_\beta}{kT}\right)$ for sorbed ions and the mass action action expressions written in terms of the neutral, $\underline{S}OH$, site and solved for the individual surface species concentrations are:

$$[\underline{S}OH_2^+] = [\underline{S}OH][H^+] \exp\left(\frac{-e\psi_0}{kT}\right) \Big/ K_{a1}^{int} \tag{16}$$

$$[\underline{S}O^-] = \frac{[\underline{S}OH]}{[H^+]} \exp\left(\frac{e\psi_0}{kT}\right) K_{a2}^{int} \tag{17}$$

$$[\underline{S}O^-Na^+] = \frac{[\underline{S}OH][Na^+]}{[H^+]} \exp\left(\frac{e\psi_0 - e\psi_\beta}{kT}\right) {}^*K_{Na}^{int} \tag{18}$$

and

$$[\underline{S}OH_2^+] = [\underline{S}OH][H^+][Cl^-] \exp\left(\frac{e\psi_\beta - e\psi_0}{kT}\right) {}^*K_{Cl}^{int} \tag{19}$$

The entire set of equations, equations (9)–(19), can be solved numerically using the modified MINEQL program. In this example known values would be specified for S_T, K_{a1}^{int}, K_{a2}^{int}, ${}^*K_{Na}^{int}$, ${}^*K_{cl}^{int}$, C_1 and C_2.

Composite Surfaces

For the system containing colloidal particles and hydrolysed aluminium it can be assumed that adsorption of hydrolysed aluminium forms a surface precipitate and hence a composite surface consisting of amphoteric base particle ($\underline{S}OH_2^+$, $\underline{S}OH$, $\underline{S}O^-$) and aluminium ($\underline{Al}OH_2^+$, $\underline{Al}OH$, $\underline{Al}O^-$) sites. In this example it was assumed that the electrolyte is KNO_3 and the surface complexes, $\underline{S}OH_2^+NO_3^-$, $\underline{S}O^-K^+$, $\underline{Al}OH_2^+NO_3$ and $\underline{Al}O^-K^+$ are formed.

The governing equations are given by the following expressions. The individual surface species concentrations are (in terms of the neutral $\underline{A}lOH$ and hypothetical base particle $\underline{S}OH$ sites):

$$[\underline{A}lOH_2{}^+] = [AlOH]\,[H^+]\,\exp\left(\frac{-e\psi_0^{Al}}{kT}\right) \Big/ K_{a1}^{int} \tag{20}$$

$$[\underline{A}lO^-] = \frac{[\underline{A}lOH]}{[H^+]}\exp\left(\frac{e\psi_0^{Al}}{kT}\right) K_{a2}^{int} \tag{21}$$

$$[\underline{A}lOH_2{}^+NO_3^-] = [\underline{A}lOH]\,[H^+]\,[NO_3^-]\,\exp\left(\frac{e\psi_\beta{}^{Al} - e\psi_0^{Al}}{kT}\right) {}^{Al}{*K}_{NO_3^-}^{int} \tag{22}$$

$$[\underline{A}l^-K^+] = \frac{[\underline{A}lOH]\,[K^+]}{[H^+]}\exp\left(\frac{e\psi_0{}^{Al} - e\psi_\beta{}^{Al}}{kT}\right) {}^{Al}{*K}_{K^+}^{int} \tag{23}$$

$$[\underline{S}OH_2^+] = [\underline{S}OH]\,[H^+]\,\exp\left(\frac{-e\psi_0^{S}}{kT}\right) K_{b1}^{int} \tag{24}$$

$$[\underline{S}O^-] = \frac{[\underline{S}OH]}{[H^+]}\exp\left(\frac{e\psi_0{}^{S}}{kT}\right) K_{b2}^{int} \tag{25}$$

$$[SOH_2^+NO_3^-] = [\underline{S}OH]\,[H^+]\,[NO_3^-]\,\exp\left(\frac{e\psi_\beta^{S} - e\psi_0^{S}}{kT}\right) \Big/ {}^{S}{*K}_{NO_3^-}^{int} \tag{26}$$

and

$$[\underline{S}O^-K^+] = \frac{[\underline{S}OH]\,[K^+]}{[H^+]}\exp\left(\frac{e\psi_0{}^{S} - e\psi_\beta^{S}}{kT}\right) {}^{S}{*K}_{K^+}^{int} \tag{27}$$

where $\psi_0{}^{Al}$, $\psi_\beta{}^{Al}$ and $\psi_0{}^{S}$, $\psi_\beta{}^{S}$ are the surface and Stern layer potentials for the aluminium hydroxide and base particle portions of the composite surface.

The total molar concentration of base particle sites, $S_t{}^S$, is assumed to be proportional to the area concentration of the uncoated part of the surface, A^S, and the surface site density, N_s^S, of the base particles, i.e.

$$S_t^S = A^S N_s^S / N_a \tag{28}$$

where N_a is Avogadro's number. Similarly, the total aluminium site concentration, S_t^{Al}, is determined by

$$S_t^{Al} = A^{Al} N_s^{Al} / N_a \tag{29}$$

where A^{Al} is the effective surface area concentration and N_s^{Al} is the effective surface site density of the adsorbed aluminium hydroxide.

The area concentration of base particle surface, A^{S}, is assumed to decrease linearly with the molar concentration of adsorbed aluminium, $M_{\mathrm{p}}^{\mathrm{Al}}$, i.e.

$$A^{\mathrm{S}} = A_{\mathrm{s}}^{\mathrm{S}} M^{\mathrm{S}} (1 - \beta M_{\mathrm{p}}^{\mathrm{Al}}) \tag{30}$$

where $A_{\mathrm{s}}^{\mathrm{S}}$ is the specific surface area and M^{S} is the mass concentration of base particles. The coefficient β is equal to the reciprocal of the molar concentration of adsorbed aluminium hydroxide required to yield complete coverage of the base particles. β is related to the surface concentration of adsorbed aluminium at complete coverage (r, mole/unit area) by the expression

$$\beta = (r A_{\mathrm{s}}^{\mathrm{S}} M^{\mathrm{S}})^{-1} \tag{31}$$

Equation (30) pertains when $M_{\mathrm{p}}^{\mathrm{Al}} \leqslant \beta^{-1}$. When $M_{\mathrm{p}}^{\mathrm{Al}} > \beta^{-1}$, $A^{\mathrm{S}} = 0$.

The area concentration of adsorbed aluminium, A^{Al}, is given by

$$A^{\mathrm{Al}} = A_{\mathrm{s}}^{\mathrm{S}} M^{\mathrm{S}} (\beta M_{\mathrm{p}}^{\mathrm{Al}}) \ . \tag{32}$$

(Equations (30) and (32) assume that when coverage of the base particle is less than complete, the adsorbed hydrolysed aluminium forms an infinitely thin, tightly bound layer of aluminium hydroxide, i.e. the total area concentration of the suspension does not change as the base particles are coated. In an earlier paper, Iyer and Letterman [22], the authors tested the appropriateness of relationships which allowed the total area concentration of the suspension to increase as the base particles are coated.)

The surface and Stern layer charge densities (σ_0 and σ_β) for the aluminium and base particles portions of the composite surface are:

$$\sigma_0^{\mathrm{Al}} = \frac{F}{A^{\mathrm{Al}}} \{[\underline{\mathrm{Al}}\,\mathrm{OH_2}^+] + [\underline{\mathrm{Al}}\mathrm{OH_2^+NO_3^-}] - [\underline{\mathrm{Al}}\mathrm{O^-}] - [\underline{\mathrm{Al}}\mathrm{O^-K^+}]\} \tag{33}$$

$$\sigma_0^{\mathrm{S}} = \frac{F}{A^{\mathrm{S}}} \{[\underline{\mathrm{S}}\mathrm{OH_2}^+] + [\underline{\mathrm{S}}\mathrm{OH_2^+NO_3^-}] - [\underline{\mathrm{S}}\mathrm{O^-}] - [\underline{\mathrm{S}}\mathrm{O^-K^+}]\} \tag{34}$$

$$\sigma_\beta^{\mathrm{Al}} = \frac{F}{A^{\mathrm{Al}}} \{[\mathrm{AlO^-K^+}] - [\underline{\mathrm{Al}}\mathrm{OH_2^+NO_3^-}]\} \tag{35}$$

and

$$\sigma_\beta^{\mathrm{S}} = \frac{F}{A^{\mathrm{S}}} \{[\underline{\mathrm{S}}\mathrm{O^-K^+}] - [\underline{\mathrm{S}}\mathrm{OH_2^+NO_3^-}]\} \tag{36}$$

The equations which apply the constraints imposed by the GCGS electrical double layer model are the same for both the aluminium hydroxide and base particle parts of the composite surface. These are given by equations (12) to (15). The effective diffuse layer charge, $\overline{\sigma}_{\mathrm{d}}$, for the composite particle is calculated using an area-weighted average of the diffuse layer charge densities for the two parts of the surface, i.e.:

$$\bar{\sigma}_d = (\sigma_d{}^S A^S + \sigma_d{}^{Al} A^{Al}) \,/\, A_s^S M^S \tag{37}$$

The effective diffuse layer potential, $\bar{\psi}_d$, is determined by substituting $\bar{\sigma}_d$ from equation (37) in the Gouy–Chapman relationship, equation (13).

To solve this set of equations (equations (12)–(15) and (20)–(37)) using the modified MINEQL program, values must be specified for the intrinsic ionization and complexation constants, the surface site densities and the electrical double layer capacitances for both the base particles and the aluminium hydroxide coating. The surface area concentration of the base particle, $M^S A_s{}^S$, and the amount of adsorbed aluminium required to coat the base particles completely, *r*, must also be specified. Table 1 lists the magnitudes used in this study.

The intrinsic ionization constants for adsorbed hydrolysed Al, K_{a1}^{int} and K_{a2}^{int}, listed in Table 1 are the same as those presented by James and Parks [24] for hydrous γ–Al_2O_3. The assumption that adsorbed aluminium hydroxide resembles an oxide particle surface is reasonable given evidence [20] that many oxide particles appear to form an outer shell of a porous, amorphous phase through hydrolysis and adsorption (possibly with surface precipitation) of hydrolysis products. It is generally observed that the isoelectric pH of hydrous Al_2O_3, 9.0 [23, 24] is similar to that of freshly precipitated aluminium hydroxide [9, 25]. Also zeta potential versus pH curves for aluminium-hydroxide coated TiO_2 particles and pure Al_2O_3 particles [23] are essentially the same for a given background electrolyte concentration.

The surface site density, $N_s{}^{Al}$, listed in Table 1 for the aluminium hydroxide coating is equal to the value listed by James and Parks [24] for γ–Al_2O_3. The intrinsic complexation constants, ${}^{Al}K_K^{int}$ and ${}^{Al}K_{NO_3^-}^{int}$, for the Al sites and K^+ and NO_3^- are equal to those listed by James and Parks [24] for Na^+ and Cl^- and an aluminium oxide surface. The assumption that K^+ and NO_3^- have affinities for aluminium sites which are similar to those of Na^+ and Cl^- is based on Table 8 in James and Parks [24]. This table lists for a given oxide similar pairs of complexation constants for KNO_3 and NaCl.

The intrinsic surface reaction constants and surface site density listed in Table 1 for TiO_2 were obtained from a paper by James *et al.* [26]. This paper describes a comprehensive analysis of the surface chemical and electrical double layer properties of the TiO_2 surface in the presence of KNO_3. The parameter values listed for SiO_2 have been discussed in a previous paper [22].

4. MIXED VERSUS PATCH SITE DISTRIBUTIONS

Initially in this study it was assumed that the surface sites associated with the adsorbed aluminium were uniformly interspersed with the base particle sites and that, therefore, the speciation of all sites was related to one set of electrical

Table 1

Parameter values used in the surface reaction part of the MINEQL program. Aluminium hydroxide–SiO_2/TiO_2 composite particles

Parameter	Symbol	Value	Reference
Intrinsic surface ionization equilibrium constants (negative log)			
Al site	pK_{a1}^{int}	5.7	James and Parks [24]
Al site	pK_{a2}^{int}	12.3	This study
Base particle sites	pK_{b1}^{int}	–1.0 (SiO_2)	Iyer and Letterman [22]
		3.0 (TiO_2)	James and Parks [24]
Base particle sites	pK_{b2}^{int}	1.0 (SiO_2)	Iyer and Letterman [22]
		8.6 (TiO_2)	James and Parks [24]
Intrinsic surface complexation equilibrium constant			
Surface site – K^+	$p^*K_K^{int}$	9.0 (Al)	James and Parks [24]
		6.8 (TiO_2)	James and Parks [24]
Surface site – NO_3^-	$p^*K_{NO_3}^{int}$	–6.9 (Al)	James and Parks [24]
		–4.8 (TiO_2)	James and Parks [24]
Surface site densities			
Al sites	N_s^{Al}	8×10^{18} sites/m^2	James and Parks [24]
Base particle sites	N_s^S	1×10^{17} sites/m^2(SiO_2)	Iyer and Letterman [22]
		8×10^{18} sites/m^2(TiO_2)	James and Parks [24]
Moles of aluminium adsorbed per unit area of base particle for complete coverage	r	9×10^{-6} mole Al/m^2(SiO_2)	Iyer and Letterman [22]
		6.5×10^{-6} mole Al/m^2(TiO_2)	Wiese and Healy [3]
		variable (TiO_2)	This study
Mass concentraton of base particles	M^S	variable	Experimental condition
Molar concentration of precipitated aluminium hydroxide	M_p^{Al}	variable	Calculated with MINEQL program (see text)
Integral capacitance for outer part of Stern layer	C_1	140 μF/cm^2	Davis *et al.* [16]
Integral capacitance for inner part of Stern layer	C_2	20 μF cm^2	Davis *et al.* [16]
Base particle specific surface area	A^S	1 m^2/g (SiO_2)	Iyer and Letterman [22]
		16 m^2/g (TiO_2)	Wiese and Healy [3]

double layer properties. In other words, instead of $\sigma_0, \psi_0, \sigma_\beta, \psi_\beta$ and σ_d, ψ_d being considered separately for the base particle and the aluminium hydroxide coating, all sites in a particular locus were used in combination. For example, in the case of the surface charge, σ_0, equations (33) and (34) are combined to yield:

$$\sigma_0 = \frac{F}{A_s^S M^S} \{[\underline{Al}OH_2^+] + [\underline{Al}OH_2^+NO_3^-] + [\underline{S}OH_2^+] + [\underline{S}OH_2^+NO_3^-] - [\underline{Al}O^-] - [\underline{Al}O^-K^+] - [\underline{S}O^-] - [\underline{S}O^-K^+]\} \ . \quad (38)$$

This 'mixed' site assumption yielded results (diffuse layer potential as a function of the fraction of the base particle surface area coated by aluminium hydrolysis products, βM_p^{Al}) which were not in agreement with experimental observations (zeta potential versus βM_p^{Al}). The 'patch' site distribution model as given by equations (20)–(37) and (12)–(15) was then derived. Figures 2 and 3

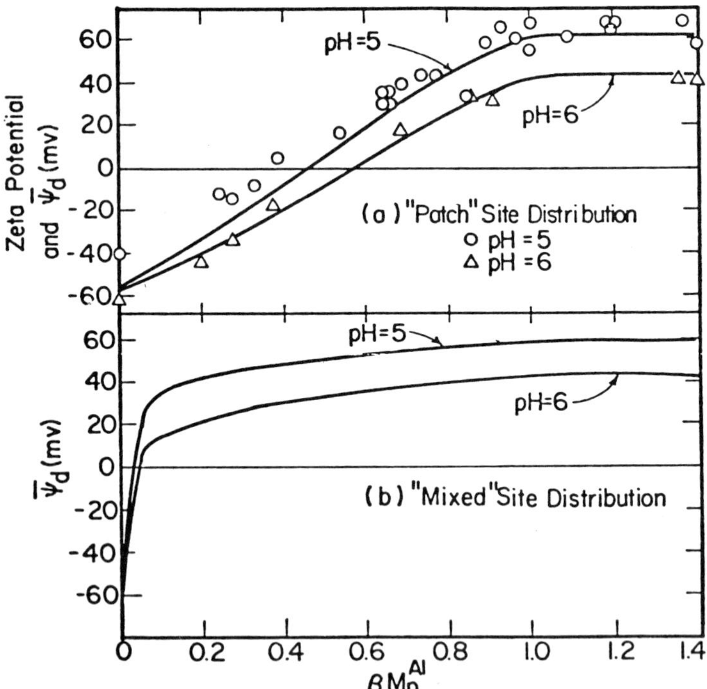

Fig. 2 – Zeta potential and model calculated diffuse layer potential versus fractional surface coverage by hydrolysed aluminium for (a) patch site distribution and (b) mixed site distribution assumptions.

illustrate the essential differences between the results obtained from the mixed and patch site assumptions.

The curves in Fig. 2(a) were plotted using the patch site model and constants given in Table 1. The data points in Fig. 2(a) are from experiments in which hydrolyzed aluminium was adsorbed on silica particles at pH 5 and pH 6. The procedures and materials used are described in a reference by Letterman *et al.* [25]. The pH 5 data are the same as those used to plot Fig. 1. In the case of silica particles a complete coating (as determined by zeta potential measure ments) is reached at a surface aluminium concentration of approximately 9×10^{-6} mole/m^2. This corresponds to $\beta M_p^{Al} = 1$. According to Fig. 2(a) the zeta potential increases almost linearly from the negative value of the uncoated silica particles to the positive value of the aluminium hydroxide coating. For pH 5; the zeta potential of the completely coated particles is approximately 64 mV and for pH 6 it is approximately 44 mV. Most of the disparity between the model calculated values of $\bar{\psi}_d$ and the measured zeta potentials is due to the MINEQL program predicting a value of $\bar{\psi}_d$ for clean silica at pH 5 of – 57 mV while the measured zeta potential in this case is approximately –40 mV.

The curves in Fig. 2(b) were plotted using the mixed site assumption and the constants for silica and aluminium hydroxide listed in Table 1. In this case the model prediction does not agree with the experimental points of Fig. 2(a). In fact, the curves can not be forced by adjustment of appropriate parameters to follow even the general trend of the data.

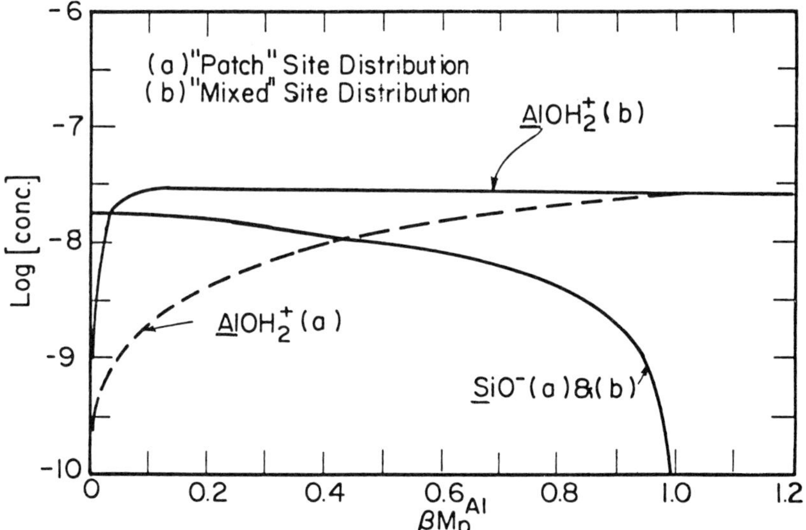

Fig. 3 – Log of the surface site concentration versus fractional surface coverage by hydrolysed aluminium for (a) patch site distribution and (b) mixed site distribution assumptions.

For a given partial surface coverage, i.e. a given value of βM_p^{Al} for $\beta M_p^{Al} <$ 1, $\bar{\psi}_d$ is consistently more positive for the mixed site assumption than for the patch site assumption. The reason for this is that when an aluminium site is interspersed with negative silica sites rather than located in a patch of aluminium sites, the net surface charge which governs site speciation will tend to be significantly negative and the quantity $-e\psi_0/kT$ will tend to be relatively large. Consequently, according to equation (20), the proportion of aluminium sites in the $\underline{Al}OH_2^+$ form will be relatively high and the adsorption of a given amount of aluminium will have a greater tendency to reverse the charge of the negative silica surface. For example, in Fig. 3 when $\beta M_p^{Al} = 0.10$, the patch site model predicts that $[\underline{Al}OH_2^+] = 2.5 \times 10^{-9}$ while the mixed site model predicts $[\underline{Al}OH_2^+] = 3 \times 10^{-8}$.

5. EVALUATION OF THE PATCH SITE DISTRIBUTION MODEL – HYDROLYSED ALUMINIUM(III) ON TiO_2 IN KNO_3 SOLUTION

The patch site distribution model was tested using experimental data obtained by Wiese and Healy [3] in studies on the adsorption of aluminium(III) at the TiO_2 – water interface. All data pertain to a colloidal TiO_2 suspension, 0.050 g/l of TiO_2 in 10^{-4} – M KNO_3, which has, according to Wiese and Healy, an available surface area concentration, $M^S A_s^S$, of 0.825 m^2/l. The aluminium, which was present in concentrations ranging from 5×10^{-7} M to 2.1×10^{-5} M, was added as aluminium nitrate.

6. RESULTS AND DISCUSSION

The variation in zeta potential with pH observed by Wiese and Healy [3] for the TiO_2 suspension with and without aluminium nitrate is shown in Fig. 4. The

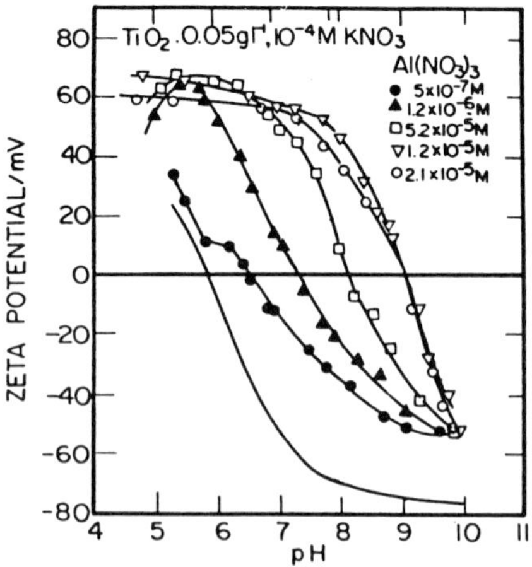

Fig. 4 – Variation of the zeta potential with pH for 0.05 g/l TiO_2 in 10^{-4}-M KNO_3 at 25 °C. From Wiese and Healy [3].

significant features of these results are consistent with the model of hydrolysed metal ion adsorption presented by James and Healy [11]. As the pH is increased with $Al(NO_3)_3$ present, the zeta potential versus pH curve changes shape abruptly at a pH somewhat below that corresponding to bulk precipitation of aluminium hydroxide. James and Healy suggested that this is due to surface nucleation of the metal hydroxide under the influence of the interfacial electric field. In the case of $Al(OH)_3$ nucleation on TiO_2, a 'CR.2' zeta potential sign reversal point as reported by James and Healy for cobalt(II) and lanthanum(III) on SiO_2 is not observed because at pH values below 5.9, the TiO_2 is positively charged without added aluminium(III). At high pH, Wiese and Healy [3] concluded that the electrokinetic properties reflect the presence of a partial or complete coating of aluminium hydroxide on the TiO_2 surface. In the case of Fig. 4 the curves for $[Al] = 1.2 \times 10^{-5}$ and 2.1×10^{-5} are essentially coincident with an isoelectric pH (labelled CR.3 in James and Healy [11]) of approximately 9 indicating that at least up to pH 9, the TiO_2 surface is completely coated by aluminium hydroxide.

Figures 5–7 show the isolelectric pH of the TiO_2 as a function of the total aluminium concentration. The solid 'experimental' curve is the same in each

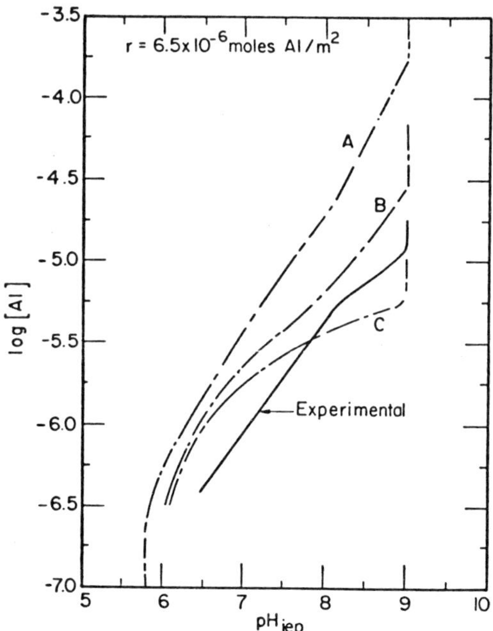

Fig. 5 – Log of the aluminium concentration versus the isoelectric pH for 0.05 g/l TiO_2 in 10^{-4}-M KNO_3 at 25 °C. The broken curves are model calculations for $r = 6.5 \times 10^{-6}$ mole Al/m^2 and aluminium hydroxide solubility products of A – $10^{-33.2}$, B – 10^{-34} and C – $10^{-36.3}$. The solid curve is from the experimental results of Wiese and Healy [3] in Fig. 4.

graph and was plotted using points derived from Fig. 4. The curves labelled A, B and C correspond to different values of the adsorbed $Al(OH)_3$ solubility product and were plotted using results obtained from MINEQL calculations with the patch site distribution model. In plotting these curves the constants listed in Table 1 were used for the TiO_2 and adsorbed aluminium hydroxide surface reactions and electrical double layer calculations. The constants listed in Table 2 were used with the solution/precipitate chemistry part of the MINEQL program to determine the speciation of the aluminium. Only monomeric hydrolysis products are assumed to exist in equilibrium with the adsorbed aluminium hydroxide precipitate. The concentration of adsorbed aluminium hydroxide, M_p^{Al}, is assumed to be equal to the concentration of $Al(OH)_3$ precipitate predicted by the MINEQL model. Figures 5, 6 and 7 correspond to aluminium

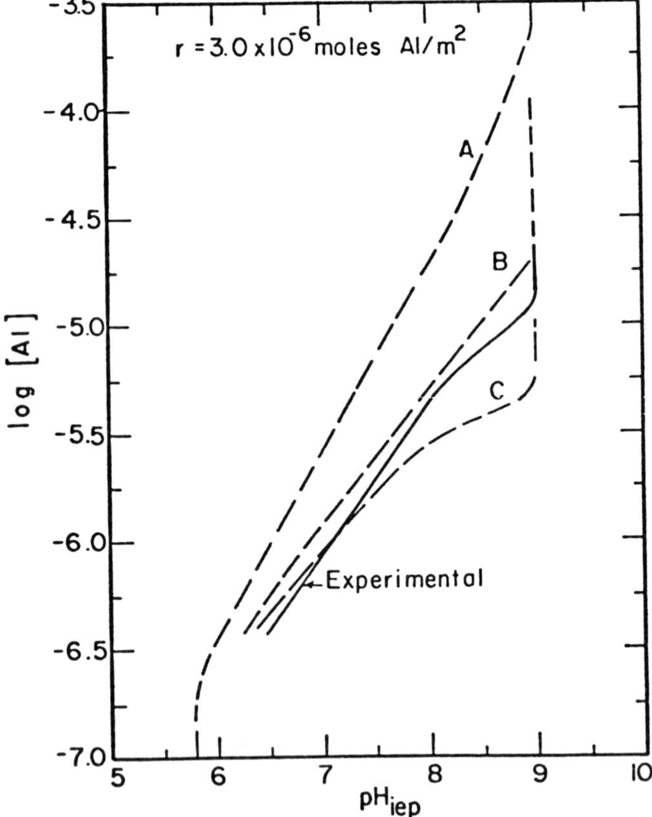

Fig. 6 – Log of the aluminium concentration versus the isoelectric pH for 0.05 g/l TiO_2 in 10^{-4}-M KNO_3 at 25 °C. The broken curves are model calculations for $r = 3.0 \times 10^{-6}$ mole Al/m^2 and aluminium hydroxide solubility products of A – $10^{-33.2}$, B – 10^{-34} and C – $10^{-36.3}$. The solid curve is from the experimental results of Wiese and Healy [3] in Fig. 4.

adsorption densities at complete coating (r) of 6.5 $\times$ 10^{-6}, 3.0 $\times$ 10^{-6} and 1.2 $\times$ 10^{-6} mole Al/m^2, respectively. Curves A, B and C correspond to $Al(OH)_3$ solubility products of $10^{-32.2}$, $10^{-34.1}$ and $10^{-36.3}$.

According to Figs 5–7, a rough fit of the experimental log [Al] – pH_{iep} coordinates is obtained with r = 3.0 $\times$ 10^{-6} mole Al/m^2 and $K_{sp} = 10^{-34.1}$. A comparison of Figs 6 and 7 suggests that a more exact fit might be obtained with a value of K_{sp} slightly less than $10^{-34.1}$ and a value of r between 1.2 $\times$ 10^{-6} and 3.0 $\times$ 10^{-6} mole Al/m^2. In Fig. 8 the model calculated diffuse layer potential,

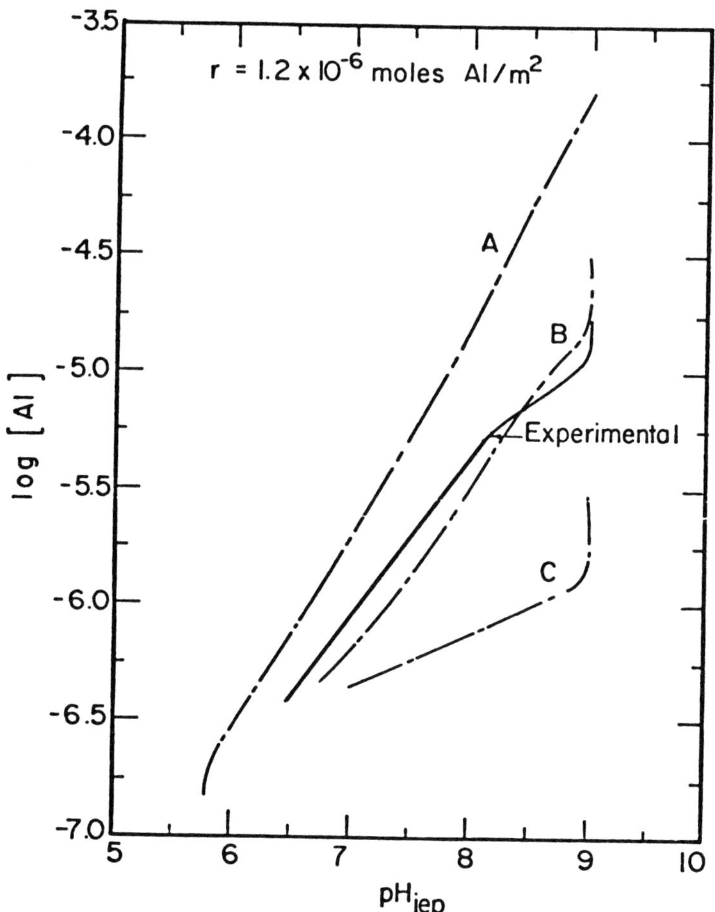

Fig. 7 – Log of the aluminium concentration versus the isoelectric pH for 0.05 g/l TiO_2 in 10^{-4}-M KNO_3 at 25 °C. The broken curves are model calculations for r = 1.2 $\times$ 10^{-6} mole Al/m^2 and aluminium hydroxide solubility products of A – $10^{-33.2}$, B – 10^{-34} and C – $10^{-36.3}$. The solid curve is from the experimental results of Wiese and Healy [3] in Fig. 4.

$\bar{\psi}_d$, has been plotted against pH for the case $r = 3 \times 10^{-6}$ mole Al/m^2 and $K_{sp} = 10^{-34.1}$. The five aluminium concentrations are the same as those used in the Wiese and Healy [3] experiments (Fig. 4).

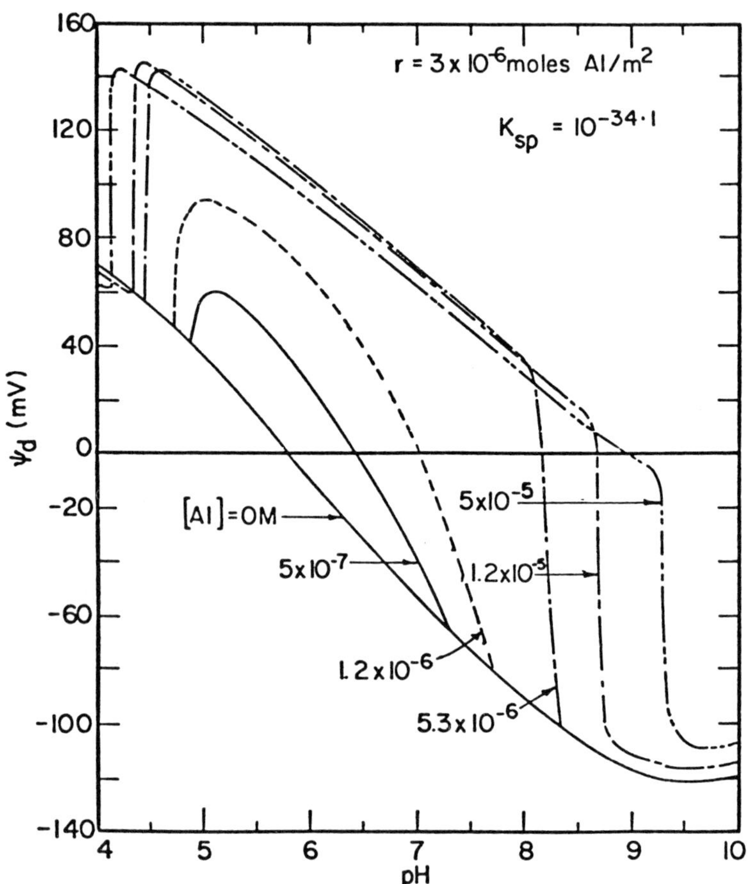

Fig. 8 – Model calculated diffuse layer potential versus pH for $r = 3 \times 10^{-6}$ mole Al/m^2, aluminium hydroxide solubility product of $K_{sp} = 10^{-34.1}$ and 0.05 g/l TiO_2 in 10^{-4}-M KNO_3.

Wiese and Healy [3], using the zeta potential results of Fig. 4 and the aluminium adsorption data replotted in Figs 9 and 10, determined that $K_{sp} = 10^{-34.1}$ and $r = 6.5 \times 10^{-6}$ mole Al/m^2 best explained their observations. The selection of $K_{sp} = 10^{-34.1}$ was based on the assumption that the abrupt increase in zeta potential at low pH (Fig. 4) is due to nucleation of freshly precipitated hydroxide in the interfacial region. The conclusion that the formation of a

Table 2

Parameter values used in the solution chemistry part of the MINEQL program

Reaction		log (equilibrium constant)
$Al^{+3} + 4H_2O$	$Al(OH)^{2+} + H^+$	−4.99
$Al^{+3} + 2H_2O$	$Al(OH)_2{}^+ + 2H^+$	−10.13
$Al^{+3} + 4H_2O$	$Al(OH)_4{}^- + 4H^+$	−21.57
	Aluminium hydroxide solubility	
	$[Al^{+3}]\ [OH^-]^3 = K_{sp}$	
Curve		$\log K_{sp}$
A		−33.2
B		−34.1
C		−36.3

complete coating occurs at 6.5×10^{-6} mole Al/m^2 was derived from the observation that in 1.2×10^{-5}-M Al, the isoelectric pH is 9.1. At pH 9.1 and 1.2×10^{-5}-M Al, the amount of aluminium adsorbed yields an aluminium adsorption density of 6.5×10^{-6} mole Al/m^2.

This result is, however, somewhat ambiguous if Fig. 4 is considered in greater detail. For example, it appears that for an aluminium concentration of 1.2×10^{-6} in the vicinity of pH 5.5 the zeta potential curve reaches the range of values (60 ± 5 mV) which is associated with a completely coated particle. If $r = 6.5 \times 10^{-6}$ mole Al/m^2 is the case, then the maximum surface coverage for [Al] = 1.2×10^{-6} would be $1.2 \times 10^{-6}/0.825 = 1.5 \times 10^{-6}$ mole Al/m^2 or $\beta M_p{}^{Al} = 0.23$, which is less than a complete coating. The value obtained from the model calculations, $r = 3 \times 10^{-6}$ mole Al/m^2, gives results which are somewhat more consistent with the measured zeta potentials of Fig. 4.

In comparing Figs. 4 and 8, it is apparent that the calculated $\bar{\psi}_d$ values can be significantly greater than the measured zeta potentials. For example, the maximum zeta potential of the $Al(OH)_3$ coated particles is between 60 and 70 mV while the maximum value of $\bar{\psi}_d$ is between 140 and 150 mV. This disparity is consistent with the significant difference usually obtained between measured zeta potentials and values of the diffuse layer potential calculated using the Gouy–Chapman–Stern–Grahame model. (For examples, see James *et al.* [24], James and Parks [23] and Anderson and Rubin [27]).

As noted in regard to the model calculations it was assumed that the amount of aluminium adsorbed is equal to the amount of aluminium hydroxide precipitate predicted by the MINEQL model. This assumption is not in complete agreement with aluminium adsorption data, e.g. Fig. 1 and Hahn and Stumm [5], and as shown in Figs 9 and 10 it is only roughly comparable to the aluminium adsorption data obtained by Wiese and Healy [3]. It is apparent in Figs 9 and 10

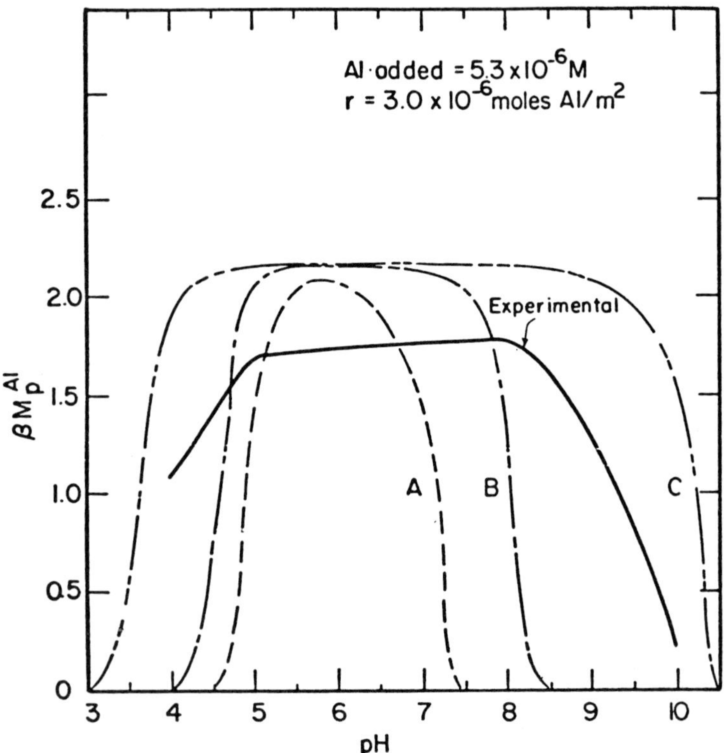

Fig. 9 – Fractional coverage by hydrolysed aluminium of the surface area in a 0.05 g/l TiO_2 suspension versus pH. Curves A, B and C refer to aluminium hydroxide solubility products of $10^{-33.2}$, $10^{-34.1}$ and $10^{-36.3}$, respectively. The experimental curve is based on results in Wiese and Healy [3]. The aluminium(III) concentration is 5.3×10^{-6} M and r is assumed to be 3×10^{-6} mole Al/m^2.

that with changing pH the model predicts significantly more abrupt changes in aluminium solubility than are observed experimentally. In general, in the region of $Al(OH)_3$ precipitation the model predicts slightly greater adsorption densities than are observed experimentally and at high pH the measured amount of aluminium adsorbed is greater than that predicted by the model. This helps to explain why at the higher aluminium concentrations in Figs. 4 and 8 the change in $\bar{\psi}_d$ with pH is much greater than the change in zeta potential with pH when pH > 8. Also as the pH increases toward pH 10, the measured zeta potentials do not approach the curve for uncoated TiO_2 particles suggesting that in Wiese and Healy's experiments [3] a significant amount of aluminium hydroxide remained

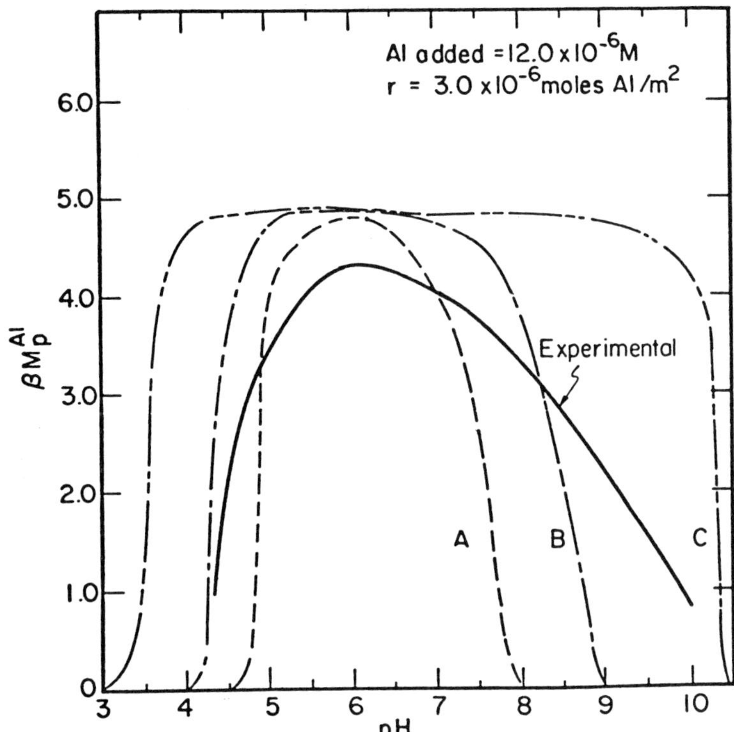

Fig. 10 – Fractional coverage by hydrolysed aluminium of the surface area in a 0.05 g/l TiO_2 suspension versus pH. Curves A, B and C refer to aluminium hydroxide solubility products of $10^{-33.2}$, $10^{-34.1}$ and $10^{-36.3}$, respectively. The experimental curve is based on results in Wiese and Healy [3]. The aluminium(III) concentration is 12 × 10^{-6} mole Al/m^2 and r is assumed to be 3 × 10^{-6} mole Al/m^2.

adsorbed on the particles at the time that the zeta potential measurements were made. This point is supported by the adsorption results plotted in Figs. 9 and 10 and is consistent with the slow aluminium(III) disolution kinetics described by Cornell, Posner and Quirk [28] for alumina-coated TiO_2 particles, and by White [12] for $Al(OH)_3$ coated polystyrene.

7. CONCLUSIONS

The application of the program MINEQL, modified with a detailed model of the electrical double layer to account for the distribution of charge and potential at the colloid–solution interface, seems to have potential as a means to predict the effect of hydrolysed Al(III) on the electrical double layer properties of aqueous colloids. In the methodology tested, the adsorption of aluminium hydrolysis products on a primary particle surface produces a new surface with chemical properties which can be modelled by assuming that ionizable and complex-forming functional groups (similar to those on insoluble oxides) are formed.

An important limitation at this time is an inability to predict accurately the amount of aluminium adsorbed on the colloidal surface. The use of equilibrium calculations and the assumption of quantitative adsorption of $Al(OH)_3$ precipitate give only approximate agreement with zeta potential measurements. Chemical kinetic and other not-well-understood factors seem to be influential, at least in the Al(III)–TiO_2 and Al(III)–SiO_2 systems. Limited results tend to justify the use of a patch site distribution model in which a partial coating of the particle by adsorbed hydrolysed aluminium leads to the formation of a composite surface. The composite surface is modelled by using two electrical double layers, one for the areas of uncoated primary particles and one for the areas covered by the aluminium hydroxide. The electrical double layers have a common dependence on the characteristics of the bulk solution and are used to calculate the net electrical double layer properties of the composite particle.

REFERENCES

[1] Brown, D. W. and Hem, J. D., *U.S. Geological Survey Paper 1827-F*, Washington, D.C. (1975).

[2] McAtee, J. L., Jr. and Wells, L. M., *J. Colloid Interface Sci.,* **24**, 203 (1967).

[3] Wiese, G. R. and Healy, T. W., *J. Colloid Interface Sci.,* **51**, 434 (1975).

[4] Matijevic, E., *Proceedings of the International Conference on Colloids and Surfaces,* San Juan, Puerto Rico, **397**, Vol. 1, Academic Press, New York (1976).

[5] Hahn, H. and Stumm, W., in *Adsorption from Aqueous Solution* (R. F. Gould, Ed.), p. 91, American Chemical Society Publications, **79** (1968); Hahn, H. and Stumm, W., *J. Colloid Interface Sci.,* **28**, 134 (1968).

[6] Rubin, A. J. and Hanna, G. P., *Env. Sci. and Technology,* **2**, 358 (1968).

[7] Mangravite, F. J. *et al., J. Am. Water Works Assoc.,* **67**, 88 (1975).

[8] Iler, R. K., *J. Amer. Ceram. Soc.,* **47**, 194 (1964).

[9] Matijevic, E., Mangravite, F. J., and Cassell, E. A., *J. Colloid Interface Sci.,* **35**, 560 (1971).

[10] Stumm, W. and Morgan, J. J., *Aquatic Chemistry,* Wiley–Interscience, New York (1981).

[11] James, R. O., and Healy, T. W., *J. Colloid Interface Sci.,* **40**, 53 (1972).

[12] White, D. W., *Ph.D. Dissertation,* Clarkson College, Potsdam, New York (1980).

[13] James, R. O., in *Adsorption of Inorganics at Solid–Liquid Interfaces,* (Anderson, M. S. and Rubin, A. J., Eds.), p. 219, Ann Arbor Science Publishers, Ann Arbor, Mich. (1981).

[14] O'Melia, C. R. and Stumm, W., *J. Colloid Interface Sci.*, **23**, 437 (1967).
[15] Westall, J. C., Zachary, J. L., and Morel, F. M. M., *MINEQL – A Computer Program for the Calculation of Chemical Equilibrium Composition of Aqueous Solutions,* Department of Civil Engineering, Massachusetts Institute of Technology, Note No. 18, EPA Grant R 803738 (1976).
[16] Davis, J. A., James, R. O., and Leckie, J. O., *J. Colloid Interface Sci.*, **63**, 480 (1978).
[17] Hohl, H., Sigg, L., and Stumm, W., in *Particulates in Water* (Kavanaugh, M. C. and Leckie, J. O., Eds.), p.1, ACS Advances in Chemistry Series, **189** (1980).
[18] Lyklema, J., *J. Electroanal. Chem.*, **18**, 341 (1968).
[19] Perram, J. W., *J. Chem. Soc., Faraday Trans. II,* **69**, 993 (1973).
[20] Perram, J. W., Hunter, R. J., and Wright, H. J. L., *Aust. J. Chem.*, **27**, 461 (1974).
[21] Grahame, D. C., *Chem. Rev.*, **41**, 441 (1947).
[22] Iyer, D. R. and Letterman, R. D., paper submitted to *Env. Sci. Technology* November (1982).
[23] Wiese, G. R. and Healy, T. W., *J. Colloid Interface Sci.*, **51**, 427 (1975).
[24] James, R. O. and Parks, G. A., in *Surface and Colloid Science,* Vol. 12, (E. Matijevic, Ed.), Plenum Publishing Corp., p. 119 (1982).
[25] Letterman, R. D., Vanderbrook, S. W., and Sricharoenchaikit, P., *J. Am. Water Works Assoc.*, **74**, 44 (1982).
[26] James, R. O., Stiglich, P. J., and Healy, T. W., in *Adsorption from Aqueous Solutions* (P. H. Tewari, Ed.), Plenum Publishing Corp. (1981).
[27] Anderson, M. S. and Rubin, A. J., Eds., *Adsorption of Inorganics at Solid-Liquid Interfaces,* Ann Arbor Science Publishers, Ann Arbor, Mich. (1981) .
[28] Cornell, R. M., Posner, A. M., and Quirk, J. P., *Colloid and Polymer Science,* **261**, 143 (1983).

LIST OF SYMBOLS

Symbol	Description	Units
A_s^S, A_s	Base particle specific surface area	m^2/gm
A^S	Base particle surface area concentration	m^2/litre
A^{Al}	Adsorbed Al surface area concentration	m^2/litre
C_1, C_2	Inner and outer layer capacitances	farad/cm^2
e	Electron charge ($= 1.6 \times 10^{-19}$)	coulomb
F	Faraday (= 96 500)	coulomb/equivalent
K^{int}	Intrinsic surface ionization equilibrium constant	–
$*K^{int}$	Intrinsic surface complexation equilibrium constant	–
k	Boltzmann constant ($= 1.4 \times 10^{-23}$)	volt coulomb/K

M, M^S	Base particle mass concentration	gm/litre
M_p^{Al}	Adsorbed $Al(OH)_3$ concentration	moles/litre
N_a	Avogadro's number ($= 6.03 \times 10^{23}$)	number/mole
N_s, N_s^S	Base particle surface site density	sites/nm²
N_s^{Al}	Adsorbed Al surface site density	sites/nm²
r	Adsorbed Al per unit base particle area at complete coverage	moles/m²
S_T, S_t^S	Base particle surface site concentration	moles/litre
S_t^{Al}	Adsorbed Al surface site concentration	moles/litre
T	Absolute temperature	K
z	Valence of ion	—
β	Reciprocal of adsorbed Al required for complete coverage	litres/mole
βM_p^{Al}	Fraction of base particle surface area covered by adsorbed Al	—
ϵ	Permittivity	farad/cm
$\sigma_0, \sigma_\beta, \sigma_d$	Inner, Stern and diffuse layer charges, respectively	coulombs/cm²
$\psi_0, \psi_\beta, \psi_d$	Inner, Stern and diffuse layer potentials, respectively	volt

CHAPTER 17

Poliovirus Adsorption on to Sand and Montmorillonite Clay

VINCENT L. VILKER, Chemical Engineering Department, University of California, Los Angeles, California 90024, USA

ABSTRACT

Separation of viruses from aqueous solutions by reversible adsorption is of interest to such diverse research areas as ion-exchange liquid chromatography and land application of virus-contaminated wastewater and sludges. Application of percolation/adsorption models for describing virus migration in soils requires experimental measurements of the equilibrium distribution of single virus particles between aqueous electrolyte solutions and suspended biomass solids in the effluent wastewater applied to land, and between percolating soil water and adsorbing solids of the soil matrix. This report describes measurements of poliovirus adsorption to activated sludge biomass, sand and clay particles. Sand and clay particles have narrow size distributions about their mean particle sizes of 123 and 1–2 μm, respectively. Adsorption to sand was carried out to high-equilibrium solution-phase virus concentration (10^9 PFU/ml) and showed saturation-limited behaviour, but low fractional coverage of sand particle surface ($f \sim 0.01$). Adsorption to clay could be made only to virus equilibrium concentrations of about 6×10^5 PFU/ml because the clay particles aggregated at higher virus concentrations. Aggregation was shown to accompany adsorption in certain ranges of clay–virus–electrolyte concentration ratios. This may have important implications for the use of isotherms determined in agitated batch electrolyte solutions for predicting virus movement in percolating soil beds where clay particles are restrained from forming aggregates. Scanning electron micrographs suggest that poliovirus adsorbs to the edges of one-micrometre montmorillonite clay particles. Below equilibrium concentrations of 6×10^5 PFU/ml, adsorption to clay was found to fit an isotherm function for which amount adsorbed per mass of clay depends on equilibrium solution-phase concentration to the 1.3 power.

1. INTRODUCTION

Enteric viruses, such as attenuated poliovirus, are the special group of viruses originating primarily from human and animal faeces. They are submicroscopic particles, several hundred Å in diameter, consiting primarily of nucleic acid and protein and act as biological macro-ions in aqueous solution.

Disappearance of *infective* enteric viruses from percolating soil water occurs by entrapment and/or sorptive interactions with soil components, inactivation of particles while still in the soil-water phase, or inactivation of particles which are associated with or retained by the soil matrix. Entrapment and sorption are most important because virus inactivation rates are usually too slow to prevent intrusion of infective virus into waters receiving the soil-water percolates [1]. Entrapment refers to the filtration by the soil matrix of suspended solids which contain embedded or adsorbed virus particles. Sorption interactions refer to the adsorption of single virus particles by specific components of the soil matrix.

A percolation/adsorption model is being developed to describe the breakthrough or concentration history of viruses in the effluent from a packed bed of soil or soil-like solid particles [1]. The applicability of this model to the establishment of site-specific criteria for the safety of landspreading of wastewater is based on the assumptions that single virus particles are the most mobile particle forms in the percolating water, and that a significant fraction of the total virus in the applied wastewater is present in the single particle form. This last assumption is examined in the experimental measurement of poliovirus adsorption to activated sludge suspended solids reported in the next section.

The modelling approach is based on the reversibility of infective virus particle association with the packed bed solids. For time t after the start of flow of the percolate containing virus at concentration C_0 to the bed, a differential virus mass balance at depth z gives:

$$\frac{\partial C}{\partial t} + u\frac{\partial C}{\partial z} + \frac{\rho_B}{\epsilon}\frac{\partial q}{\partial t} = 0 \qquad (1)$$

where $C(t, z)$ is liquid-phase virus concentration at time t and depth z, $q(t, z)$ is solid-phase virus concentration, ρ_B is mass of solid per unit mixed volume (solid + fluid), ϵ is bed void fraction and u is pore fluid velocity. The dispersion term, $(\partial^2 C/\partial z^2)$, is not included due to the small diffusivity of virus in solution, $\sim 10^{-8}$ cm^2/s. Virus inactivation rates are assumed to be negligible relative to the time scale of migration through the bed. Volumetric flow rate applied to the top of the bed is $\epsilon u A_c$ where A_c is bed cross-sectional area. The rate of adsorption at any z is given by

$$\rho_B\frac{\partial q}{\partial t} = k_{f,c}(C - C^*) \qquad (2)$$

where $k_{f,c}$ is an adsorption rate parameter which accounts for the diffusive transport of virus from the bulk solution to the surface of each solid particle and C^* is virus concentration in the liquid in immediate contact with this surface.

In order to solve equations (1) and (2), initial conditions $C(0, z)$ and $q(0, z)$ and the boundary condition $C(t, 0)$ must be specified, the value of the rate parameter $k_{f,c}$ which depends on bed permeability must be known, and an isotherm relationship must be available for $q(C^*)$. At present, these isotherm relationships must be measured experimentally for particular viruses adsorbing to specific solid adsorbents. The principal focuses of this paper are our experimental studies directed toward the determination of these isotherms for the adsorption of poliovirus from electrolyte solutions to well-characterized size fractions of sand and montmorillonite clay. These two soil components have a dominant influence on soil bed permeability and virus adsorption activity.

2. POLIOVIRUS ADSORPTION TO ACTIVATED SLUDGE SOLIDS

This study of the adsorption of poliovirus to suspended solids from the activated sludge unit of a local treatment plant was undertaken to define adsorption capacity over a wide range of solids concentration [2]. The results permit calculation of virus distribution between liquid and adsorbed phases over the entire range of solids concentration levels ranging from the high loadings typical for an activated sludge mixed liquor to the low loadings typical of a treated wastewater effluent.

2.1 Materials and Methods

Samples of biomass suspended solids were collected from an activated sludge unit that consisted of a typical complete-mix aeration basin and multistage clarifier. The biomass concentration of the aeration basin mixed liquor is about 5000 g/m^3. Suspended solids concentration at the clarifier overflow weirs is typically less than 10 g/m^3. Most samples consisted of solids obtained by filtering clarifier liquor through 20-μm mesh cloth screen. Filtered solids from the clarifier overflow and unfiltered samples of solids in the aeration basin mixed liquor were also included in the study. The solids were disinfected either by autoclaving at 120 °C for 15 minutes or by chlorination with sodium hypochlorite followed by neutralization with sodium thiosulphate. The method of disinfection did not appear to affect adsorption results and no further examination of biomass particle morphology was attempted. Disinfected solids were resuspended to desired concentrations in either 0.15-M saline or autoclaved, filtered effluent water. The pH of all suspensions was approximately 7.5. Final solids concentrations were determined by the filtration/gravimetric technique described in *Standard Methods* [3].

Poliovirus type 1 (CHAT strain, 10^8 PFU/ml stock titre, Microbiological Associates, Los Angeles) diluted to desired initial concentration in phosphate-buffered saline was used in this study. In all experiments, virus concentrations were determined by the plaque assay method. Solution aliquots (0.3 ml) for virus assay were inoculated into primary African green monkey kidney cell cultures in 25 cm^2 plastic bottles. Replicates of four bottles were used for each sample. After 2 h adsorption, they were overlayed with a maintenance medium containing agarose and neutral red. Observations for plaque-forming units (PFU) were made daily for 14 days. Virus concentration of the inoculuum corresponds to the number of PFU at the end of this period. Further details on the tissue culture and plaquing methods used here can be found in Hoskins [4] and Lennette and Schmidt [5].

Adsorption experiments were performed by mixing 24 ml of freshly prepared solids suspensions and 1 ml of poliovirus solution of appropriate concentration to give the desired initial titre (10^2 to 10^5 PFU/ml) in 100-ml screw cap bottles. After brief swirling to mix the solution, samples were withdrawn and inoculated immediately for enumeration of time zero virus concentration. The bottles were then shaken continuously at about 180 oscillations per minute in a gyratory waterbath (22 °C) for a standard 30 minutes. Preliminary kinetic studies showed that virus adsorption equilibrium was achieved in less than 10 minutes. Following the 30-minute contact period, 10 ml aliquots from each bottle were placed in centrifuge tubes. Solids were separated from solution by centrifugation at 14,700 g for 15 minutes. For these centrifugation conditions, we estimate that only virus, biomass, soluble intracellular macromolecules and cellular debris smaller than three to four times the virus (30 nm equivalent spherical diameter) remain in the resulting supernatant. The supernatant samples were inoculated to determine liquid-phase equilibrium virus titre, C_e. Control samples consisting of virus solution in solids-free saline or filtered wastewater were shaken and centrifuged at the same conditions as the solids samples in order to account for titre loss by inactivation or adsorption to container walls. This fractional loss in infectivity was applied to the solids sample time zero concentration for each experiment to give total virus concentration, C_i. The average loss in virus concentration from all control samples was about 12% of initial titre.

2.2 Results and Conclusions

Virus distribution between the liquid and solid phases from measurements of poliovirus uptake by a known mass of suspended solids, w mg/ml, can be correlated with a relationship such as the Freundlich isotherm:

$$q = \frac{C_i - C_e}{w} = uC_e^n \, , \tag{3}$$

where u and n are coefficients determined by linear regression after writing equation (3) in logarithmic form. Typical measurements for adsorption to activated sludge solids are shown in Fig. 1.

These results suggest $n = 1$ but that u depends on solids concentration. These results, along with measurements at several additional solids concentrations were

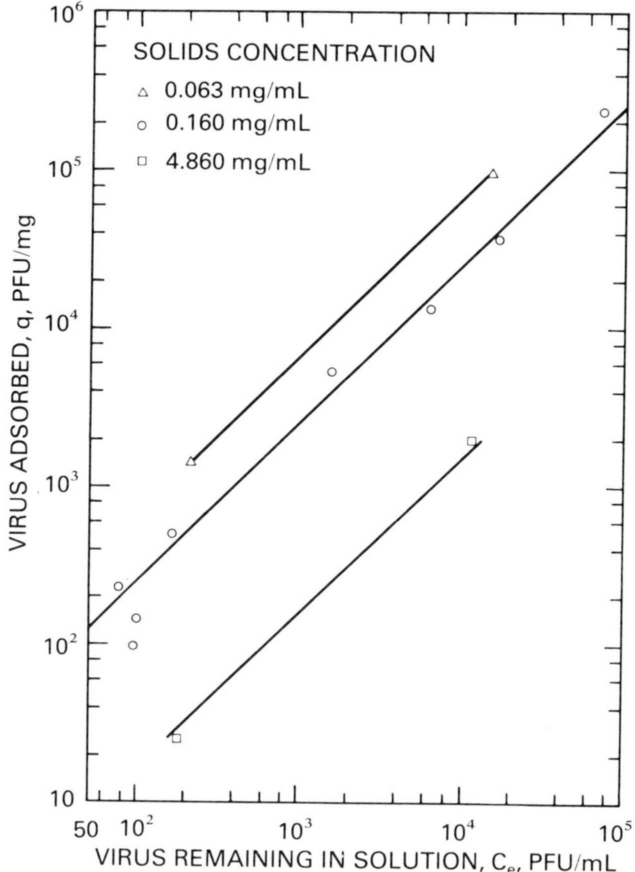

Fig. 1 – Adsorption of poliovirus to activated sludge solids. Reprinted with permission from Gordon & Breach. V. L. Vilker *et al. Chem. Engr. Comm.*, **4**, 569 (1980).

correlated using linear least squares regression analysis to give the isotherm relationship:

$$q = 0.63 w^{-0.81} C_e \ . \tag{4}$$

The correlation coefficient for this result is 0.80 and the confidence limits of the regression constants are 0.63 ± 0.10 and 0.81 ± 0.07.

The correlation result shows a relatively strong decrease in capacity for virus adsorption with increasing solids concentration. We interpret this effect to be the result of competitive adsorption between biomass particles for virus and for each other since biomass solids are primarily wasted cells and cellular debris which have a strong tendency to aggregate.

Equations (3) and (4) may be rearranged to find the ratio of free unassociated to total virus, C_e/C_i, as a function of biomass solids concentration. This calculation shows that at typical water treatment effluent solids loadings of about 10 mg/l, almost 80% of the total virus in the effluent will be in the free unassociated state. This supports the contention that when such effluent is disposed by landspreading, the potential exists for the presence of a large number of free single virus particles that will percolate into the subsurface soil.

3. POLIOVIRUS ADSORPTION TO OTTAWA SAND

In order to obtain unambiguous isotherm functions and parameters that can be used to characterize virus-adsorbent interactions, it is necessary that adsorption data be of high precision and extended over a large range of virus solution-phase concentration. These experiments are usually performed by mixing separate suspensions of virus and adsorbent particles in agitated batch systems under constant temperature control. After sufficient contact time has elapsed to ensure equilibrium between virus particles in the free and adsorbed state, adsorbent particles are separated from the suspension, usually by centrifugation, and the equilibrium solution-phase concentration of infective virus, C_e, is determined. Note that C^* of equation (2) and C_e are both liquid-phase equilibrium concentrations (with q), but the former is conventionally used to designate the interfacial concentration in a packed bed. The amount adsorbed is calculated indirectly from the *solution-phase* concentration difference, $(C_i - C_e) \times$ suspension volume, where C_i is measured at the moment before absorption begins. This change in solution infective virus concentration can occur due to multiple processes including: (1) natural or induced inactivation and adsorption to extraneous surfaces like shaker vessel walls, (2) changes in the state of aggregation of virus particles, (3) changes in the state of coagulation of adsorbent particles which may alter the adsorbent surface properties during the measurement time or remove virus from solution by entrapment and (4) true adsorption of single virus particles to individual adsorbent particles. The adsorption isotherm will be reasonably precise only when the fourth process dominates the concentration change $(C_i - C_e)$ and the other processes are suppressed or account is taken of their effects. Inactivation and losses to extraneous surfaces are usually accounted for by adjustment of the concentration difference $(C_i - C_e)$ based on the concentration changes that take place in control solutions which are subjected to the same batch adsorption experimental protocol but without adsorbent present. Virus aggregation and adsorbent coagulation can be eliminated, or at least controlled by regulation of electrolyte composition and pH, but this does restrict generality of the resulting isotherm.

Uniformity of adsorbent particle size also influences the precision of measured isotherm parameters when adsorption is normalized to per unit mass adsorbent as shown in equation (3). Although normalization on a per unit mass basis is not strictly proper, since the correct extensive variable is total adsorbent surface area, this choice is usually made because of lack of knowledge concerning the constitution and extent of total adsorbent surface that is active for virus adsorption. Failure to find correlations between nitrogen gas adsorption [6] and virus adsorption simply indicates that surface properties and mechanisms that lead to N_2 or virus adsorption are not the same. Experimentalists can probably do no better at present than to correlate adsorption results with adsorbent particle physical size and therefore in our measurements, effort was made to insure homogeneity of particle sizes among the adsorbent mass measurements, w, that make up the isotherm data points $q(C_e)$. Sand and clay represent particle size extremes in soil, although differences in the chemistry of their surfaces are undoubtedly also important for virus adsorption.

3.1 Materials and Methods

Finely crushed Ottawa sand was first dry sieved according to ASTM Standard D 422-63. This was followed by wet sieving with deionized water in order to eliminate the fine materials which passed through No. 325 mesh screen. The final sand samples were air dried and their particle size determined by photomicrography [7]. The particle size distribution was found to be normal within a range of 56–200 μm (Feret's diameter), and had an average size of 123 μm with a standard deviation of 24 μm. These samples were sterilized and suspended in phosphate-buffered saline (PBS) until use in the adsorption experiments.

In order to obtain solutions of high virus concentration, new attenuated poliovirus type 1 (Sabin strain) was propagated in secondary African green monkey kidney cell cultures. The solutions obtained after harvesting these cultures had a virus titre of about 10^8 PFU/ml. This virus was concentrated and purified to contain only single virus particles by ultracentrifugation and caesium chloride density gradient sedimentation [7]. The final preparation contained 2×10^9 PFU/ml in PBS buffer and was stored at -20 °C until use.

The adsorption studies were performed in 5 ml total volumes of PBS solutions (0.15 M, pH 7.3) at 20 °C. Each experiment was started by adding small aliquots of the high titre poliovirus solution to slightly less than 5 ml PBS solution to give the initial desired virus concentration. Virus solutions were then placed in sterile, round-bottom, 10-ml boiling flasks which were placed in a rotary shaker bath at 20 °C. Samples were taken for assay of the initial virus concentration and 0.5 ml of sand suspension added to each of the flasks (except those used for controls) to give sand concentrations of 0.1 mg sand/l. Control and adsorption flasks were agitated in the shaker bath at 20 °C for 1 hour. Separate kinetic studies showed that about 15 minutes was sufficient time to ensure that adsorption equilibrium had been established. The entire contents of these flasks were then transferred to polycarbonate centrifuge tubes and the solid adsorbent particles separated from the suspensions using a Damon IEC HT centrifuge. Sand particles were separated at 3000 g for 5 minutes. This centrifugation condition gave complete sand particle separation and minimal inactivation losses as shown by the controls. The supernatant solutions were analysed by the plaque assay method to determine the equilibrium solution-phase virus concentration, C_e.

3.2 Results and Conclusions

Nine adsorption measurements were made over the range of $C_e = 2.7 \times 10^3$ to 1.8×10^9 PFU/ml. The five data points at highest C_e are shown on a rectilinear scale in Fig. 2.

Errors shown in the tabulated data were calculated from a propagation of error analysis performed for each experiment. This analysis includes estimates of errors associated with making serial dilutions for the virus assay and for errors that are inherent in plaque counting. The results suggest that sand has a saturation limit for virus adsorption. It is noteworthy that no straight line can be passed through all these points even when the effects of measurement errors are included.

Saturation limited adsorption is often fit by the Langmuir equation,

$$\frac{C_e}{q} = \frac{1}{QK_L} + \frac{1}{Q} C_e \ . \tag{5}$$

When the conditions of the Langmuir adsorption reaction are applicable, the constants Q and K_L are equal to the maximum number of adsorption sites per mass of adsorbent and the adsorption reaction equilibrium constant, respectively. Linear regression analysis using equation (5) for all sand adsorption data gives $Q = 5.27 \times 10^8$ sites/mg and $K_L = 4.81 \times 10^{-9}$ ml/PFU with a correlation coefficient $r = 0.927$. This correlation is shown by the solid curve in Fig. 2 (also the curve passing through the open circles of Fig. 5 shown later). The dashed curve of Fig. 2 results from linear regression analysis for only the five data points shown in the figure. Values of the isotherm parameters for this function are $Q = 6.44 \times 10^8$ sites/mg and $K_L = 1.85 \times 10^{-9}$ ml/PFU and the correlation coefficient is improved to $r = 0.937$. From these values for the isotherm parameter Q, and the calculated surface area per unit mass of 123 μm sand particles, it can be shown that the virus saturation limit corresponds to low fractional surface coverage, about 2.6% of the surface area of an individual sand particle.

4. POLIOVIRUS ADSORPTION TO MONTMORILLONITE CLAY

Measurements of poliovirus adsorption to clay particles are complicated by the effects of clay particle aggregation [8] and by inactivation of virus under the relatively severe centrifugation conditions required to sediment all of the clay from a clay-virus–electrolyte equilibrium mixture [7]. In addition, a fundamental question about the nature of the virus–clay adsorption force arises since both particles bear net negative charges under the solution conditions used in the laboratory experiments and under the commonly encountered soil–water conditions found in the field. In this section, results from studies of the particle

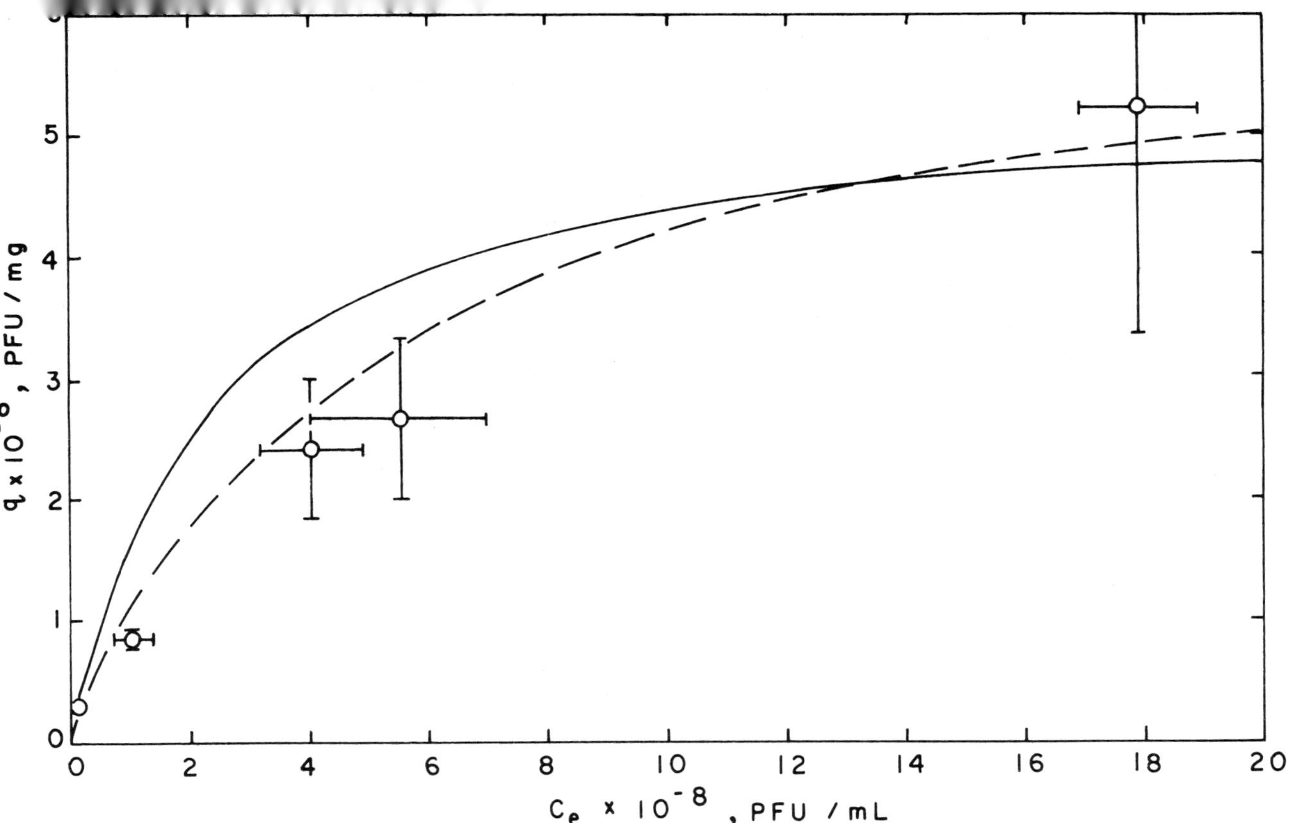

Fig. 2 – Langmuir isotherms applied to poliovirus adsorption on Ottawa sand (123 μm particle size). Regression analysis applied to all data, gives solid isotherm; regression applied only to five highest titre data points shown in figure gives dashed isotherm. Reprinted with permission from Academic Press. V. L. Vilker *et al.*, *J. Colloid Interface Sci.*, **92**, 422 (1983).

interactions in poliovirus–clay–electrolyte solutions using bright field and scanning electron micrography are reported. These studies provided the guidance that was needed to make the experimental measurements of a true, unambiguous poliovirus–clay adsorption isotherm.

4.1 Materials and Methods

Poliovirus preparations and plaque assay methods for measurement of virus concentration were the same as those used in the sand adsorption studies.

Na^+-montmorillonite clay particles were prepared from purchased bentonite (Ward's Natural Science Establishment, No. 46W 0435 PAC, Wyoming bentonite) by a series of chemical purification and size fractionation steps. Raw clay was suspended in water, dispersed in an ultrasonic bath and fractionated by gravity sedimentation to eliminate agglomerates larger than 20 μm from the suspension. Clay particles in the supernatant were centrifuged and the pellet washed in a series of sodium acetate, hydrogen peroxide and sodium citrate/sodium bicarbonate solutions in order to remove soluble ions, trace organic material and free iron oxides, respectively. The 1–2 μm fraction of the chemically treated clay was isolated by a series of low speed centrifugations. Final yield of the 1–2 μm particles was estimated to be about one percent of the initial bentonite mass. Clay particles were sterilized and stored in sodium carbonate solution at pH 9.5 and 4 °C. Monodisperse, uniformly sized particles were readily obtained from this preparation by mild sonication before withdrawing an aliquot for adsorption or coagulation studies.

Clay particle size and state of dispersion were checked throughout these studies using a Zeiss model 20T microscope fitted with a Polaroid 4 $\times$ 5 Land film holder. Viewing samples were prepared by placing a single drop of clay suspension on a standard glass microscope slide. A stabilized thin film was formed by placing a coverglass on the drop for 1 minute. The coverglass was removed before viewing in order to improve image sharpness and focus. Manipulations for viewing and photographing the specimens were done quickly before evaporation caused dissolved salts to precipitate in the thin film. The size of the clay particles was determined using a calibrated eyepiece micrometer.

Scanning electron microscopy (SEM) was used to inspect single clay particles that were isolated from equilibrated virus–clay suspensions under conditions where clay aggregation was absent. Jakubowski reported an earlier unsuccessful attempt to examine poliovirus type 1 absorption to clay minerals in seawater by electron microscopy [9]. We eliminated the problem of salt crystal interference in SEM sample preparation by sedimenting the poliovirus from PBS electrolyte and resuspending it in a solution of the volatile salt ammonium bicarbonate [10]. A small drop of an equilibrium mixture of virus and clay in NH_4HCO_3 solution (ionic strength 0.078 M, pH 8.0) was placed on a prepared SEM sample holder and dried overnight in a vacuum desiccator. The samples were next coated with thin conductive layers of either gold, gold–palladium or carbon. Electron micrograph quality varied with the several different coating materials. Gold and gold–palladium coatings gave the brightest images but the least reproducible coating thickness. Carbon coating gave darker images but more reproducible coating thickness as determined by coating, and subsequent measurement of the images of polystyrene latex spheres of known diameter (850 ± 55 Å).

Adsorption measurements for the determination of the poliovirus–clay isotherm were made by the same procedures described above for the sand studies with the exception of the centrifugation conditions. The separation of the 1–2 μm clay particles from the equilibrated clay–virus suspension required that the centrifugal force and time be sufficient to sediment all clay particles and none of the virus, and that virus inactivation losses be minimized. These effects were studied in separate experiments and it was found that the optimum centrifugation condition was 11,000 g for 10 minutes [7].

4.2 Results of Poliovirus–Clay Particle Interaction Studies

The interactions between 1–2 μm clay particles and virus particles were first investigated by mixing solutions of clay, virus and PBS electrolyte in various ratios and examining small drops of the suspensions at 640 X magnification with bright-field illumination. The result of the study in which poliovirus concentration at 0.026 M was held constant is shown in Fig. 3.

A transition from nearly monodisperse clay particles to aggregates occurs as clay concentration is increased from 0.028 to 0.080 mg/ml. In the absence of virus, the clay particles remain unaggregated to at least 0.08 mg/ml at these conditions. Lowering virus concentration or raising PBS concentration relative to the conditions for Fig. 3B also produced greater aggregation. In other studies at 0.15-M PBS, we found that 1–2 μm clay particle suspensions remain monodisperse to concentrations of 0.4 mg clay/ml with no virus present, or for clay concentration held constant at 0.04 mg clay/ml, the suspensions remained monodisperse to virus concentrations of about 10^7 PFU/ml.

Figure 4 is a reproduction of a scanning electron micrograph of a particle that was isolated from an equilibrated monodisperse clay–virus solution. The figure shows a carbon-coated image of a spherical particle attached to the edge of one or more overlapping plate-like particles.

Examination of the shape and size of these particles (by estimation of the thickness of carbon coating from measured coating thicknesses applied at the same time to polystyrene latex particles) gives strong indication that these are images of the virus adsorbed to the edge of a clay particle. The width of the clay particle is 0.9 μm and the diameter of the spherical particle is 308 ± 128 Å. This diameter of the spherical particle agrees closely with the diameter of poliovirus, 280–300 Å [11]. Similar inspections of 144 plate-like particles from three different clay–virus–electrolyte preparations showed an average of one virus-sized spherical particle adsorbed for every 6½ clay particles. In each case the spherical particle was attached to an edge and in only one case was more

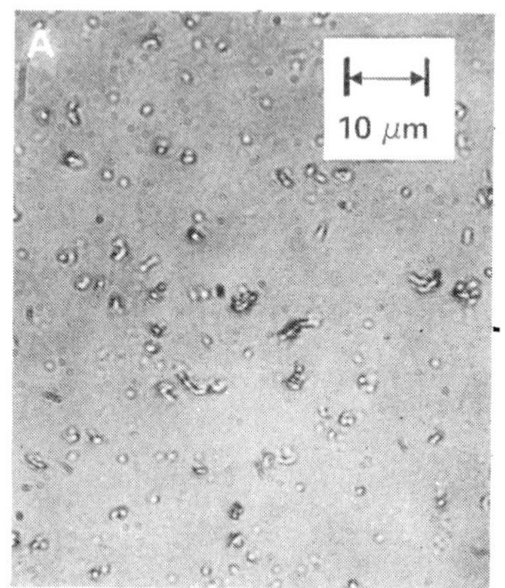

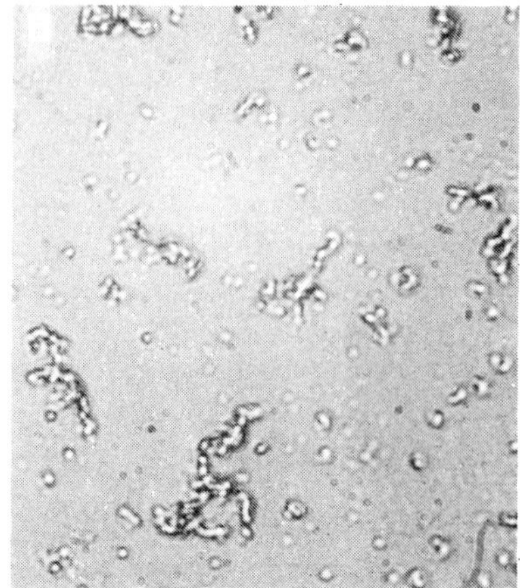

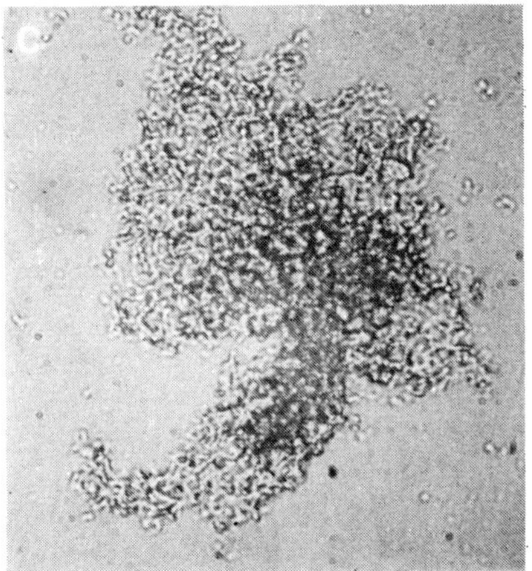

Fig. 3 – Photomicrographs of 1–2 μm montmorillonite clay particles (single particle size in A) suspended in solutions of phosphate buffered saline (0.026 M, pH 7) and poliovirus (8×10^7 virus/ml) showing increasing aggregation with increasing clay concentration: A, 0.028 mg clay/ml; B, 0.055 mg clay/ml, C, 0.080 mg clay/ml.

than one spherical particle found on a clay particle. This adsorbed virus-to-clay particle ratio gives an estimate of $q = 3 \times 10^7$ virus/mg which is within the range of the isotherm data in Fig. 5 below for the comparable value of C_e ($\sim 3 \times 10^4$ PFU/ml).

4.3 Results of Poliovirus–Clay Adsorption Isotherm Studies

Data for adsorption of poliovirus to monodisperse 1–2 μm clay particles are presented in Fig. 5.

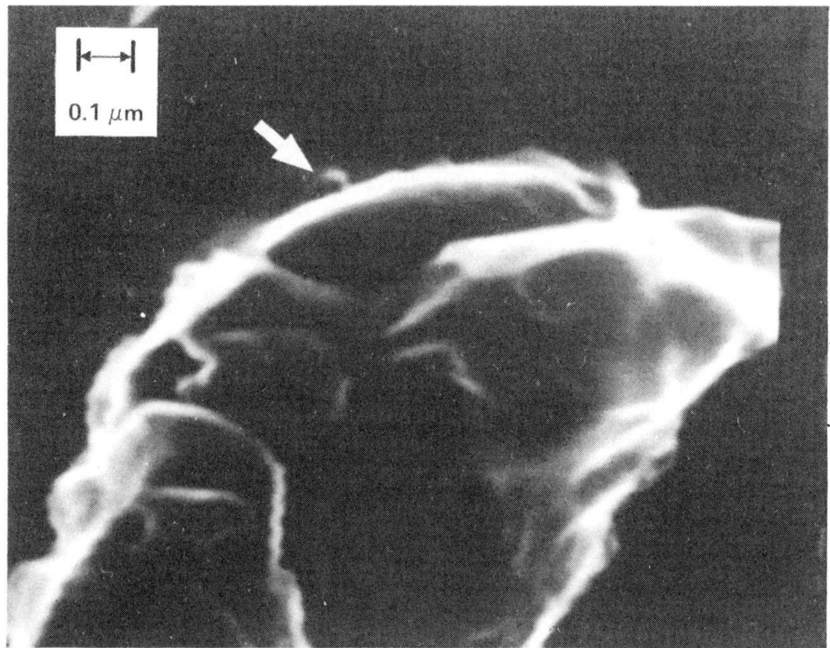

Fig. 4 – Scanning electron micrograph (80,000 ×) of spherical particle adsorbed to edge of probable clay particle. Particles were isolated from solution of 1 μm clay particles (0.31 mg/ml), 0.078-M NH_4NCO_3 (pH 8) and purified poliovirus (3.2×10^4 virus/ml). Carbon coating thickness estimated to be 65 Å from other electromicrographs of coated latex spheres of known diameter. Spherical particle diameter is 308 ± 128 Å and clay particle width is 0.9 μm, after correction for carbon coating.

The isotherm function which best fits the data is of the Freundlich type like equation (3). The log–log form of this equation was used to find the isotherm parameters u and n by linear regression analysis. We found $u = 4.54$ and $n = 1.345$ with C_e and q having the dimensions PFU/ml and PFU/mg clay, respectively. The correlation coefficient is $r = 0.995$. This isotherm is shown as the solid line passing through the clay adsorption data (squares) in Fig. 5.

5. COMPARISON OF POLIOVIRUS ADSORPTION ISOTHERMS

Our data and isotherm functions are plotted as log q *vs* log C_e on Fig. 5. The isotherm shown for sand adsorption is weighted more heavily by the data at larger values of C_e. Recent adsorption measurements of Moore *et al.* [6] for poliovirus adsorbing to sand from solutions of 0.006 ionic strength and pH 7.0 are also presented. The sand used in this study was prepared by dry sieving commercial Ottawa sand to give particles in the stated size range of 600–800 μm. These investigators fit their low virus concentration data (open triangles) and high virus concentration data (solid triangles) to separate isotherm functions. No physical or experimental reason is offered in support of this separation of the adsorption data into two different populations.

Both our results and those of Moore *et al.* show that poliovirus adsorption to sand approaches a saturation limit and that the virus saturation limits are

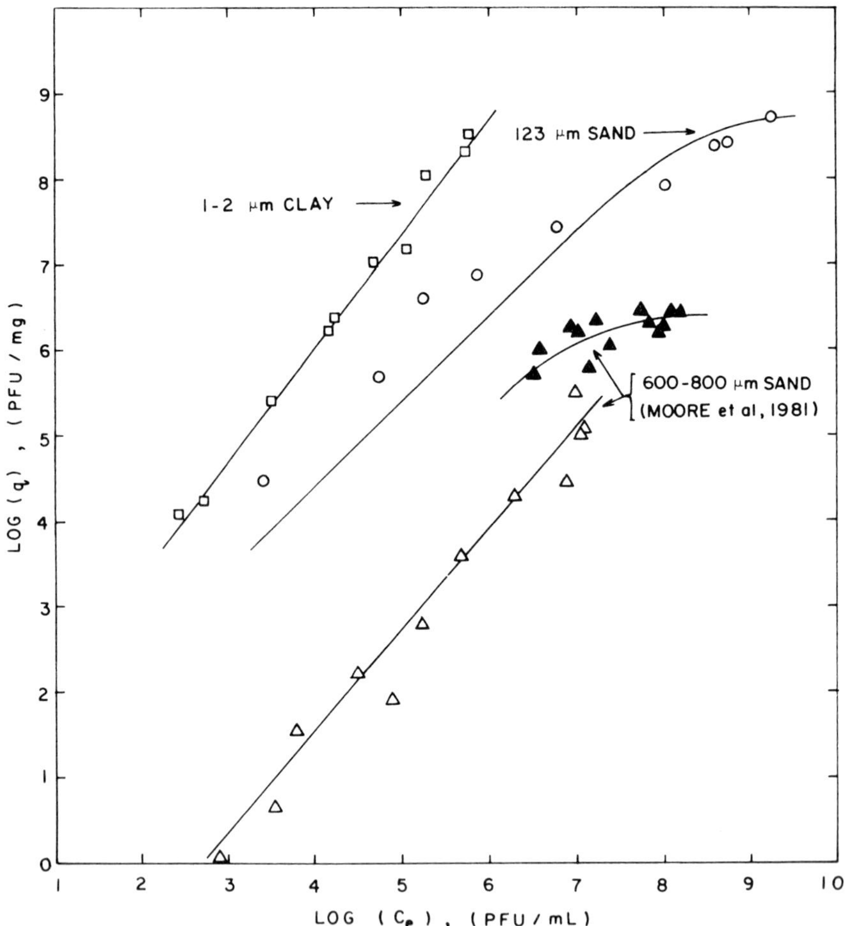

Fig. 5 – Comparison of data and isotherms for poliovirus in PBS solution ($I = 0.15$, pH 7.3) adsorbing to 123 μm Ottawa sand particles (○), 1–2 μm Na-montmorillonite particles (□) and the data and isotherms of Moore *et al.* [6] for poliovirus in synthetic freshwater ($I = 0.006$, pH 7.0) adsorbing to 0.7-mm Ottawa sand particles (△▲). Reprinted with permission from Academic Press. V. L. Vilker *et al.* *J. Colloid Interface Sci.*, **92**, 422 (1983).

found to correspond to low fractional surface coverage: 0.026 for our results and 0.001 for the results of Moore *et al.* The apparent discrepancy between our adsorption data to sand and that of Moore *et al.* can probably be attributed to differences in sorbent surface area and solution ionic strength, although several additional experimental variables differed between these two studies. The larger particle size of Moore *et al.* gives an average surface area per unit mass sorbent which is about 32 times smaller than our 123-μm particle size preparation. Therefore, virus adsorption per unit sorbent surface area will compare favourably for the two studies at $C_e > 10^6$ PFU/ml. Their adsorption at low C_e is brought into closer agreement with our results when both results are normalized by sorbent surface area, but would still be less by 1–2 orders of magnitude. This difference may be due in part to the higher ionic strength (0.15 *vs* 0.006) used in our studies. Qualitative increases of adsorption with increasing ionic strength have been previously reported for both proteins [12] and viruses [1, 13]. The effect is presumably caused by ionic double layer compression.

Comparison of the isotherm functions derived for our sand and clay studies show that q is about 20–200 times larger for clay in the range of C_e equal to $10^3 - 10^6$ PFU/ml. The power of C_e for the dependence of adsorption to clay was found to be statistically greater than unity: $n = 1.345 \pm 0.106$ at the 95% confidence level. We do not know of a mechanistic interpretation for this dependence but note that this observation was part of the motivation to perform our clay aggregation studies. Although photomicrographs such as Fig. 3A suggest that large scale aggregates are not present in the adsorption studies, undetected smaller scale aggregates might still be a possible explanation for $n > 1$.

ACKNOWLEDGEMENT

This research was supported in part by the University of California, Water Resources Center, Project UCAL-WRC-W-597 and by NSF Grant CPE-82-12227.

REFERENCES

[1] Vilker, V. L. in *Modelling Wastewater Renovation-Landtreatment,* (I. K. Iskandar, Ed.), Chapter 9, John Wiley & Sons; New York (1981).

[2] Vilker, V. L., Kamdar, R. S., and Frommhagen, L. H. *Chem. Engr. Commun.,* **4**, 569 (1980).

[3] American Public Health Association, *Standard Methods for the Examination of Water and Wastewater,* 13th ed., APHA, New York (1971).

[4] Hoskins, J. M., *Virological Procedures,* p. 31, Butterworth, London (1967).

[5] Lennette, E. and Schmidt, N., Eds. *Procedure for Viral and Rickettsial Infections;* 4th ed., p. 115, American Public Health Association, New York (1969).

[6] Moore, R. S., Taylor, D. H., Sturman, L. S., Reddy, M. M., and Fuhs, G. W., *Appl. Environ. Microbiol.*, **42**, 963 (1981).
[7] Vilker, V. L., Fong, J. C., and Seyyed-Hoseyni, M., *J. Colloid Interface Sci.*, **92**, 422 (1983).
[8] Vilker, V. L., Meronek, G. L., and Butler, P. C., *Environ. Sci. Tech.*, accepted for publication (1983).
[9] Jakubowski, W., *Bacteriol. Proc.*, V198 (1969).
[10] Schwerdt, C. E. and Fogh, J., *Virology*, p. 1283, **4**, 41 (1957).
[11] Davis, B. D., *Microbiology*, 2nd ed., Harper and Row (1973).
[12] Feder, J. and Giaever, I., *J. Colloid Interface Sci.*, **78**, 144 (1980).
[13] Taylor, D. H., Moore, R. S., and Sturman, L. S., *Appl. Environ. Microbiol.*, **42**, 976 (1981).

CHAPTER 18

Practical and Theoretical Aspects of Constant Pressure and Constant Rate Filtration

A. RUSHTON and C. KATSOULAS, UMIST, PO Box 80, Manchester M60 1QD, UK

1. INTRODUCTION

Most industrial solid–liquid separations proceed in conditions of constant pressure differential, constant pumping rate or, when using centrifugal pumps, with variable flow and pressure. Some separations may involve a mixed mode, e.g., where the initial conditions are at constant rate, followed by a period of constant pressure, at a preselected upper limiting pressure.

Elementary theory may be used to obtain an expression relating pressure differential, flowrate and volume filtered:

$$\Delta P = (2K_1 V + K_2)\, q \tag{1}$$

in which ΔP is the pressure differential at time t; q is the instantaneous flowrate, V is the volume filtered up to time t; K_1 is a factor related to cake and liquor properties and K_2 involves the fluid flow resistance of the filter medium. Thus

$$K_1 = (\alpha \mu C)/2A^2 \tag{2}$$

and

$$K_2 = \mu R/A \tag{3}$$

Attempts have been made [1] [2] to use the above expression, in integrated form, in developing process design equations for use with rotary and horizontal belt filters, pressure filters and centrifugal filters [3]. Great care has to be taken in the interpretation of pilot plant work, particularly where processes of cake drainage or washing follow the separation. An overall intention in the development of theory is the provision of a basis for the estimation of process time. Varying periods of time are required to filter, dewater and wash a cake and each period has an effect on the overall productivity and quality of the filtered product.

Equation may be developed for drainage and wash times leading to overall optimization of batch cycles in pressure filter [4].

Integration of equation (1) in conditions of (a) constant pressure differential and (b) constant rate leads to the conclusion that the time required for constant rate processes is twice that for constant pressure, assuming the same volume filtered and the same final pressure differential.

Such elementary conclusions are, of course, dependent upon the assumption that the cake properties (e.g. specific resistance) are unaffected by the separation mode.

Certainly the importance of cake properties can be demonstrated if the essential compressible nature of most filter cakes is recognized. Assuming a relationship of the form:

$$\bar{\alpha} = \alpha_0 \, \Delta p^n \tag{4}$$

to describe the effect of pressure differential increase on cake specific resistance α, equation (1) may be integrated as follows.

Assuming negligible cloth resistance

$$q = (\Delta P A^2)/(\bar{\alpha} \mu c V) \tag{5}$$

from (4)

$$q = \Delta P^{(1-n)} A^2/(\alpha_0 \mu c V) \tag{6}$$

For constant ΔP:

$$\frac{\mu \alpha_0 c}{A^2} \frac{V^2}{t} = 2 \, \Delta P^{(1-n)} \tag{7}$$

In constant rate mode, with maximum differential pressure ΔP_m

$$\frac{\alpha_0 \mu c}{A^2} \frac{V^2}{t} = \Delta P_m{}^{(1-n)} \tag{8}$$

Hence for equivalent amounts filtered

$$(\Delta P_m/\Delta P) = 2^{1/(1-n)} \tag{9}$$

Thus for incompressible materials ($n = 0$) a pressure ratio of 2 may be expected; for highly compressible solids ($n > 0.75$) the ratio is sixteen. Constant rate processes are, therefore, sometimes considered unsuitable for compressible materials.

Again this conclusion depends on the constancy of α_0 in the two modes.

Ruth [5], by contrast, found that the specific resistance takes different values in constant pressure and constant rate filtration of the same slurry. On the other hand Carman [6] stated that the property is identical in both operations. Walker *et al.* [7] also contradicted Ruth stating that values of α and n agreed reasonably well in the two modes.

More recently the work of Heertjes [8] points to the importance of flow-rate in its effect on the packing of particles. Open structures with low specific resistance were obtained at high flowrates. Later work [9] has highlighted the combined role of flow and concentration in cake formation and cloth-pore bridging processes.

The study reported below was undertaken to collect experimental evidence on the specific resistance of suspensions filtered in the same pressure filter, under constant rate and constant pressure conditons. Considering the vital importance of constant rate processes in industry it is somewhat surprising to note the relative dearth of information in this area.

2. EXPERIMENTAL

Slurries of calcium silicate and calcium carbonate were studied. The latter material is essentially incompressible in most of the practical conditions used; calcium silicate on the other hand exhibits a high degree of compressibility, as may be observed in the various graphs below.

Equation (1) leads to the following relationships between volume filtered and time:

Constant pressure $$\Delta Pt = K_1 V^2 + K_2 V \quad (10)$$

Constant rate $$\Delta P = 2K_1 K_3^2 t + K_2 K_3 \quad (11)$$

In Equation (11), K_3 is the rate of flow. These expressions suggest linear relationships between $(t/V, V)$ for constant pressure filtration and $(\Delta P, t)$ for constant rate in those circumstances where K_1 and K_2 remain constant during the separation.

Suspensions in the concentration range 0.5–8 wt.% of solids were filtered at pressures up to 2.1×10^5 N/m^2 (30 psig) in one frame of a small pilot scale pressure filter. The area available for filtration was 0.043 m^2. Filtrate rates and volume were measured by direct weighing using a balance with the maximum capacity of 60 kg. The properties of the particles used are listed in Table 1. Other experimental conditions and arrangements are reported elsewhere [10].

Table 1

Particulate properties

Solid	Size Range μm	Mean Size μm	Density kg.m^{-3}
Calcium carbonate	4–50	8.6	2490
Calcium silicate	4–30	6.5	1750

A multifilament, mixed-fibre cloth (supplied by P & S Textiles (Neotex 1188)) was used in the studies. This cloth was found to produce clear filtrates, without excessive cloth binding.

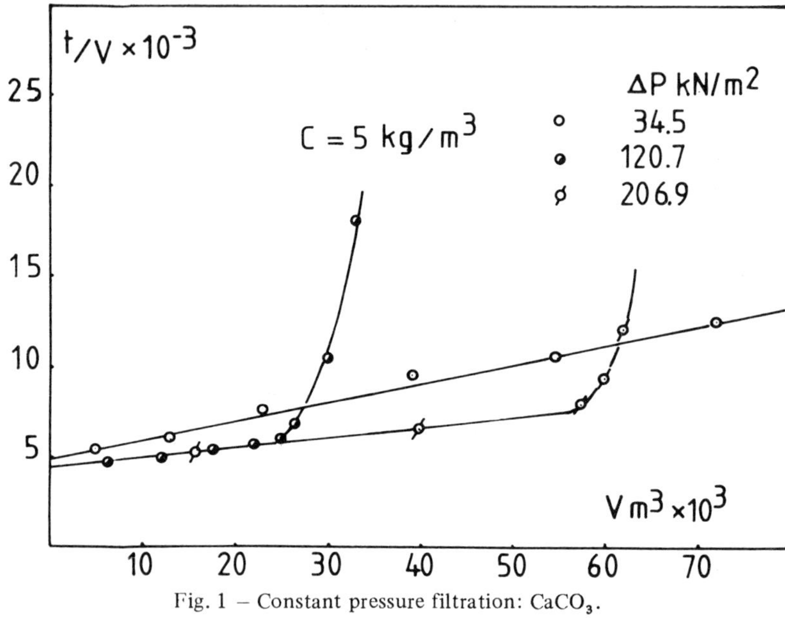

Fig. 1 – Constant pressure filtration: $CaCO_3$.

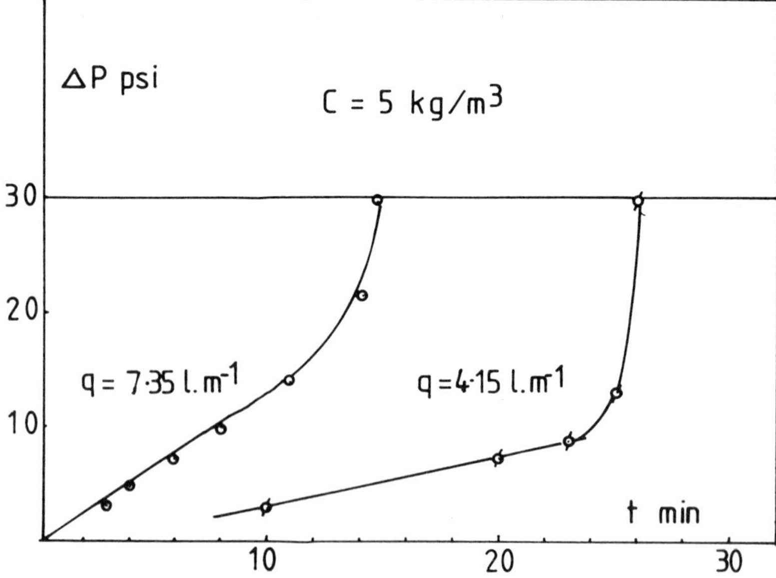

Fig. 2 – Constant rate filtration: $CaCO_3$.

Graphical representation of some of the results are presented in Figs. 1–4. The collected data indicate that the calcium carbonate used exhibited weak compressibility at the lower concentration (5 kg/m^3); the constant rate data on the latter also indicates a weak effect of flow rate on $\bar{\alpha}_{CR}$. At higher concentrations up to 80 kg/m^3, the calcium carbonate appears incompressible and unaffected by flowrate. The latter effect is demonstrated in Fig. 2 where linear (ΔP, t) graphs are depicted.

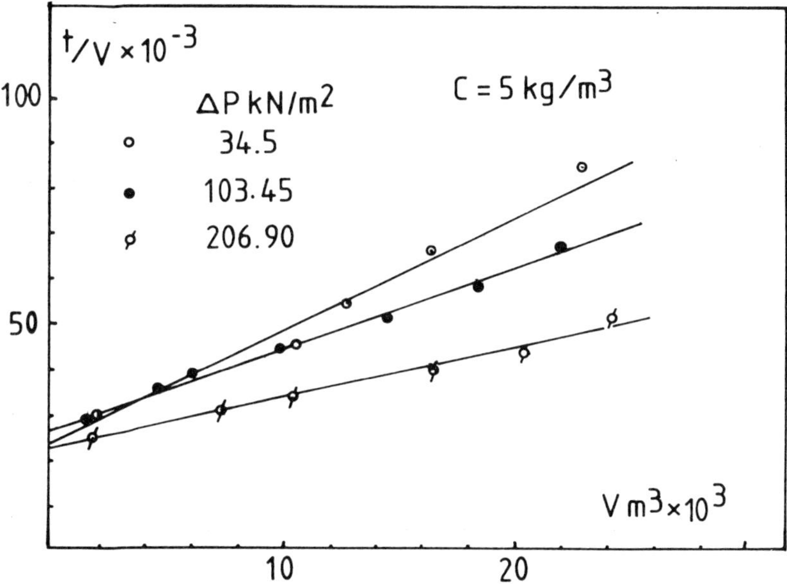

Fig. 3 – Constant pressure filtration: $CaSiO_3$.

On the other hand calcium silicate exhibits strong compressibility ($n = 0.6$ for pressures up to 206 kN/m (30 psig). The curves obtained in constant rate tests, Fig. 4, also indicate the compressible nature of this material. The results here have been examined in order to estimate the compressibility index n in constant rate conditions. Here the filtration data were linearized by substituting for α, using equation (4) to produce:

$$\Delta P = K_4 \alpha_0 \Delta p^n t + K_2 K_3 \tag{12}$$

where $K_4 = 2K_1 K_3{}^2/\alpha = \mu c q^2/A$.

If the initial resistance to flow cannot be neglected, the constants α_0 and n may be obtained interactively by:

(1) assuming a value for n,
(2) for each (ΔP, t) pair of data, calculate $\Delta P^n y$;

(3) plot ΔP v. $(\Delta P^n t)$;
(4) perform linear regression analysis on (3);
(5) calculate the correlation coefficient for the data;
(6) repeat, from Step 1, for a different value of n;
(7) continue until a satisfactory value of the correlation coefficient is obtained.

The results of this analysis are presented in the next section.

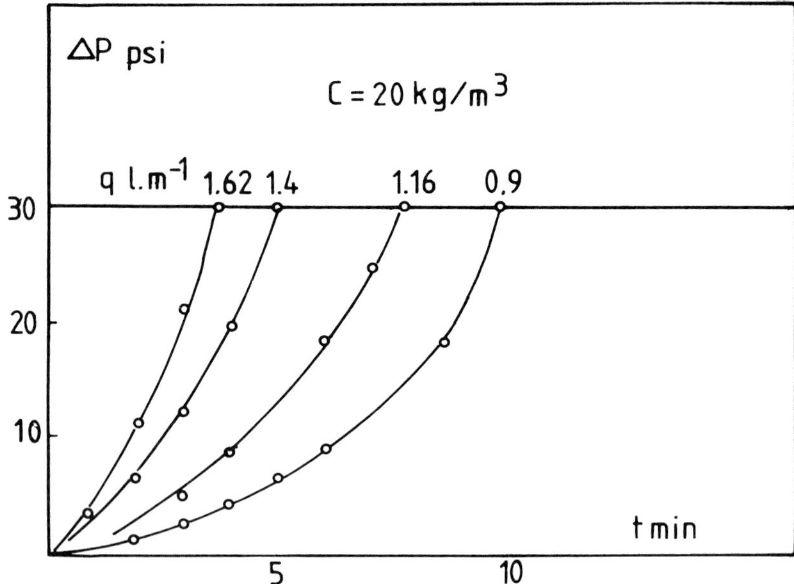

Fig. 4 – Constant rate filtration: $CaSiO_3$.

3. ANALYSIS OF RESULTS

3.1 Constant Rate Filtration of Calcium Silicate

The data collected for calcium silicate depicted in Fig. 4 were analysed by plotting ΔP v. $\Delta P^n t$ using various assumed values of n. Some of the results obtained are presented in Table 2. This information is plotted in Fig. 5 where it may be noted that both α_0 and n are apparently influenced by the flow conditions. The trends indicate that high flowrates are associated with high α_0 values, but with a decrease in the associated compressibility coefficient. The effect of flowrate on n makes the conclusions from Equation (9) quite dubious. The above information can be used to calculate average resistance by means of the expression:

$$\alpha = \Delta P / \int_0^{\Delta P} \Delta P/\alpha \qquad (13)$$

Table 2

Constant rate experimental results

Run	Flow $(m^3/s) \times 10^5$	Concentration kg/m^3	n	α_0 $(m/kg) \times 10^{-8}$
63	1.538	10	0.68	4.17
64	1.953	10	0.40	9.55
65	2.333	10	0.40	9.28
66	2.713	10	0.35	16.90
67	1.510	20	0.77	1.70
68	1.933	20	0.48	3.14
69	2.333	20	0.46	4.24
70	2.700	20	0.40	8.56

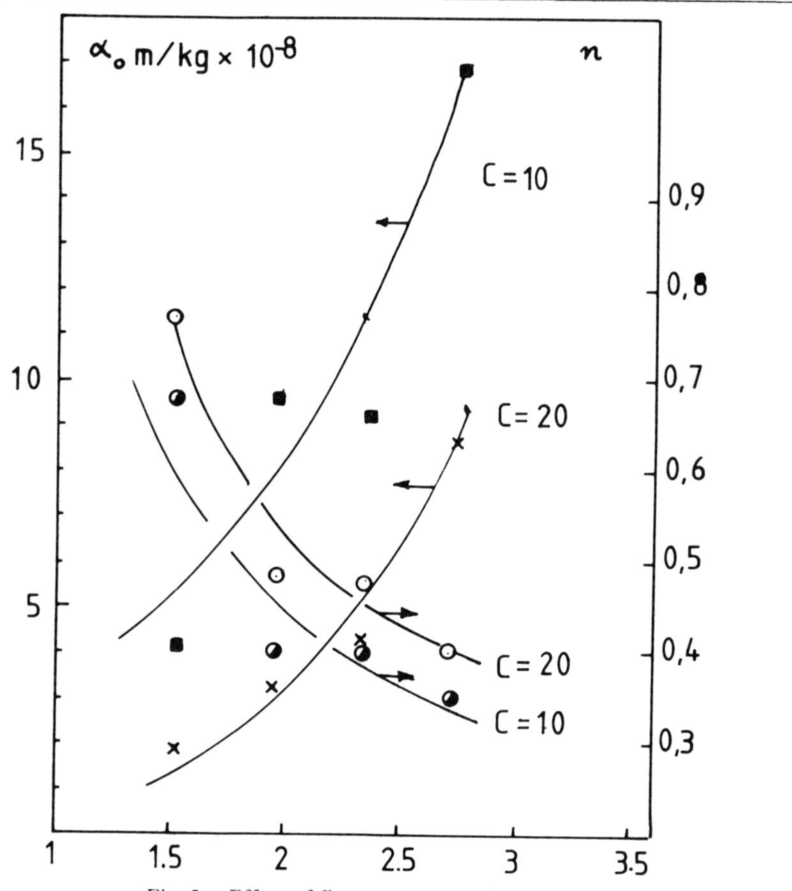

Fig. 5 – Effect of flowrate on α_0 and n.

which for systems describable by:

$$\alpha = \alpha_p \, \Delta p^n$$

reduces to:

$$\alpha = \alpha_0 \, (1 - n) \, \Delta p^n \tag{14}$$

and leads to the conclusion that high flowrates will be associated with lower, overall specific resistances. This is in agreement with the findings of Heertjes [8].

An obvious interest lies in the possible use of constant rate data to predict filterabilities in other modes. Data taken from constant pressure separation of calcium silicate, using identical slurries, were compared using the average resistances from the C. P. run and those predicted from the C.R. data and equation (14) (see Table 3). These calculations demonstrate a serious divergence between

Table 3

Calculated and experimental specific resistances

ΔP kN.m^{-2}	αC.P. m.kg^{-1} X 10^{-10}	q_0 m.^{3}s^{-1} X 10^{-5}	α C.R. m. kg^{-1} X 10^{-10}
34.48	6.38	2.90	6.42
68.96	9.17	2.60	5.26
103.45	10.98	3.00	9.49
137.93	12.39	3.30	5.73
172.41	14.75	3.66	5.97

predicted resistances (using constant rate data) and those obtained in constant pressure conditions. The effect of increasing pressure in C.P. filtration is underestimated when data from C.R. are used for comparison. It may be concluded that specific resistances produced by the two modes are different and, in view of the effects of flow on cake structure, data collected in one mode cannot readily be used to predict flowrates in the other. It is, therefore, quite spurious to make any conclusions about the relative merits of the two modes, if identical α values are used in the argument.

3.2 Practical Comparisons

The inconclusive nature of the results above leads to a simpler method of comparison between the two modes. As reported above, assumption of identical αs leads to the result that, at the same final ΔP and V, the time for filtration in C.R. conditions will be *twice* that for C.P. processing. This concept has been tested using practical data on slurries of the same composition; it has been found that the above result is quite wrong in that C.R. filtration times required to reach a particular (ΔP, V) point can be *shorter* than the corresponding C.P. separation. Again if we consider the energy expended in the separation as the product of the pressure differential and the volume filtered, C. R. filtration involves less energy loss by friction. Of course the total energy consumed must take into account kinetic effects and other factors.

These points are recorded in Table 4 where productivities have been estimated for the two modes. The productivity is calculated from the volume filtered (when the ΔP_{CP} and ΔP_{CR}) divided by a cycle time t_c which is the sum of the filtration time t_f and a 'down-time' t_d. In Table 4 the latter is taken as constant at 900 s.; it is interesting to note the difference between t_f for C.R. and C.P., the data for the latter is taken from (V, t) data at constant pressure and t_f for C.R. is calculated from V^*/q. Here V^* is the volume filtered when the ΔPs equalize (see Fig. 6) and q is the flowrate.

Table 4

Practical comparison of constant rate and constant pressure

Calcium silicate slurry: 10 kg/m³							
Material Concn. kg/m³	q m³/s × 10⁵	ΔP kN/m²	V^* m³ × 10³	t_{fCR} s	t_{fCP} s	β_{CR}/β_{CP}	E_{CR}/E_{CP}
$CaSiO_3$ 10	1.538	34.48	7.3	475	480	1.003	0.339
		172.41	15.7	1020	860	0.917	0.283
$CaSiO_3$ 10	1.953	34.48	5.0	256	242	0.986	0.409
		172.41	13.6	696	720	1.017	0.419
$CaSiO_3$ 10	2.333	34.48	4.2	180	190	1.010	0.399
		172.41	11.3	484	560	1.055	0.404
$CaSiO_3$ 10	2.713	34.48	3.0	110	115	1.002	0.508
		172.41	9.0	331	420	1.071	0.399
$CaSiO_3$ 10	1.953	68.96	7.75	356	470	1.089	0.481
		103.45	10.0	512	590	1.055	0.437
		137.92	12.0	614	660	1.031	0.412
$CaCO_3$ 40	9.660	34.48	6.5	67	80	1.013	0.458
		68.96	12.1	125	230	1.102	0.437

The calculations indicate that as the flowrate used is increased the productivity ratio is affected by the C.P. level chosen and by the nature of the slurry as evidenced in the $CaCO_3$ data, where a 10% productivity improvement is noted.

Figure 6 shows the (ΔP, V) relations involved. It is well known that an optimum volume filtered can be specified for C.P. conditions. Such a point is indicated on the horizontal pressure line. It is suggested that a C.R. curve could be specified which would result in a higher productivity. It remains to be seen whether the total energy requirements for the separation in the two modes,

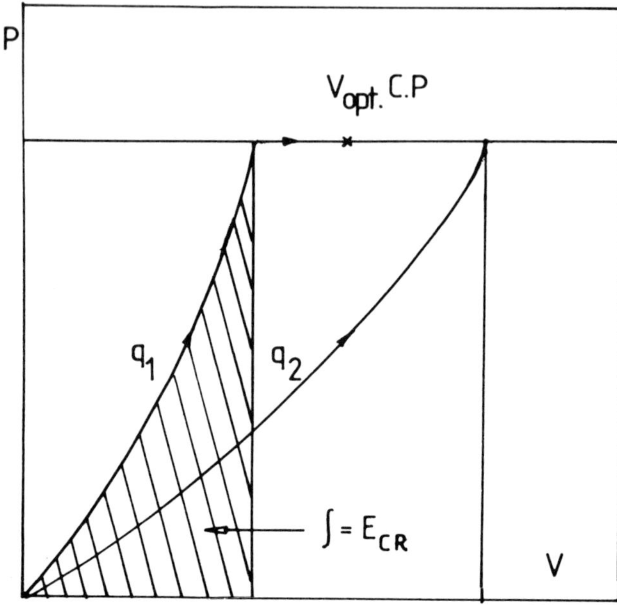

Fig. 6 – $\Delta P, V$ relations in filtration.

taking into account other frictional losses, kinetic energy, etc., will lead to a clear preference for a particular mode. However, the data presented do point to the overly simple economic conclusions which follow the assumption of equal specific resistances. The $\sqrt{2}$ 'law' is merely a result of calculus and has little bearing on the physical situation obtaining in the filtration of suspensions.

CONCLUSIONS

1. The specific resistance of filter cakes produced in conditions of constant pressure and constant rate can be quite different, an effect which is attributed to (a) the effect of flowrate on the 'constant' α_0 and (b) the reduction in compressibility of the cake.
2. The $\sqrt{2}$ 'law' has been demonstrated to be erroneous for the slurries filtered.
3. Constant rate filtration has been shown to produce improvement in productivity terms and possibly in the energy consumed in solid–fluid separation.

NOMENCLATURE

A	Area	m^2
C	Concentration	$kg.m^{-3}$
E	Energy	J

E_{CR}	Energy consumed in constant rate filtration	J
E_{CP}	Energy consumed in constant pressure filtration	J
K_1	Cake 'constant': $(\alpha\mu C/2A^2)$	$N.s.m^{-8}$
K_2	Medium 'constant': $(\mu R/A)$	$N.s.m^{-3}$
K_3	Flowrate	$m^3.s^{-1}$
K_4	Constant: $2K_1K_3{}^2/\alpha$	$N.kg.\ s^{-1}m^{-3}$
n	Compressibility coefficient	—
P	Pressure	$N.m^{-2}$
ΔP	Pressure differential	$N.m^{-2}$
ΔP_m	Pressure drop (maximum)	$N.m^{-2}$
ΔP_{CP}	Pressure differential in constant pressure filtration	$N.m^{-2}$
ΔP_{CR}	Pressure differential in constant rate filtration	$N.m^{-2}$
q	Flowrate	$m^3.s^{-1}$
q_0	Initial flowrate	$m^3.s^{-1}$
R	Medium resistance	m^{-1}
t	Time	s
t_f	Filtration time	s
t_c	Cycle time	s
t_d	Down time	s
t_{fCR}	Filtration time in constant rate filtration	s
t_{fCP}	Filtration time in constant pressure filtration	s
V	Volume	m^3
V^*	Volume filtered when $\Delta P_{CP} = \Delta P_{CR}$	m^3
α	Specific resistance	$m.\ kg^{-1}$
α_0	Specific resistance at unit pressure differential	$m.\ kg^{-1}$
α	Average specific resistance	$m.\ kg^{-1}$
α_{CR}	Specific resistance calculated from C.R. data	$m.\ kg^{-1}$
α_{CP}	Specific resistance calculated from C.P. data	$m.\ kg^{-1}$
β	Productivity	$m^3\ s^{-1}$
μ	Viscosity	$N.s.m^{-2}$

REFERENCES

[1] Tiller, F. M., *Chem. Eng. Progr.*, **51**, 282 (1955).

[2] Rushton, A., International Symp. Hydrometallurgy, AIME, 112th Annual Meeting, Atlanta, Georgia, March 6 (1983).

[3] Rushton, A. and Wakeman, R. J., *Powder and Bulk Solids Tech.*, **1**, 58 (1977).

[4] Rushton, A., I.Chem.E. Jubilee Symposium, E. F.CE Pub. Series No. 21. Event No. 263. London 6/8 April (1982).

[5] Ruth, B. F., *I.E.C.*, **27**, 708 and 806 (1935).

[6] Carman, P. C., **J.**

[6] Carman, P. C., *J. Soc. Chem. Ind.*, **53** 159T, 301T (1934).

[7] Walker, W. H. *et al. Principles of Chemical Engineering,* 3rd edn, McGraw-Hill (1937).
[8] Heertjes, P. M., *Chem. Eng. Sci.,* **6**, 269 (1957).
[9] Rushton, A., Hosseini, M. and Hassan, I., *J. Separ. Proc. Technol.,* **1** (3), 35–41 (1980).
[10] Katsoulas, C. V., M.Sc. Thesis, Manchester University (1982).

CHAPTER 19

Deep Bed Filtration in a Network of Random Tubes

M. LEITZELEMENT, P. MAJ, J. A. DODDS and J. L. GREFFE, Laboratoire des Sciences du Genie Chimique CNRS–ENSIC, 1 rue Grandville, 54042, Nancy, France

ABSTRACT

A network of tubes with a range of sizes, distributed randomly in a two dimensional square array, is used as a model of a deep bed filter. The importance of the structure as an interconnected network on the pattern of fluid flow and residence time distribution is emphasized and it is shown that average pore sizes cannot be used to calculate flow when there is a distribution in pore sizes.

The model is used to simulate deep bed filtration and the variation of flow pattern with progressive clogging is shown as ending with the establishement of preferential paths through the bed similar to the 'worm holes' mentioned in the literature.

1. INTRODUCTION

Deep bed filtration is a long-established engineering process in which large volumes of liquids containing small concentrations of fine particles in suspension are clarified by passage through beds of granular material such as sand.

The pores in filter beds are larger than the size of the particles to be retained so as to avoid cake formation on the inlet surface and the particles are thus retained inside the pores of the filter and accumulate on the surfaces of the filter grains. Filters of this type are normally made with granular media since the retained solids have little value and can be removed by backwashing either to clean the grains as in municipal water filters, or to discharge the filter as in precoat filtration.

Deep bed filtration is never a steady state operation as the progressive accumulation of deposited particles changes the nature of the collector surfaces and the structure of the pore space. Therefore as the operation progresses the filtration efficiency changes and, if a constant pressure is maintained, the rate of flow decreases. Filtration is stopped when the bed is exhausted, that is either the effectiveness of filtration falls below a limiting value or the resistance to flow exceeds a set value. Ideally, in an optimally designed filter, these two limits should occur at the same time.

To design a deep bed filter it is therefore necessary to know the efficiency of filtration and the flow characteristics of the bed as well as how these vary with the progress of the filtration. These two parameters are both dependent upon the structure of the pore space in the bed and the fundamental problem is to describe how the structure of the bed changes in response to the accumulation of deposited solids both for the local alterations in geometry and in the hydrodynamic conditions in the filter pores.

2. THEORIES OF DEEP BED FILTRATION

Recent reviews [1, 2] class theories of deep bed filtration into two families, macroscopic and microscopic descriptions. In the former the dynamic operation of a deep bed filter is described by partial differential equations based on semi-empirical functions derived from bench scale experiments. This theory is used in design and has been developed to the point of being able to give an optimum construction for the full cycle of operation of multimedia filters including the backwash phase.

The microscopic theory gives a mathematical formulation based on an analysis of the interaction between the filter grains and the particles in suspension in a hydrodynamic flow field. The particles in suspension are brought to capture sites by direct interception, inertia, diffusion, gravity or hydrodynamic forces, and when they encounter the collector surfaces they are retained by electrostatic, double layer, or van der Waals forces.

These theories have obtained a high degree of sophistication and can successfully predict the initial filtration period in a clean filter bed. Nevertheless, as has been mentioned [3] the major part of practical filtration takes place after this initial period and in a bed whose porous structure is continuously changing, both geometrically and physico-chemically, due to the accumulation of deposits.

The microscopic theory has mainly dealt with the first of the two fundamental parameters of deep bed filtration, namely the efficiency of collection, and has paid little attention to other fundamental parameter, i.e. the change in flow rate/pressure drop characteristics. Relatively little work has been done on analysing the effects of deposits on the action of filter beds despite the fact that this is of fundamental importance for extending the scope of capture theory after the initial clean filter period.

Work has been done in observing deposits in beds [4–7] and tracer methods [8, 9] have been used to follow structural changes during filtration. However, as in the macroscopic theory, this work is usually based on a description of flow in porous media using the Kozeny–Carman equation which is inadequate in this case. The structure of the bed and its influence of on fluid flow must be examined in more detail.

3. FLUID FLOW IN PACKED BEDS

The flow of fluids in packed beds is described by Darcy's law

$$\nu = \frac{K}{\eta}\frac{\Delta P}{L} \tag{1}$$

which introduces the notion of the permeability of a porous medium as a structural parameter. The Kozeny–Carman equation gives this structural parameter in terms of the porosity, and the specific surface of the porous medium together with a factor called the Kozeny constant:

$$K = \frac{\epsilon^3}{h_\mathrm{k}\,(1-\epsilon)^2}\,\frac{1}{S_0{}^2}\ . \tag{2}$$

Derivation shows that the Kozeny constant h_k includes a pore shape factor and a tortuosity of the pores in the bed:

$$h_\mathrm{k} = k_0\,\tau^2 \approx 5.0\ . \tag{3}$$

There are several ways of deriving the Kozeny–Carman equation from first principles. The original derivation by Kozeny was by assuming that the packed bed was one large channel of complex shape which retards flow by friction at its surface. Carman's derivation was based on an empirical correlation for flow in non-circular channels using the hydraulic radius, which for a porous medium can be given in terms of ϵ and S_0. It is also possible to derive the Kozeny–Carman equation from Poiseuilles law for flow in a capillary tube applied to a bundle of equal sized capillary tubes.

The Kozeny–Carman equation is a remarkably simple relationship for a complex phenomenon and which gives good results for many different materials once the Kozeny constant has been determined. Nevertheless it is based essentially on a description of a porous medium by average pore size. It seems to owe a lot of its success to the fact that most packed beds have a similar structure and that it is impossible to vary S_0 and ϵ independently over a sufficient range to test the equation. This is demonstrated by the excessive range of h_k values which must be introduced for such extreme cases as fibre beds, flocculated filter cakes or consolidated porous media.

As Carman himself mentioned [10] there are important conditions which must be observed for the equation to be valid. Amongst these there are:

(1) No pores must be sealed off as this would affect the porosity available to flow, but not the absolute porosity.

(2) The particles forming the bed must be randomly distributed and be reasonably uniform.
(3) The porosity must not be either extremely large or extremely small.

None of these conditions obtain in a filter bed during clogging. In a classic paper Childs and Collis-George [11] showed that surface area can be altered almost independently of permeability, since large pores contribute most to permeability but small pores contribute most to specific surface. They also tried to derive the Kozeny–Carman equation for a bundle of capillary tubes with a distribution of sizes and showed that this was only valid if all the tubes were the same size.

It seems, therefore, that the success of the Kozeny–Carman equation only reflects the essential similarity of most porous media. But, over the course of a filtration run the bed undergoes large changes in structure, permeability, porosity, specific surface and, consequently, overall pressure drop.

An analogous problem is in describing two-fluid (such as oil and water) flow in the same porous medium. In this case at any given saturation, capillary forces act to structure the distribution of the two fluid phases, wetting phase in small pores, non-wetting phase in large pores. Thus the oil phase flows through a porous medium composed of the solid phase plus the distributed water phase. And the water phase flows through a porous medium formed by the solid phase and the distributed oil phase. If the respective proportions of the oil and water phases are varied the permeability of each phase can vary from 0 to the full single phase permeability. It is found in petroleum engineering literature that these variation in permeability in response to 'structural' changes can only be modelled by including (1) the pore size distributions and (2) the inter connection of pore space.

The inescapable conclusion is that a description of the effects of deposits in the structure and flow in a deep bed filter must also include these two effects. Furthermore any model should be able to give a relationship between the variation in the permeability and the structure of pore space related to measurable parameters such as, for example, residence time distribution and the way the permeability of the bed varies.

4. FLUID FLOW IN A NETWORK OF RANDOM SIZED TUBES

As a first approach to developing a model of deep bed filtration to take into account structural changes in the pore texture as filtration progresses, we use a two-dimensional network of capillary tubes with a range of sizes in a regular square lattice. This is a minimum for introducing the two fundamental parameters of pore size distribution and pore interconnectivity. For the moment we neglect the finding of Payatakes *et al.* [12] that capillary tubes give a poor representation of collector efficiency and concentrate on the study of fluid flow and structural changes during filtration in this simple model.

This model of porous media was first proposed by Fatt [13] to model capillary pressure and relative permeability curves in two phase flow in porous media. There has recently been a renewal of interest due to the application of percolation theory to this sort of problem.

An example of the network used in this study is shown in Fig. 1.

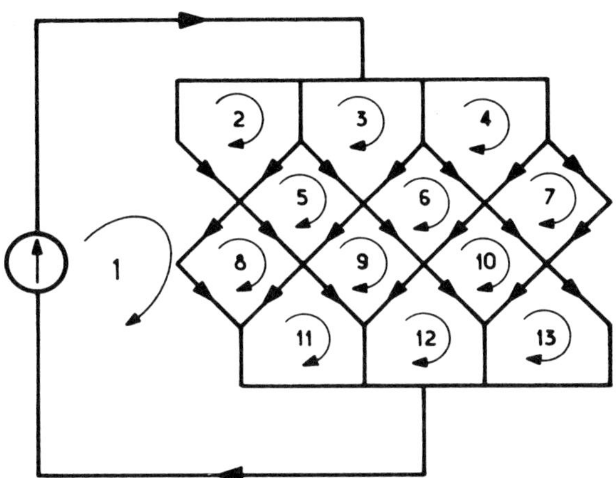

Fig. 1 – Example of a network of capillary tubes showing assumed loop flows and tube flows.

Fluid flow in such a network can be calculated using a matrix expression for Darcy's law derived from electrical network theory as described in Appendix 1.

$$[Q] = [Z]^{-1}\,[P] \tag{4}$$

$[Z]^{-1}$ is the inverse of $[Z]$ called the mesh resistance matrix. This is an operator which acts as the resistance of the whole network. In the same way as the permeability in equation (1) it is a structural parameter and consists of the connection matrix of the network and a diagonal matrix containing the resistances to flow in the capillary tubes:

$$[Z] = [C]_t\,[L/R^4]\,[C]\ . \tag{5}$$

A scalar multiplier of $(8\eta/\pi)$ should be included in this expression but is neglected here to be lumped into a constant multiplying factor to the final results. In a further paper we will show that the $[L/R^4]$ matrix can be a reduced pore size distribution multiplied by a scalar operator giving the hydraulic radius term in the Kozeny–Carman equation. We may then assimilate the connection matrix to a sort of tortuosity factor. It should however be remembered that $|Z|$ is not an identifiable entity in itself and can only be used as an operator on other matrices. Therefore the overall resistance or permeability of the network can only be obtained from the overall pressure drop and the total flow in the network.

In general the pressure drop in the network is imposed in loop 1 only and the mesh resistance matrix is used to calculate the flow in all the loops of the network. Thus the $[P]$ matrix is a column vector containing the imposed pressure drop as the first element with all the other elements being zero. We may take the imposed pressure drop to be unity. The problem is therefore to find the inverse of the mesh resistance matrix $[Z]$.

If we assume the inverse of $[Z]$ to be say $[X]$ then (4) becomes:

$$[Z]^{-1} \begin{bmatrix} 1 \\ 0 \\ \vdots \\ 0 \end{bmatrix} = \begin{bmatrix} X_{11} & X_{12} & \cdots \\ & & \\ & & \\ & \cdots & X_{nn} \end{bmatrix} \begin{bmatrix} 1 \\ 0 \\ \vdots \\ 0 \end{bmatrix} = \begin{bmatrix} X_{11} \\ X_{21} \\ \vdots \\ X_{n1} \end{bmatrix} = \begin{bmatrix} Q_1 \\ Q_2 \\ \vdots \\ Q_n \end{bmatrix} . \tag{6}$$

It can be seen from (6) that only the first column of the inverse of $[Z]$ is required which, from the definition of the inverse, is equivalent to the solution of a set of n simultaneous equations with the first column of the unit matrix as coefficients where n is the number of loops in the network.

$$\begin{bmatrix} Z_{11} & Z_{12} & \cdots \\ Z_{11} & & \\ \vdots & & \\ \cdot & & Z_{nn} \end{bmatrix} \begin{bmatrix} X_{11} \\ X_{21} \\ \vdots \\ X_{n1} \end{bmatrix} = \begin{bmatrix} 1 \\ 0 \\ \vdots \\ \cdot \end{bmatrix} . \tag{7}$$

The total flow in the network caused by unit pressure drop is equal to Q_1 and is therefore equivalent to the permeability.

Going further the column vector $[Q]$ contains the flows in all n loops in the network and by using equation (A2) from the appendix

$$[q] = [C]\ [Q] . \tag{8}$$

That is multiplying $[Q]$ by the connection matrix we obtain a column vector $[q]$ containing the flows in all the tubes in the network. We can therefore calculate the total flow Q_1 and the individual flows $[q]$ in the whole network.

5. CALCULATED FLOWS IN A NETWORK OF TUBES

We have calculated the flows in a network of the form shown in Fig. 2. This has 12 rows of 6 loops per row giving a total of 73 loops including the first or external loop. There are 13 tubes in each row for a total of 143 tubes in the network. Calculations of flows in this network require the solution of a set of

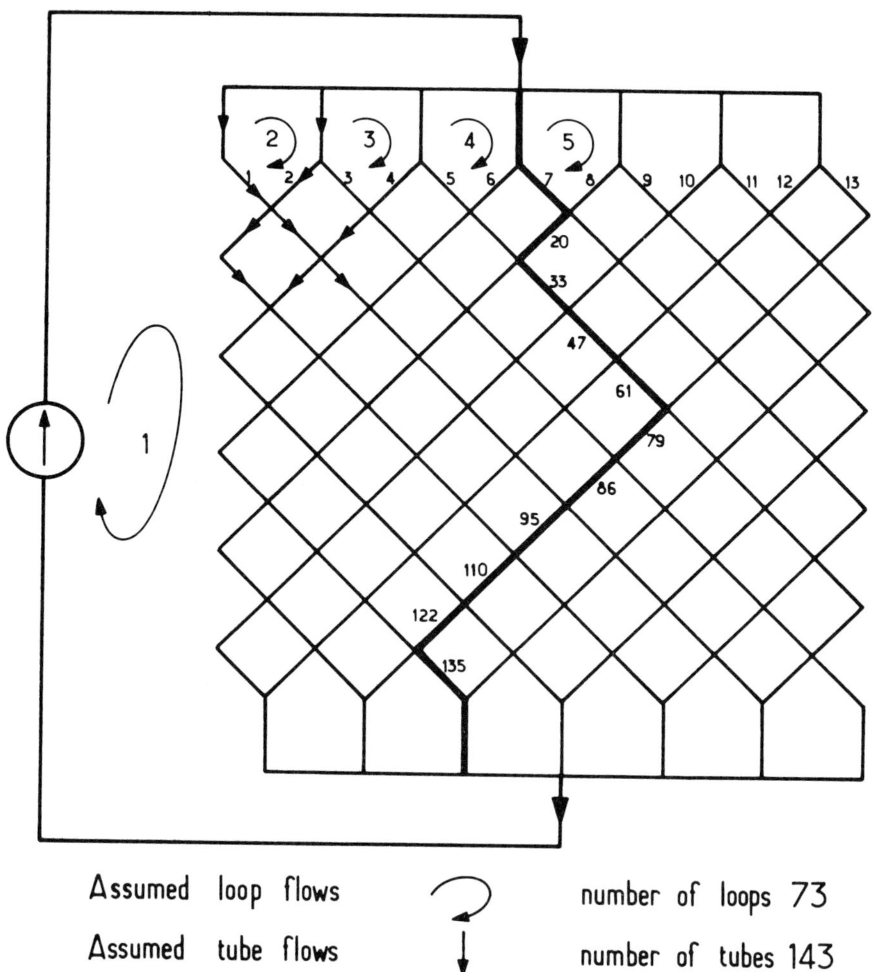

Fig. 2 – Network of 73 loops and 143 tubes showing one path through the network.

73 simultaneous equations and the multiplication of the resulting 73 element column vector by a connection matrix of 143 × 73. This was the largest network which could be treated with the computer available.

Tube radii from a gaussian distribution were fitted randomly to positions in the array of Fig. 2 using the IBM subroutine GAUSS. This allowed filling the network with a distribution of tubes of the same mean radius (1.0) and with different standard deviations (σ = 0.12, 0.23, 0.35) in increments of 0.1 of a radius.

The tube size distributions are given in Fig. 3.

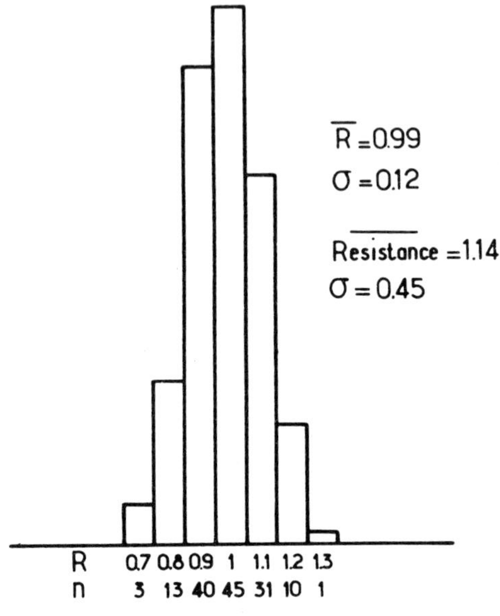

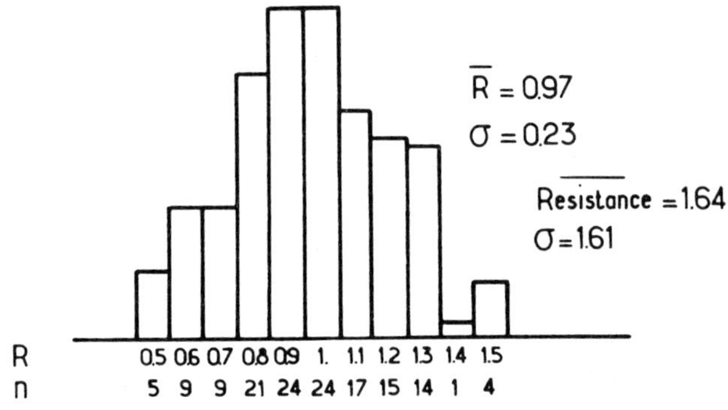

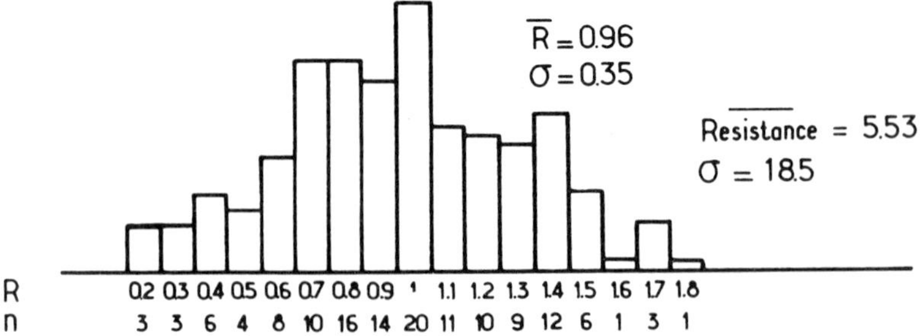

Fig. 3 – Tube radius distribution used in the calculations.

To calculate flows in the network we have assumed that the length of each tube is proportional to its radius. Thus wide tubes are taken to be long and narrow tubes taken to be short. The resistance to flow in a tube is therefore proportional to the reciprocal of the tube radius.

Flow resistance is $\propto L/R^4$ (9a)
and since L is $\propto R$ (9b)
we have flow resistance $\propto 1/R^3$ (9c)

Table 1 gives the results of the calculations for the total flow in three networks with the same mean tube radius and the different distributions shown in Fig. 3. Values are given relative to the total flow of 100 in a network with of uniform tubes of a uniform size of 1.0

Table 1

The total flow in a network as a function of the tube radius distribution

Network	A	B	C	D
σ	0	0.12	0.23	0.35
Total flow	100	93.5	85.8	71.9
$\bar{R}$	1	0.99	0.97	0.96
$\overline{\text{Resistance}}$	1	1.14	1.64	5.53
$(\overline{\text{Resistance}})_{2000}$	1	(1.07)	(1.35)	(3.37)
Average flow in pores of size 1.0	7.7	8.1	8.4	8.8

It can be seen that the permeability of a network (that is the flow under unit pressure drop) for the same mean pore size depends to a marked degree on the spread of the distribution. This is obviously associated with the fact that the mean flow resistance is not related to the mean tube radius in a simple fashion. That is:

$$1/(\bar{R})^3 \neq \overline{(1/R^3)} \qquad (10)$$

However, these results cannot be accounted for even using the mean flow resistance calculated from the tube size distribution. As can be seen from Table 1, as the standard deviation of the distribution goes from 0 to 0.35 the total flow is reduced by about 30% at the same time as the mean flow resistance is increased by a factor of 5.5. The network used in these calculations is quite small and 143 tubes is not sufficient to give a good gaussian distribution. However the second last line of Table 1 gives the mean flow resistances calculated from a distribution of 2000 tubes to give a better approach to a true gaussian. There is still a large discrepancy.

The clear conclusion is that the configuration of the network must be included in the averaging procedure. A few large resistances in a network are sufficient to change the mean resistance but flow finds other paths through the network and reduces their effects. For porous media with a distribution of pore sizes there is no way of correctly deducing the permeability without some knowledge of the structure of the pore space. Such an evaluation can perhaps be made using the effective medium theory.

Figure 4 shows the tube size distribution in network C with $\sigma = 0.23$ together

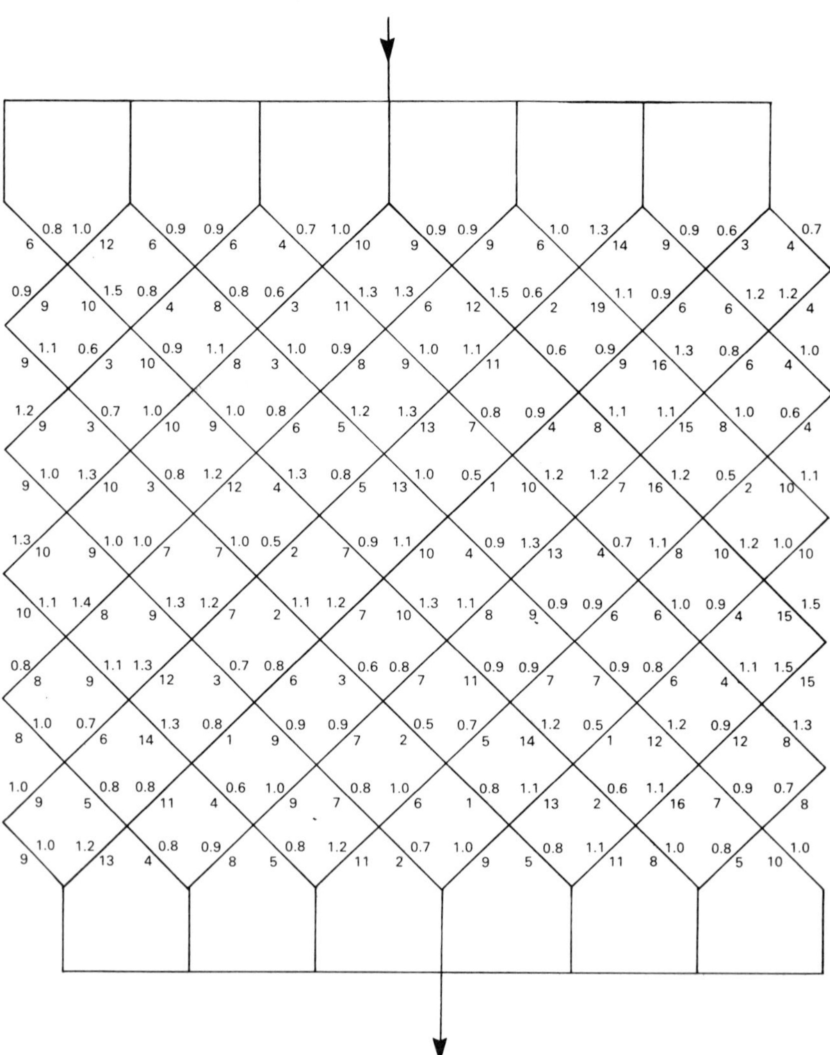

Fig. 4 – Flows in the tubes in network C ($\sigma = 0.23$).

with the flows in each tube given as a percentage of the total flow in the network. These results are reproduced in Table 2 where the flows in each tube size class are presented together. The mean flow in the tubes of the mean radius are given in Table 1 for each of the three networks.

Table 2

Flows in network *C* with mean tube radius 1.0 and standard deviation 0.23

Tube Radius	Tube Resistance	Mean Flow	Flows in each tube
0.5	8.00	1.4	0.7 1.1 1.5 1.6 2.2
0.6	4.63	3.1	2.1 2.4 2.7 2.9 3.0 3.3 3.5 3.6 4.1
0.7	2.92	4.4	2.1 3.2 3.4 3.6 3.7 4.4 5.1 6.2 7.7
0.8	1.95	5.7	1.0 1.4 3.1 3.7 3.8 4.8 5.1 5.3 5.5 5.5 6.1 6.1 6.2 6.4 6.4 6.7 6.7 7.0 8.2 8.3 11.3
0.9	1.37	7.7	3.7 4.1 4.4 5.6 5.9 5.9 6.1 7.2 7.3 7.4 7.5 7.8 8.5 8.6 8.6 8.7 9.0 9.0 9.2 9.2 10.7 11.6 11.8
1.0	1.0	8.4	3.1 3.6 5.7 5.9 6.5 7.3 7.3 7.9 8.0 8.3 8.7 8.8 8.9 9.0 9.1 9.1 9.1 9.4 9.7 10.0 10.1 10.5 12.4 13.1
1.1	0.75	10.0	2.2 4.1 7.7 8.0 8.4 8.5 8.5 8.8 9.6 9.8 10.1 10.7 10.8 12.7 14.5 16.3 18.7
1.2	0.58	9.4	4.4 5.5 5.6 6.7 7.0 7.3 8.7 9.6 9.6 10.8 11.6 12.2 13.0 14.0 15.7
1.3	0.46	10.7	3.6 6.3 7.7 8.6 9.6 9.8 10.1 11.1 11.9 13.0 13.3 14.3 14.4 16.1
1.4	0.36	7.5	7.5
1.5	0.30	13.2	10.2 12.2 15.2 15.2

It can be seen that in general, as could be expected, the high flow rates are in the large tubes and the low flow rates are in the small tubes, and also that the mean flow rate in a given tube class is roughly proportional to the resistance of that class. However there are large variations within this general framework of average values and in networks with the widest distribution we even obtain negative flows, that is in an opposite direction to that which is assumed in Fig. 2. Thus in any tube size we can find flows from half to almost twice the mean flow rate in that tube class. This may, once again, be interpreted as an effect brought on by the structure of the network, or more precisely, the flow in a given tube depends not only on the size of the tube considered but also on the size of the tubes around it. This conclusion would seem to be important for the application of collector theory to deep bed filtration.

6. DISPERSION DUE TO FLOW IN A NETWORK OF TUBES

The main conclusion from the previous sections is that the structure of the network is an important parameter. The only really practical way of studying the structure of pore space in porous media is by residence time distribution experiments where the dispersion due to flow is followed by injecting tracers.

The phenomena causing dispersion of a tracer in flow through a porous medium are not well established but four contributions may be expected:

- –Molecular diffusion due to concentration gradients.
- –Taylor diffusion due to the interplay of velocity profiles and concentration gradients.
- –Backmixing effects in the voids in the pore space.
- –Dispersion due to the continual division, recombination and redivision of the flow streams.

In laminar flow, which is not too slow as is usually the case under practical operations conditions for deep bed filtration, we may expect the last of these four effects to be predominant. Furthermore this effect is the fundamental contribution of the structure of the porous medium to dispersion and the other effects may be considered to be superimposed on it. We have calculated dispersion due to this mechanism in the network C.

At each node of the network shown in Fig. 2 we have two fluid flows entering and two fluid flows leaving. This ignores possible recirculations found in networks with a wide tube-size distribution.

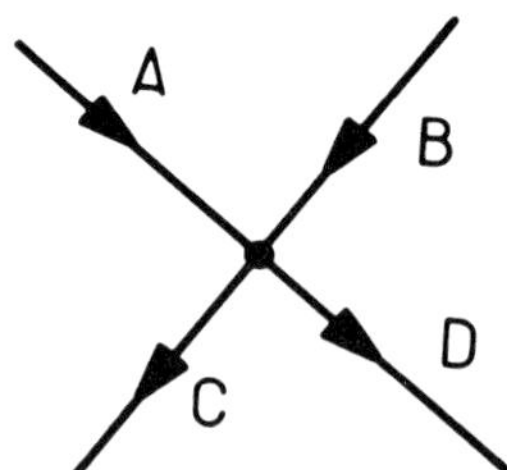

Fig. 5 – Flows in a node in the network.

If these flows are called q_A q_B q_C q_D then a given element of fluid entering the node in either stream A or B must leave in stream C or D and the probability of it leaving in either of them will be, respectively:

$$p_C = \frac{q_C}{q_C + q_D} \tag{11}$$

$$p_B = \frac{q_D}{q_C + q_D} . \tag{12}$$

In the previous section we showed how to calculate the flows in all the branches of the network. We may now use this information to determine how the fluid streams divide at each node. Furthermore knowing the volume and the flow in each tube we may calculate the passage time in each tube in Fig. 5.

$$t_C = \frac{\text{volume tube C}}{\text{flow in tube C}} . \tag{13}$$

Examining the network shown in Fig. 2 we can see that an element of fluid entering the network will pass through 10 nodes or 11 tubes before leaving and at each node it will be divided in the proportions given in equations (11) and (12). Thus each element of fluid will be subjected to a series of divisions at each node and the overall path through the network will be a tree of tubes to be recombined at the exit.

For the case of a network of uniform sized tubes, say of size 1.0, the result is obvious. The total flow in the network, taken to be 100, will be divided equally amongst the 13 tubes in each row to give a flow of 7.7 in each. The time of passage in each take will be 1/7.7 and each node will divide the entering flows in the proportion 1 to 1. Thus each element of fluid will follow a tree of tubes with 10 nodes and 11 identical tubes which will have the same total passage time and will be recombined at the outlet when they all appear simultaneously at the total passage time of 11 × 1/7.7 = 1.43. Consequently there will be no dispersion. A step injection at the inlet will emerge as a perfect step response at the outlet.

However, when there is a distribution of tube sizes and hence different flow rates and passage times in each branch of the network, the problem is more complicated. For example one possible path through the network in Fig. 2 is shown to be through the tubes numbered: 7 20 33 47 61 79 86 95 110 172 135.

Fluid using this path will emerge at the outlet after a time of

$$t_7 + t_{20} + t_{33} + t_{47} + \ldots + \text{etc.} = \Sigma\, t_i \tag{14a}$$

and the amount of fluid, or the concentration of tracer following this path will be

$$p_7 \times p_{20} \times p_{33} \times \ldots \times \text{etc.} = \Pi p_i . \tag{14b}$$

We must therefore make an inventory of all the possible paths through the network then piece together the response of the flow dispersion.

In the network shown in Fig. 2 there are 9816 different paths as made by an inventory using a pile storage alogorithm since analysis as a Markov process required too much computer storage space. We present dispersion results in Fig. 6 with respect to a step function in tracer by plotting the cumulative contribution 1 each flow path in the order of increasing passage times. The step function presentation is preferred as the small size of the network induces spurious effects in a impulse response version of these results for the wide take size distributions.

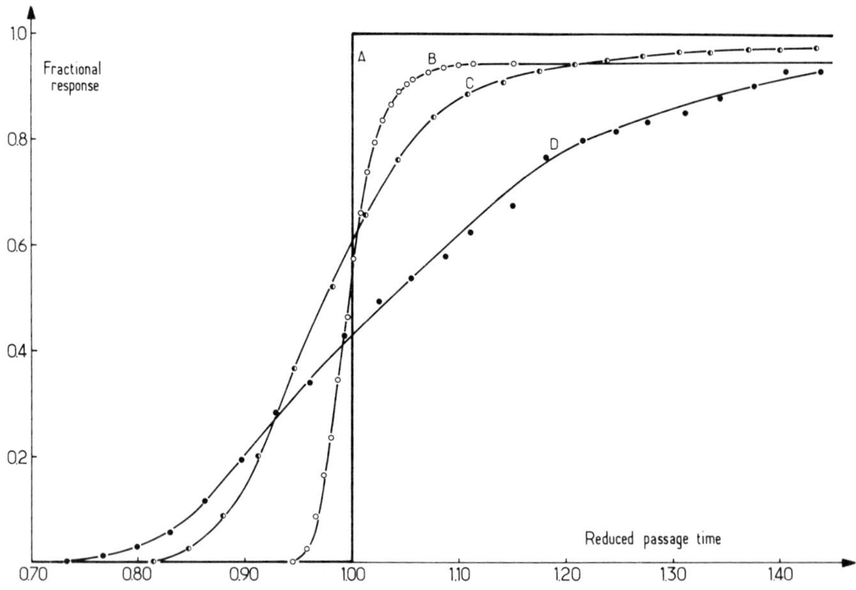

Fig. 6 – Residence time distributions in networks A, B, C, D.

We find the mean passage time of flow in the networks varies with the width of the distribution since as was mentioned previously the average pore volume $(\bar{R^3})$ is not the same as $(\bar{R})^3$. We have therefore normalized each curve by the mean volume and plotted them together on the basis of a reduced mean passage time of 1.0.

It can be seen that, as expected, there is more dispersion in networks with a wide tube size distribution. Also the curves become more skewed as the distribution is widened, that is they have a longer tail which is not compensated by a corresponding increase in the leading part of the curves. This result is found with badly packed chromatographic columns which tend to produce skewed peaks.

7. THE EFFECTS OF PROGRESSIVE CLOGGING ON FLUID FLOW IN A NETWORK OF TUBES

To simulate deep bed filtration in a network of tubes we have used an incremental procedure and have assumed that in a unit of time during which flow occurs a deposit is formed in each tube which reduces the radius in the next unit of time. The new resistance to flow in the ith tube for the $(n + 1)$ time unit is given by

$$S_{in+1} = \frac{\alpha \,.\, S_{in}}{\beta - \gamma\, q_{in} \cdot S_{in} \cdot Q_{tn}} \tag{15}$$

in which

α, β and γ are constants,
S_{in} is the flow resistance in tube i during the time n,
Q_{tn} is the total flow in the network during time unit n,
q_{in} is the fraction of the total flow passing through the tube i in time unit n.

Equation (15) is derived by assuming that the volume of deposit in a tube is proportional to the volume of fluid flowing through it, which implies that the efficiency of filtration does not depend on the evolution of the deposit nor on the flow velocity; that the volume of a tube is proportional to R^3; and that the resistance to flow is proportional to $1/R^3$. This is almost the same as saying that the deposit is formed uniformly on the inside surface of the tube but in addition we must also allow that the length of the tube remains proportional to the radius even as the radius is reduced by deposition on the walls. A rigorous expression analogous to equation (15) where the original length of each tube is conserved could be used but the equation is much more complicated and requires memorizing the length of each tube in the network as well as its radius. The extra complication to accommodate what is in fact a very questionable filtration mechanism did not seem justified.

We have used the methods described in the previous sections to calculate the flows in a network together with the residence time distribution. Once the flows and therefore the filtration deposit in each tube are known we use equation (15) to calculate the modified tube radii and hence the new flow resistances and we repeat the calculations. When a tube radius is reduced to below 0.05 it is considered to be saturated, it continues to participate in flow with its high resistance, but does not continue to filter.

We have performed this calculation procedure on network C ($\sigma = 0.23$) used in the previous section and have used 15 filtration stages. This results in 16 out of the 143 tubes becoming saturated, the mean tube size being reduced from 0.97 to 0.67, and the total flow rate under unit pressure drop being reduced from 1 to 0.002. These results are given in Tables 3, 4, 5 and in Figs 7, 8 and 9.

8. DISCUSSION OF THE CLOGGING RESULTS

The overall results for the variation of the total flow rate and the average values of the network parameters as a function of the 15 stages of filtration are given in

Table 3. It can be seen that the variation in flow rate bears no relation to the variation in the mean tube size. As the mean tube size goes from 0.97 to 0.67 the total flow rate is reduced by a factor of 500. This is shown graphically on Fig. 7 where we show the variation in total flow rate over the 15 stages and the

Table 3

Variation of flow and average network parameters over the filtration stages 1 to 15

	(Q_t/Q_{t1})	$\bar{R}$	$\bar{R}^3$	$\overline{1/R^3}$	No. tubes effective	$(Q_t/Q_{t1}).\dfrac{\bar{1}}{R^3}$
1	1	0.971	1.073	1.65	143	1.65
2	0.86	0.928	0.956	1.98	143	1.70
3	0.70	0.887	0.859	2.43	143	1.70
4	0.57	0.850	0.779	3.05	143	1.74
5	0.45	0.815	0.714	4.01	143	1.80
6	0.34	0.783	0.662	5.86	143	1.99
7	0.25	0.754	0.623	11.7	143	2.93
8	0.17	0.728	0.594	129	143	21.9
9	0.12	0.708	0.573	307	138	36.8
10	0.070	0.692	0.539	1287	137	90.1
11	0.034	0.677	0.549	772	131	26.2
12	0.017	0.672	0.544	778	131	13.4
13	0.007	0.664	0.542	793	131	2.34
14	0.003	0.666	0.541	813	129	2.44
15	0.002	0.665	0.540	890	127	1.78

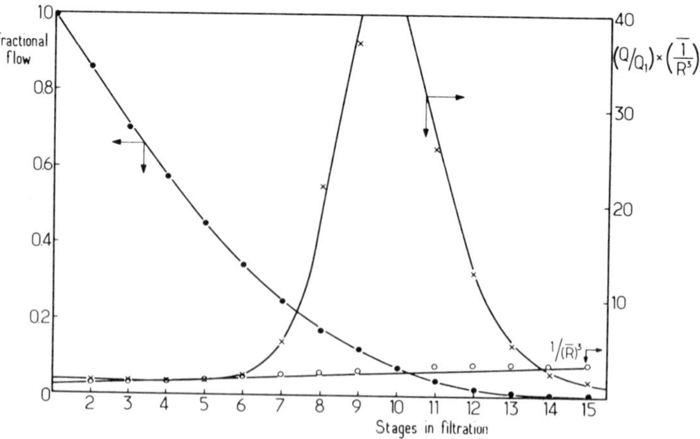

Fig. 7 – Variation in flow, as the filter becomes clogged.

mean resistance to flow calculated from the mean tube size. Also on this graph and in Table 3 we show the product of the flow and the mean resistance $(\overline{1/R^3})$. Whilst the mean resistance to flow $(\overline{1/R^3})$ bears no relation to the resistance to flow calculated from the mean pore size $(1/\bar{R}^3)$ we could have hoped for some relationship with the total flow rate but it can be seen that this is not simple. For stages 1 to 6 the product of flow and mean resistance is roughly constant. For Stages 7 to 14 it shows a very large variation with a peak at stage 10 to reduce in the last stage to almost its initial value. This re-emphasizes the conclusions drawn previously as to the irrelevance of average pore sizes for describing permeability in structures with distributed parameters.

Table 4

Variation in the tube size distribution during filtration stages 1 to 15

Stages	Tube radii																Q_t/Q_{t1}
	0	0.1	0.2	0.3	0.4	0.5	0.6	0.7	0.8	0.9	1.0	1.1	1.2	1.3	1.4	1.5	
1						5	9	9	21	24	24	17	15	14	1	4	1
2					5	9	10	20	24	24	17	15	14	1	4		0.86
3					8	10	15	23	17	23	14	14	14	1	4		0.70
4					15	8	18	24	23	13	13	14	10	1	4		0.57
5				9	10	12	20	18	24	15	10	14	6	3	2		0.45
6			5	8	11	15	21	18	17	15	9	15	4	4	1		0.34
7		4	2	11	13	16	19	22	9	14	10	15	3	4	1		0.25
8	3	3	4	13	11	23	10	21	10	12	12	13	3	4	1		0.17
9	5	2	6	13	12	21	9	22	10	11	11	13	3	4	1		0.12
10	6	5	6	12	14	18	10	14	12	10	10	13	3	4	1		0.070
11	12	0	8	10	17	14	13	16	13	9	11	12	3	4	1		0.034
12	12	1	9	10	17	14	11	16	13	10	10	12	3	4	1		0.017
13	12	3	9	9	17	13	12	15	13	10	10	12	3	4	1		0.007
14	14	3	7	9	17	13	12	15	13	10	10	12	3	4	1		0.003
15	16	1	7	9	17	13	12	15	13	10	10	12	3	4	1		0.002

Table 4 gives more detail of the changes in the network structure with progressive blocking and shows how the tube sizes vary as filtration proceeds. An obvious finding is that the small tubes are more affected by filtration than the large tubes. Thus from stages 1 to 7 we have a reduction in the mean tube radius (Table 3) which is accompanied by an increase in the spread of the distribution. This is due to the relative stability of the sizes of the large tubes which require more deposit to reduce their radii.

After stage 8 we start to have tubes which become saturated and this gradually leads to a stabilization of the spread of the distribution as it is necessarily limited between 0.1 and 1.5. These three phases seem to correspond to the three phases found in the variation of the shape of the curve of flow × mean resistance as shown in Fig. 7. However remembering the effects of the width of the tube size distribution on the flow in networks there is little hope of this three-phase evolution being accounted for in a simple fashion.

In Fig. 8 we show the results of the calculation of the residence time distribution by the method presented in the previous sections. These results are only given for filtration stages 1 to 11 as afterwards there are very pronounced back flows in some of the tubes which cannot be handled by our rather simple analysis. Work is in progress on this point. Table 5 gives an analysis of the variation in the shape of the curves as filtration progresses by noting the times at which

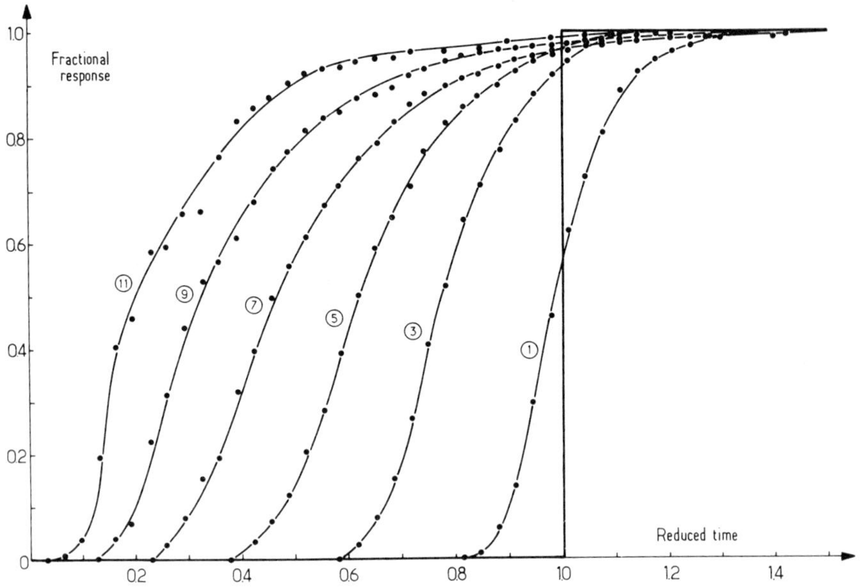

Fig. 8 – Changes in residence time distribution as the filter becomes clogged.

Table 5

Variation in the shape of the residence time distribution curves with filtration

Stage	Point				
	0.1	0.5	0.9	Δtotal	$\Delta b/\Delta f$
1	0.900	0.986	1.13	0.230	1.67
3	0.666	0.775	0.965	0.299	1.74
5	0.476	0.620	0.875	0.399	1.77
7	0.310	0.470	0.790	0.480	2.00
9	0.196	0.325	0.665	0.469	2.64
11	0.120	0.200	0.500	0.380	3.75

responses 0.1, 0.5 and 0.9 leave the bed. With these we have calculated the variation in the total width of the curves together with the relative proportions of the front and back parts of the curves. It is found that the width of the curves increases to a maximum at stage 8 then decreases and that the 'skewness', measured by the ratio of the front width of the curves to the back width, increases progressively. This supports the other results reported above.

To emphasize these points further, we give the flows in the various tubes at stage 11 in Fig. 9. The tubes which are saturated are marked as are tubes which

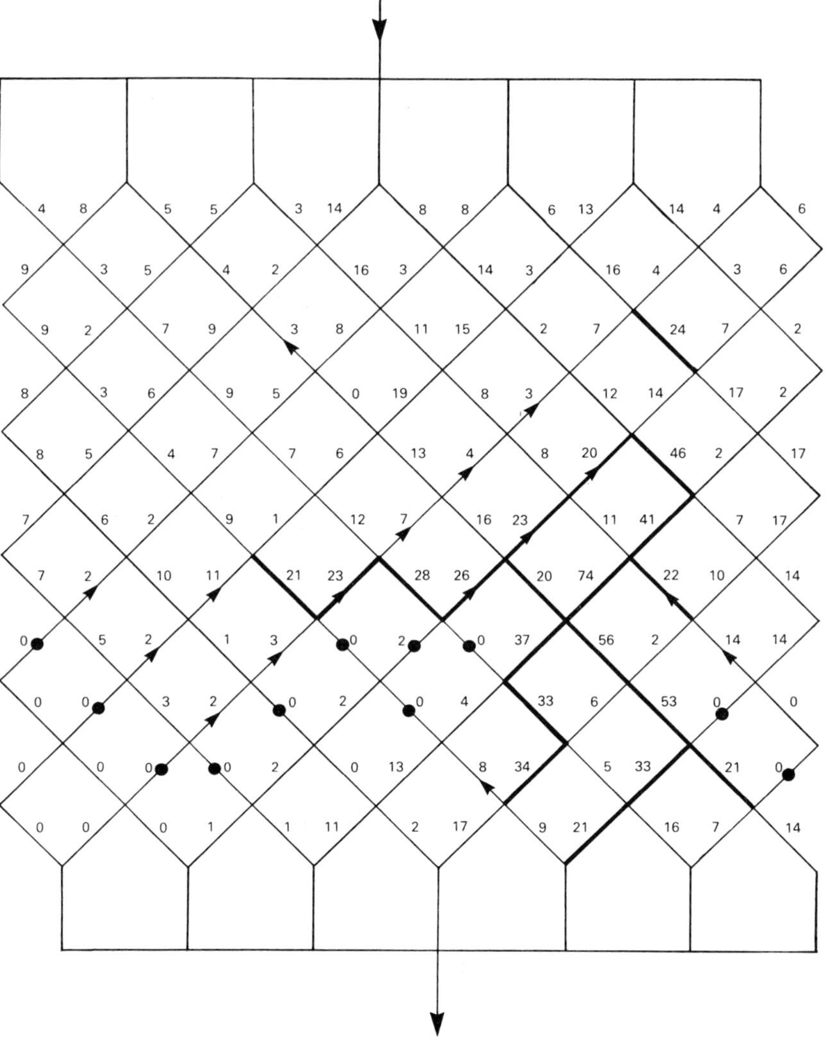

Fig. 9 – Flows in the tubes in network C after 11 stages of filtration.

have a back flow. The tubes which have a flow of more than 20% are shown in heavy lines and it can be seen that there are preferential paths through the filter of the sort of 'worm holes' mentioned by Baumann [14]. This figure also gives an explanation of the shape of the residence time distribution for stage 11 as shown in Fig. 8. The preferential path will tend to have a very short residence time and corresponds to the rapid early rise in the step response Fig. 8. Most of the other tubes have very low flow rates and will correspond to the long tail. In addition, and which is not accounted for in the calculations, the many recirculation paths will tend to give a much longer tail.

9. CONCLUSION

In the model of porous media used here, which involves a relatively small two dimensional regular square array of tubes, we have shown that the effects due to the distribution in pore sizes, which is inevitable in practice, cannot be accounted for by average pore parameters. As would be expected the residence time distribution of the model is strongly affected by the pore size distribution. These conclusions are reconfirmed by a simulated clogging of the network model.

However the question remains open as to whether these findings correspond to the reality of flow in real three dimensional porous media which only experiments, can decide.

ACKNOWLEDGEMENT

We would like to thank Alain Lemaitre for help with programming and especially for his elegant solution to the inventory of flow paths in the network.

NOMENCLATURE

$[C]$	Connectivity matrix
e	Voltage
h_k	Kozeny constant
i	Current
K_0	Pore shape factor
K	Permeability
L	Length
P	Pressure
p	Probability of passage
Q_i	Flow in loop i
Q_t	Total flow in network i.e. flow in loop 1
q_i	Flow in tube i
R	Tube radius
r	Electrical resistance

S	Resistance to flow
S_0	Specific surface
t	Passage time
v	Flow velocity
$[Z]$	Mesh resistance matrix
α, β, γ	Constants
ϵ	Porosity
η	Viscosity
σ	Standard deviation
τ	Tortuosity

Superscript

$(\bar{\ })$	Average value

APPENDIX 1 MATRIX EXPRESSION OF KIRCHHOFF'S LAWS FOR NETWORKS

The following treatment is based on electrical network theory such as given in Stigant [15].

A network A1(a) may be decomposed in to the so called 'primitive' network A1(b) and the relation between the two systems can be represented by a connection matrix.

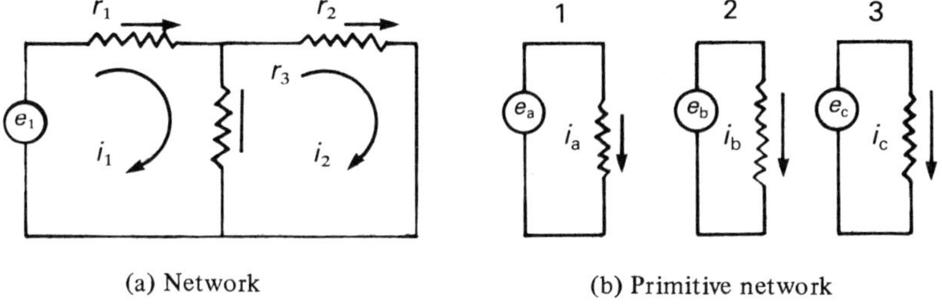

Fig. A1

In Fig. A1(b) Ohm's law may be expressed as

$$\begin{bmatrix} e_a \\ e_b \\ e_c \end{bmatrix} = \begin{bmatrix} r_1 & 0 & 0 \\ 0 & r_2 & 0 \\ 0 & 0 & r_3 \end{bmatrix} \begin{bmatrix} i_a \\ i_b \\ i_c \end{bmatrix}$$

$$\begin{bmatrix} e' \end{bmatrix} = \begin{bmatrix} r \end{bmatrix} \begin{bmatrix} i' \end{bmatrix} \tag{A1}$$

But from the real network in Fig. A1(a)

$$i_a = i_1$$

$$i_b = i_2$$

$$i_c = i_1 = i_2$$

which may be expressed

$$\begin{bmatrix} i_a \\ i_b \\ i_c \end{bmatrix} = \begin{bmatrix} 1 & 0 \\ 0 & 1 \\ 1 & -1 \end{bmatrix} \begin{bmatrix} i_1 \\ i_2 \end{bmatrix}$$

$$[i'] = [c]\ [i] \tag{A2}$$

where $[C]$ is the connection matrix and provides the relationship between the currents in the actual network and the primitive network.

Similarly the relationship between the voltages in (a) and (b) may be given in terms of a matrix which is the transpose of $[C]$

$$e_1 = e_a \qquad + e_c$$

$$e_2 = \qquad e_b \quad - e_c$$

$$[e] = [C]_t\ [e'] \tag{A3}$$

The connection matrix is a rectangular matrix with i rows and j columns where i is the number of elements and j is the number of loops in the real network. To define the connection matrix for a network, the elements and the loops in the network are numbered and the directions of current flow arbitrarily assigned. Here loop currents are taken as being clockwise and currents in elements taken as either left to right for horizontal elements or top to bottom for vertical elements. The elements of the matrix are given by:

$C_{ij} = 0$ If element i is not part of loop j;
$C_{ij} = +1$ If the assumed direction of current flow in element i coincides with the assumed loop current flow in j;
$C_{ij} = -1$ If the assumed direction of current flow in element i is opposite to the assumed loop current flow in j.

Combining (A1) (A2) (A3) gives

$$[e] = [C]_t\ [r]\ [C]\ [i] \tag{A4}$$

Ohm's law for the network may be written

$$[e] = [Z]\,[i] \tag{A5}$$

where
$$[Z] = [C]_t\,[r]\,[C] \tag{A6}$$

In electrical terminology $[Z]$ is called the 'mesh impedance matrix' and is an operator which acts as the resistance of the whole network.

In the network shown in Fig. A1(a)

$$[Z] = \begin{bmatrix} 1 & 0 & 1 \\ 0 & 1 & -1 \end{bmatrix} \begin{bmatrix} r_1 & 0 & 0 \\ 0 & r_2 & 0 \\ 0 & 0 & r_3 \end{bmatrix} \begin{bmatrix} 1 & 0 \\ 0 & 1 \\ 1 & -1 \end{bmatrix}$$

$$[Z] = \begin{bmatrix} (r_1 + r_3) & -r_3 \\ -r_3 & (r_2 + r_3) \end{bmatrix}$$

$[Z]$ is a symmetrical square matrix with its order given by the number of loops in the network, in this case 2 × 2 and may also be defined directly by a set of rules:

Z_{ij} = the sum of resistances in loop i
$Z_{ij} = Z_{ji} = -$ (the sum of resistances common to loops i and j) .

Where in general the potential drop across an array of resistances is known, the mesh impedance matrix can be used to calculate the loop current matrix from

$$[Z]^{-1}\,[e] = [i] \tag{A7}$$

Therefore knowing the potential drop across the network and the current it causes, the total effective resistance of the network may be calculated. In general therefore the problem is one of finding the inverse of the mesh impedance matrix.

To transpose these expressions to flow in networks of capillary tubes we make use of the analogy between Ohm's law and Poiseuille's law.

Ohm's law
$$e = r\,i \tag{A8}$$

Poiseuille's law
$$p = \frac{8\eta}{\pi}\frac{L}{R^4}\,Q \tag{A9}$$

Therefore we replace

$[e]$ by $[P]$ $\qquad$ $[i']$ by $[q]$

$[i]$ by $[Q]$ $\qquad$ $[e']$ by $[p]$

$[r]$ $\qquad$ $[L/R^4]$

and an equivalent for Darcy's law for flow in the network is then

$$[Q] = [Z]^{-1} [P] \tag{A10}$$

REFERENCES

[1] Tien, C. and Payatakes, A. C., *AIChE J.*, **25**, 737–59 (1979).
[2] Ives, K. J., NATO-ASI *Mathematical Models and Design Methods in Solid Liquid Separation,* Lagos (1982).
[3] Pendse, H. and Tien, C., *AIChE J.*, **28**, 677–86 (1982).
[4] Ison, C. R. and Ives, K. J., *Chem. Eng. Sci.*, **24**, 717–29 (1969).
[5] Beizaie, M., Wang, C. S., and Tien, C., *Chem. Eng. Comm.*, **13**, 153–80 (1981).
[6] Cleasby, J. L., and Baumann, E. R., *J. Am. WWA*, **54**, 579–602 (1962).
[7] Maroudas, A., and Eisenklam, P., *Chem. Eng. Sci.*, **20**, 867-73 (1964).
[8] Pendse, H., Tien, C., Turian, R. M., and Rajagopalan, R., *AiChE J.*, **24**, 473–85 (1978).
[9] Coad, M. A., and Ives, K. J., Filtration Society Conference, *Filtration and Separation Equipment for Optimum Results,* London (1981).
[10] Carman, P. C., *The Flow of Gases through Porous Media,* Butterworth, London (1956).
[11] Childs, E. C., and Collis-George, N. C., *Proc. Roy. Soc., A*, **201**, 392–405 (1950).
[12] Payatakes, A. C., Rajagopalan, R., and Tien, C., *Canad. J. Chem. Eng.*, **52**, 722–31 (1974).
[13] Fatt, I., *Trans. A.I.M.E. Pet. Div.*, **207**, 144–59; 160–3; 164–81 (1956).
[14] Baumann, E. R., NATO ASI, *Mathematical Models and Design Methods in Solid–Liquid Separation,* Lagos (1982).
[15] Stigant, S. A., *Matrix and Tensor Analysis in Electrical Network Theory,* MacDonald, London (1964).

CHAPTER 20

Transient Behaviour of Deep Bed Filters

HSU-WEN CHIANG and CHI TIEN, Department of Chemical Engineering and Materials Science, Syracuse University, Syracuse, NY 13210, USA

ABSTRACT

Experimental data on the filtrate quality of and pressure drop across experimental filters were obtained over extended periods of time. These measurements were conducted under well-defined and controlled conditions using nearly mono-sized filter grains and mono-dispersed suspended particles. The results were used to test existing theories of deep-bed filtration.

1. INTRODUCTION

In deep-bed filters, suspensions are clarified by passing them through filter media, with the retention of the suspended particles taking place through the entire media. As the deposited particles accumulate within the interstices of the granular media, the geometry, structure, and possibly the surface condition of the media undergo continuous changes. These changes, in turn, affect the passage of the suspension through the filter bed as well as the ability of the medium to capture suspended particles. In essence, the dynamic behaviour of deep-bed filtration is characterized by the time-dependent behaviour of the filtrate quality and the pressure drop across the bed.

It is obvious that a quantitative prediction of the dynamic behaviour of deep-bed filters forms the cornerstone of this rational design, control, and optimization of filtration. In spite of several attempts, researchers have been unable to achieve such a predictive capability. There are two main reasons for this failure: the absence, until very recently, of an adequate theoretical framework which can be used to interpret experimental data properly, and the lack of experimental data systematically collected under well-defined and controlled conditions. In regard to the latter, it should be noted that limited accuracy in particle counting instrumentation has been a major obstacle in the experimental work.

Developments during recent years have allowed formulation of a number of hypotheses on the nature and mechanisms of particle capture within granular media and the resulting changes in the media and structure. Based on these hypotheses, models can be constructed for predicting filter performance in which the effect of particle retention may be significant.

In the present work, experiments in deep-bed filtration were conducted. The data obtained were the histories of filtrate quality and pressure drop. It was not the purpose of this work to collect massive data upon which empirical correlations could be obtained. Rather, these results were used to test current theories of deep-bed filtration.

The following sections briefly outline the current deep-bed filtration theories, then describe the experimental work performed, the treatment of the experimental data, and the comparison between experiments and theories.

2. THEORIES OF DEEP-BED FILTRATION

As previously stated, the dynamic behaviour of deep-bed filtration is characterized by the time-dependent behaviour of filtrate quality and pressure. Since the filtrate quality results directly from the extent of particle retention, which, in turn, can be related to the local value of the filter coefficient, λ, the practical use of filtration theories is to predict the variation of the filter coefficient and the pressure gradient necessary to maintain a given flow rate as functions of the local state of the filter media, in other words, the identification and evaluation of the functional form

$$\frac{\lambda}{\lambda_0} = F_1(\alpha, \sigma) \tag{1}$$

$$\frac{\partial p/\partial z}{(\partial p/\partial z)_0} = F_2(\beta\ \sigma) \tag{2}$$

where λ and $(\partial p/\partial z)$ are the filter coefficient and pressure gradient, respectively. The subscript '0' denotes the clean filter or the initial filter state. σ is known as the specific deposit, defined as the number of particles deposited per unit volume volume of filter bed. α and β are the parameter vectors characteristic of the relevant phenomena.

Although the functional form of F_1 and F_2 can be obtained empirically from experimental data, the empirical approach, in general, does not yield sufficiently accurate results [1]. On the other hand, F_1 (or F_2) can be derived or developed with the use of different assumptions. A brief outline of the various approaches which can be used to obtain F_1 and F_2 is given below.

2.1 Approach Based on Prior Assumptions Concerning Deposit Morphology

If the morphology of the deposits formed by the captured particles is known and the change in the geometry of the filter grains is moderate, then variation of the filter coefficient (and the pressure gradient) as a function of the extent of particle deposition can be readily obtained. A typical example of utilizing this approach is the work of Tien *et al.* [2]. Here, the deposition of particles within a filter is viewed on the basis of the following assumptions, as consisting of two stages:

(a) During the initial stage, particle deposition occurs primarily through the direct adhesion of individual particles to filter grains. The consequence of this mode of deposition is the formation of a relatively smooth layer of deposits on the outside of the filter grains. This first stage will continue until σ reaches a specific value, σ_{tran}.

(b) The second stage of the deposition is dominated by the formation of particle aggregates through deposition, the re-entrainment of these aggregates, and, subsequently, the re-deposition of these aggregates into pore constrictions. Thus, the consequence of the processes prevailing in this second step is the blocking of certain parts of the filter bed.

Based on these assumptions, expressions were derived for F_1 and F_2 in each stage of deposition. These expressions are given in ref. [2].

The initial (or clean) unit collector efficiency, η_0, can be obtained by evaluating η at conditions corresponding to those existing with clean filter grains by the method of Rajagopalan and Tien [3].

2.2 Approach Based on Trajectory Calculation

The traditional trajectory calculation estimates the rate of particle collection by determining the so-called 'limiting trajectory' [4] for a given collector geometry. This same concept can be applied to determine the instantaneous unit collector efficiency (or filter coefficient) if a relationship between the change in the collector geometry and the rate of particle deposition can be established. Under such conditions the investigation of the dynamics of deep-bed filtration is reduced to the study of the formation and growth of particle deposits outside filter grains. Incrementally, one may estimate the effect of the change of collector geometry on the flow field around (or within) the collector, which, in turn, causes changes in the rate of particle collection and in the pressure drop.

The applicability of this approach rests upon the assumption that deposited particles are contiguous. Consider a surface element δA on a collector. The change of the local deposit thickness (Δ) can be related to the flux in number of particles locally deposited, J_i, by the following expression:

$$\frac{\mathrm{d}}{\mathrm{d}t}[(1-\epsilon_{\mathrm{d}})(\delta A)(\Delta)] = (\delta A)J_i\nu \qquad (3)$$

where ϵ_{d} is the porosity of the deposits and ν the volume of a single suspended particle.

As a special case, if one considers that the dominant mechanism for particle collection is interception, J_i can be readily determined from the value of the stream function associated with the flow. Consider a collector of the constricted-tube type. The increase in deposit thickness over a time interval $t_{i-1} < t < t_i$ on a dimensionless basis becomes

$$\delta(\Delta^*) = \int_{t_i^*}^{t_{i+1}^*} \frac{\mathrm{d}\left(\psi^*\Big|_{y^* = \frac{N_R}{2}}\right)}{\mathrm{d}x^*} \frac{\nu^*}{p_w^*(1-\epsilon_d)} \, \mathrm{d}t^* \; . \tag{4}$$

The physical situation that this expression described is depicted in Fig. 1.

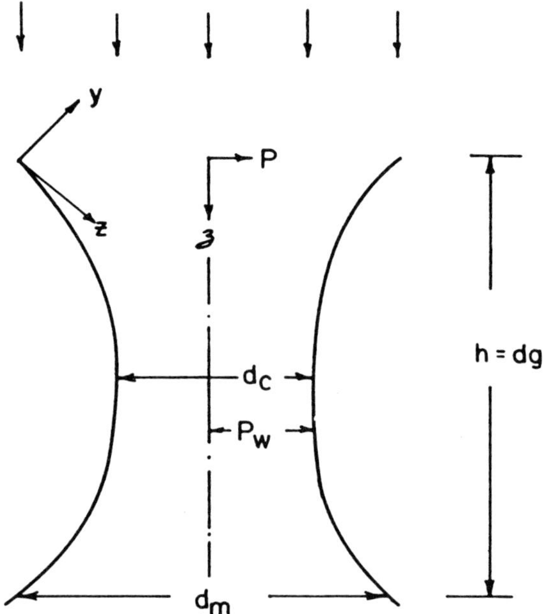

Fig. 1 – Coordinate system for a constricted tube.

The stream function, ψ^*, present in equation (4) changes with time as the dimension of the constricted tube alters as a result of increased particle collection. Thus, with the knowledge of p_w^* (as a function of x^*) for any instant, ψ^* can be evaluated from the general solution given by Chow and Soda [5] for flow through constricted tubes. Similarly, the problem can also be solved if the initial collector geometry is assumed to be spherical. In that case, the solution of Happel and Brenner [6] for flow over deformed spheres can be used to assess the change in the collector geometry resulting from deposition.

A detailed discussion of this aspect of deep-bed filtration theory can be found in Chiang's dissertation [7].

2.3 Approach Based on Simulation by Stochastic Modelling

The use of a simulation technique which considers particle deposition by examining sequentially each particle as it passes through the filter media is, at least conceptually, the simplest and most straightforward way of studying filtration. This methodology was first formulated by Tien and coworkers [8, 9] and subsequently adopted by Pendse and Tien [10] for aerosol filtration in granular beds. In their study, Pendse and Tien employed the constricted-tube model for filter media representation. Consequently, filtration was considered in terms of the flow of a suspension through a constricted tube of appropriate geometry and the subsequent deposition of particles from the suspension. The suspended particles were assumed to enter into the tube at a time (i.e. singular behaviour), and deposition is assumed if a particle's trajectory comes within a distance of a_p from the wall of the tube or the surface of previously deposited particles, where a_p is the radius of the particle.

The method that Pendse and Tien [10] developed can be readily applied to the filtration of liquid suspension. In view of the importance of particle adhesion, however, in determining the outcome of deposition in liquid filtration [11], the assumption that particle deposition results automatically from contact with a collector or previously deposited particles requires modification. Instead, for an approaching particle, upon its making contact with the surface of the collector or previously deposited particle, the question of whether or not deposition will result is determined by its adhesion probability. Specifically, assuming that this adhesion probability is γ, a random number (between 0 and 1) is generated. If the value of the number is equal to or less than γ, the particle is considered to be collected. Otherwise, it is considered to have escaped.

The adhesion probability, γ, can be estimated from the standard particle adhesion force measurements [12–14]. Briefly, these measurements are made by subjecting a given number of particles placed on a flat surface to a well-defined shear flow. The results obtained are in the form of the fraction of particles remaining on the surface against the drag force to which the particles were subjected (see Fig. 2 for example). For a single particle, these results can be interpreted as the adhesion probability corresponding to a given value of the drag force. To estimate the drag force acting on the approaching particles, one can readily apply the method developed recently [15].

3. EXPERIMENTAL WORK

The results obtained from the experimental work are the time-dependent behaviour of effluent quality and the pressure drop across the filter bed. Since a large number of variables are involved, direct correlations of experimental results with respect to these variables become impractical. Accordingly, as stated earlier, the purpose of this experimental work is to gather a relatively small amount of data under controlled conditions, which can then be used to test the different filtration theories.

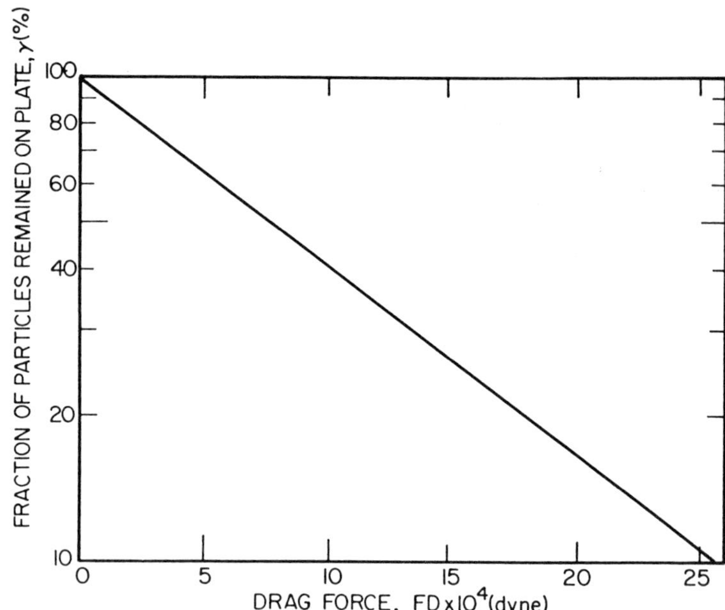

Fig. 2 – Fraction of particles remained on plate *vs* drag force acting on particles. From Johannisson [13].

3.1 Apparatus

The schematic diagram of the experimental apparatus is shown in Fig. 3. The main components of the apparatus included a suspension preparation tank; a pump; an overflow constant-head tank; and an experimental filter, which was made of a plexiglass cylinder 1 inch in diameter. A divergent funnel preceded the inlet to the filter to ensure that the plug-like velocity profile within the filter could be realized. Also, a screen was placed at the lower section of the filter as a filter grain support. Two pressure taps were located, respectively, at 2 cm below the divergent funnel and 3.5 cm below the screen support. Sampling tubes, one influent and one effluent, were placed at 3 cm below the funnel and 10 cm below the support screen, respectively. The distance between the influent sampling tube and the screen was 5 cm, representing the maximum height of the filter media used in this study.

3.2 Materials

The suspended particles used were lycopodium and ragweed. The lycopodium particles were supplied by BDH Chemicals Ltd. and distributed in the United

States through the Galland-Schlesinger Chemical Manufacturing Corp. (Carle Place NY). These particles were nearly monodispersed, with 90% (by volume) of the particles in the range of 22.5 μm–30.5 μm and a mean diameter of 26 μm. The ragweed particles were obtained from Sigma Chemical Co. (St. Louis, MO). Ninety per cent of them were in the size range of 18 μm–23 μm with a mean diameter of 19.5 μm. Glass beads were used as filter grains. The average size of the glass beads was 505 μm, obtained by sieving (between US No. 30 and 40 sieves). The ratio of the filter bed diameter to filter grain diameter was more than 50, thereby avoiding wall effects.

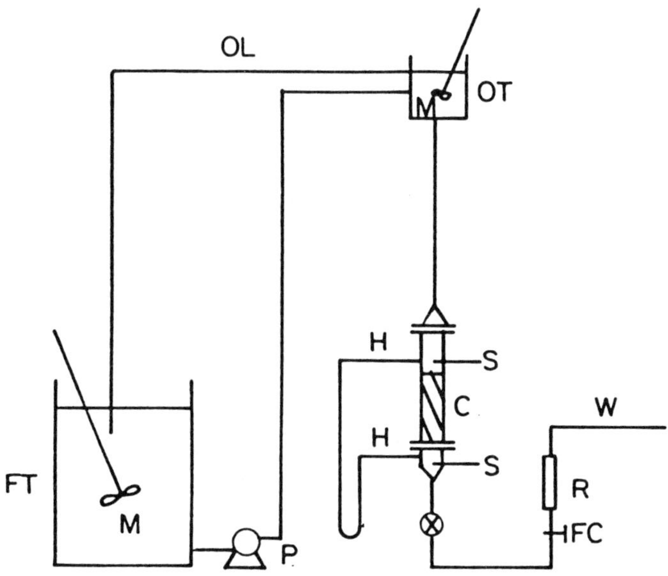

C - FILTER COLUMN
FC-FLOW RATE CONTROLLER
FT-SUSPENSION MIXING TANK
H - PRESSURE TAP
M- MECHANICAL MIXER
OL- OVERFLOW LINE
OT- OVERFLOW TANK
P- PUMP
R- ROTAMETER
S- SAMPLING TAP
W- WASTE LINE

Fig. 3 – Schematic diagram of the experimental apparatus.

3.3 Procedures

To prepare the test suspension, suitable amounts of suspended particles, together with distilled water, were added to a 200 ml flask. The flask was thoroughly shaken and then placed in an ultrasonic cleaner (Heat Systems-Ultrasonic, Inc., Plainview, NY) to insure the absence of flocculation. The suspension was then introduced into the mixing tank with distilled water added to obtain the desired particle concentration. An ultrasonic disintegrator (Model 150, Artek Systems Corp., Farmingdale, NY) and a mechanical mixer were used to keep the particles uniformly suspended.

To start an experiment, the suspension was prepared in the mixing tank, pumped to the constant-head tank, and passed into the filter packed with glass beads of specified height. The flow rate was kept constant by adjusting a valve placed before the rotameter. Influent and effluent samples were taken at specific intervals beginning a few minutes after the start of the experiment. This time lapse was necessary in order to assure a total displacement by the test suspension of the clean water initially present in the filter.

The particle concentration and size distribution of the samples taken were determined using a Coulter Counter (Model ZB, Coulter Electronics, Inc., Hialeah, FL) equipped with a multichannel analyser and an Apple II PLUS computer.

A typical set of data obtained from one experimental run consisted of the effluent concentration and pressure drop at various times as shown in Fig. 4.

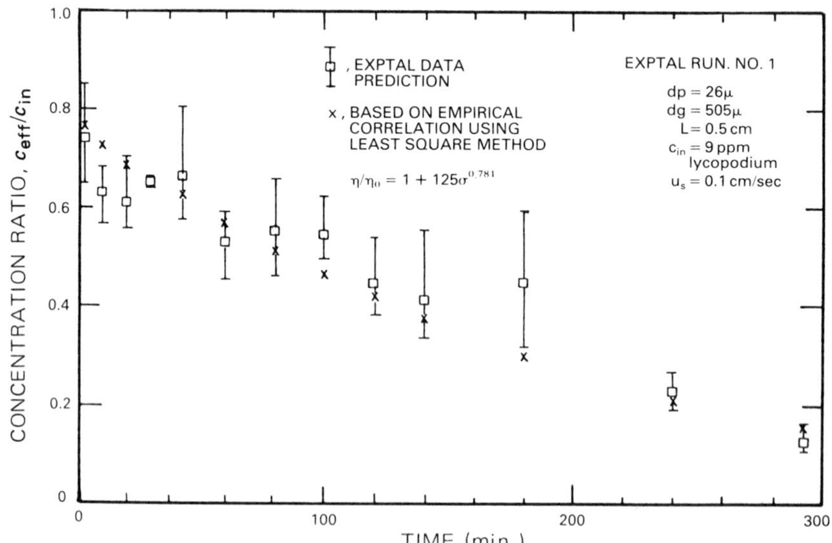

Fig. 4 – Ratio of effluent concentration to influent concentration *vs* filtration time (experimental run no. 1).

4. INTERPRETATION OF EXPERIMENTAL RESULTS

The experimental data obtained are in the form of the effluent concentration and pressure drop, Δp, across the bed. The procedures used to gather information concerning the effect of particle deposition on filter performance is briefly outlined below.

4.1 Unit Collector Efficiency

If a filter bed is viewed as an assembly of unit collectors, the filter's capability for particle retention can be expressed by the efficiency of the collectors. Furthermore, as shown previously, the unit collector efficiency, η, and the filter coefficient, λ, are related by the expression

$$\lambda = \frac{1}{l} \ln \left(\frac{1}{1-\eta} \right) \approx \frac{\eta}{l} \text{ if } \eta \ll 1 \tag{5}$$

Accordingly,

$$F_1(\alpha, \sigma) = \frac{\lambda}{\lambda_0} = \frac{\eta}{\eta_0} \; . \tag{6}$$

The axial dimension of a unit collector, l, is given as

$$l = \left(\frac{\pi}{6} \frac{1}{1-\epsilon} \right)^{1/3} d_g \; . \tag{7}$$

A filter bed of depth L may be considered to be composed of $N(=L/l)$ unit collectors connected in series. Let c_i denote the particle concentrations of the suspensions leaving the ith unit collector. The ratio of the effluent to the influent concentration can be expressed as

$$\frac{c_{eff}}{c_{in}} = \prod_{i=1}^{N} \frac{c_i}{c_{i-1}} = \prod_{i=1}^{N} [1 - \eta_i] = \prod_{i=1}^{N} [1 - \eta_0 F_i(\alpha, \sigma_i]) \tag{8}$$

where η_0 is the clean collector efficiency, or the unit collector effeciency at its initial state. For a homogeneous filter bed, η_0 is the same throughout the bed.

The extent of particle deposition is described by σ, the specific deposit. If σ_n denotes the specific deposit of the nth unit collector, the values of σ_n can be found by solving the following equations:

$$\frac{d\sigma_n}{dt} = \frac{u_s}{l} c_{in} \eta_0 F_1 (\alpha, \sigma_n) \prod_{i=1}^{n-1} [1 - \eta_0 F_1 (\alpha, \sigma_i)] \tag{9}$$

$$\text{for } n = 1, 2, \ldots, N$$

with $\sigma_n = 0, \quad t = 0 \; .$ (10)

From the experimental results, the concentration ratio, c_{eff}/c_{in}, at various times is known, and $N = L/l$ is specified. The problem is to determine the functional form of F_1. Toward this end, an objective function, ϕ, is defined as

$$\phi = \sum_{m=1}^{M} w_m \left[(c_{eff}/c_{in})_{exp,m} - (c_{eff}/c_{in})_{pre,m}\right]^2 \tag{11}$$

where M is the number of data points. w_m's are the weighting factors associated with the mth data point. w_m may be assigned to $1/\delta_m^2$ where δ_m is the experimental error (%) associated with each data point. This choice was made on the grounds that the weight assigned to each datum in determining F_1 should be inversely proportional to its accuracy.

The determination of F_1 from experimental data is reduced to an optimization-search problem. In other words, F_1 is found on the condition which gives ϕ a minimum. In actual practice, a given functional form of F_1 is specified first. For example, as one possibility, F_1 may be assumed to be

$$F_1 = 1 + a\sigma^b \ . \tag{12}$$

The coefficient and exponent are to be determined on the basis which yields a minimum of the objective function ϕ. The search was performed using the algorithm OR GLS (A General FORTRAN Least Square Program, Oak Ridge National Laboratory, Oak Ridge, TN). The calculated values of c_{eff}/c_{in} were obtained from the solution of equations (8) and (9). The integration of equation (9) was made using the algorithms EPISODE (Effective Package for the Integration of Systems of Ordinary Differential Equations) which is based on Gear's method for solving the ordinary differential equation.

4.2 Pressure Gradient

The procedures used to obtain the effect of deposition on the pressure gradient necessary to maintain a given flow rate were similar to those described in section 4.1. Again based on the premise that a filter bed is represented by a numer of unit collectors connected in series and from the definition of F_2 [i.e. equation (2)], one has

$$\frac{(\Delta p/L)}{(\Delta p/L)_0} = \frac{1}{N} \sum_{i=1}^{N} F_2(\beta, \sigma_i) \tag{13}$$

where the subscript 0 denotes the initial or clean collector state; σ_i, the specific deposit of the ith unit collector, can be found from the solution of equation (9). The objective function, ϕ_p, becomes

$$\phi_p = \sum_{m=1}^{M} W_m \left[\left\{\frac{(\Delta p/L)}{(\Delta p/L)}\right\}_{exp,m} - \left\{\frac{(\Delta p/L)}{(\Delta p/L)}\right\}_{pre,m} \right]^2 \tag{14}$$

In other words, the function F_2 is determined by minimizing the objective function ϕ_p from the pressure data.

5. RESULTS AND DISCUSSIONS

The discussions presented below are based on four sets of experiments conducted under the conditions listed in Table 1. The first three experiments were intended

Table 1
Experimental conditions

Run Numbers	1	2	3	4
Filter medium	Glass	Glass	Glass	Glass
Particles in suspension	Lycopodium	Lycopodium	Ragweed	Lycopodium
Bed porosity	0.41	0.41	0.41	0.41
Superficial velocity, u_s (cm/sec)	0.1	0.2	0.1	0.1
Grain diameter, dg(cm)	0.0505	0.0505	0.0505	0.0505
Particle diameter, dp(cm)	0.0026	0.0026	0.00195	0.0026
Bed height, L(cm)	0.5	0.5	0.5	1.0
Feed concentration, c_{in} (p.p.m.)	9.0	5.0	12.0	6.0

to provide comparisons with various filtration theories. The experimental data were obtained in the form of c_{eff}/c_{in} *vs* time as shown in Fig. 4. Applying the optimization–search procedure, the values of λ/λ_0(or η/η_0) and $(dp/dz)/(dp/dz)_0$ (or F_1 and F_2) were obtained as functions of specific deposits, σ. These results are shown as Figs. 5(a)–6(c).

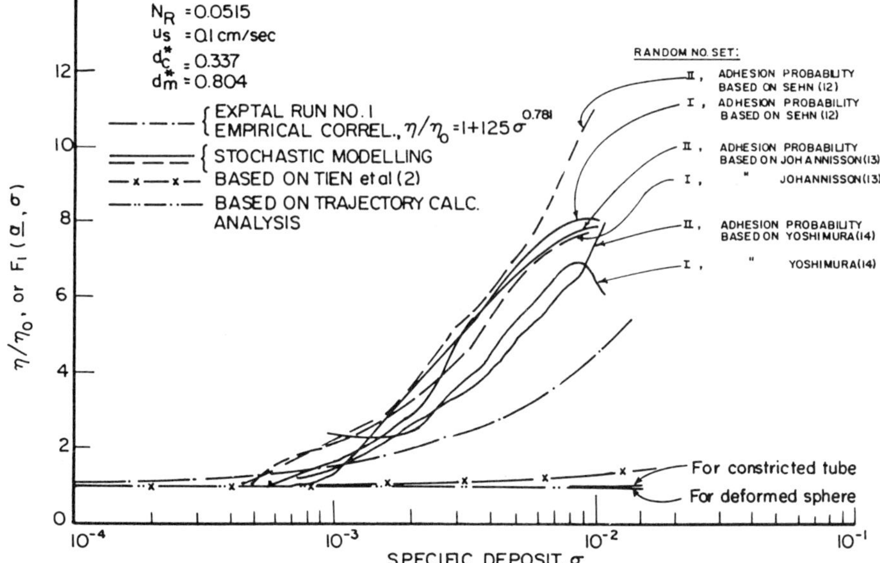

Fig. 5(a) – Comparison of the predicted relationship of F_1 (α, σ) *vs* σ with experimental data. Experimental run no. 1.

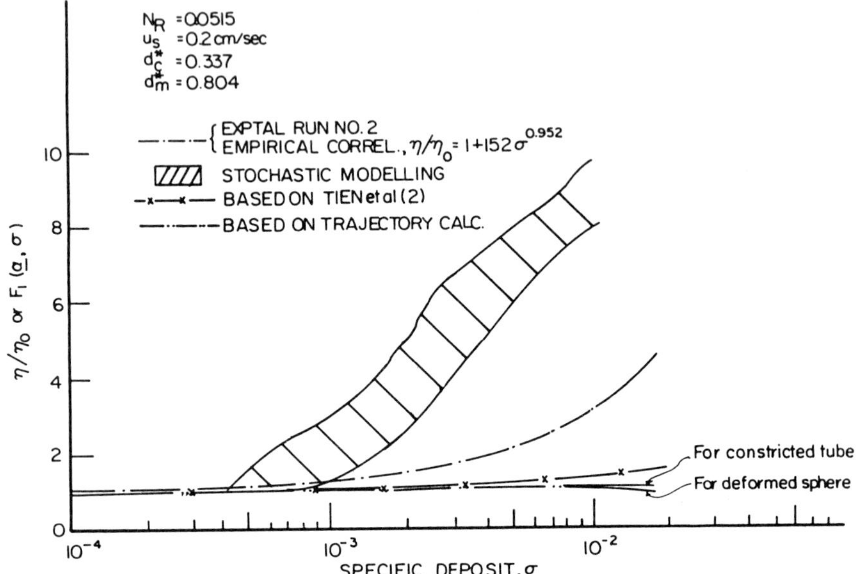

Fig. 5(b) – Comparison of the predicted relationship of F_1 (α, σ) *vs* σ with experimental run no. 2.

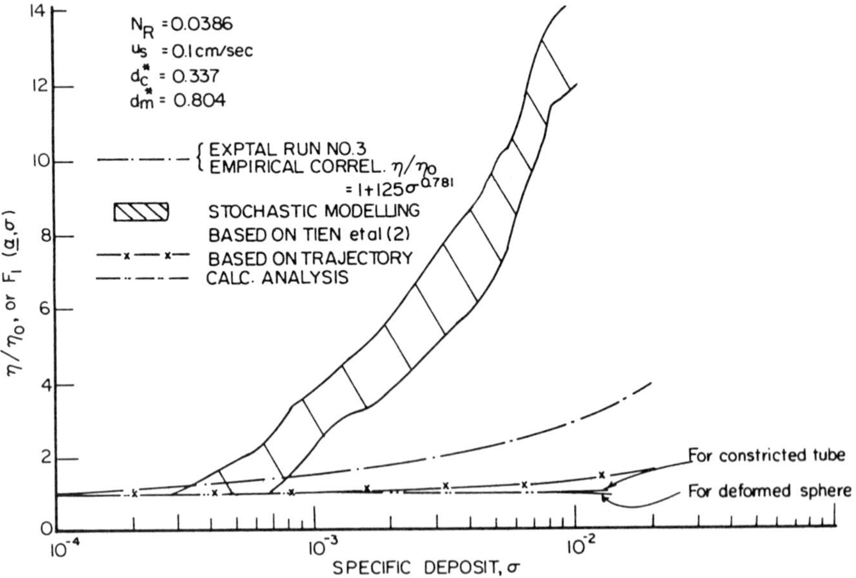

Fig. 5(c) – Comparison of the predicted relationship of F_1 (α, σ) *vs* σ with experimnetal data. Experimental run no. 3.

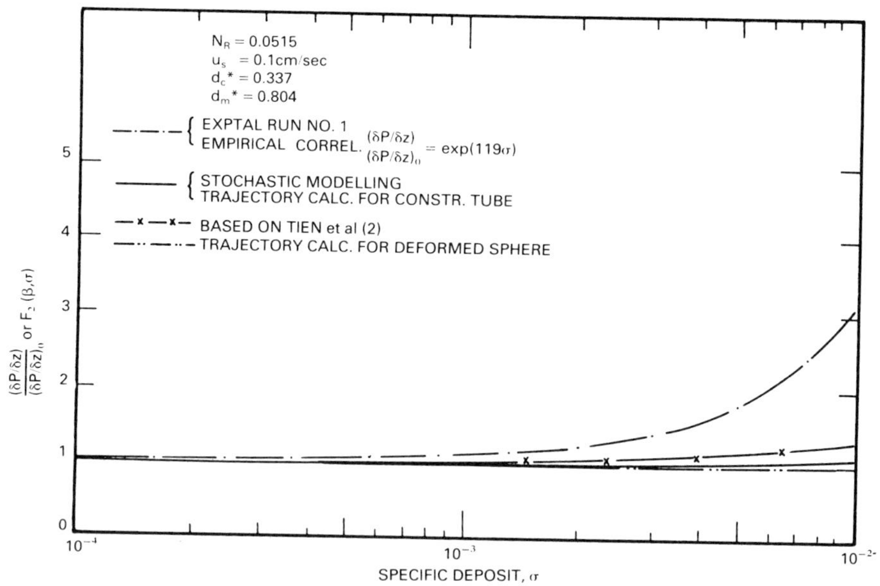

Fig. 6(a) – Comparison of predicted relationship of $F_2(\beta, \sigma)$ and σ with experimental data. Experimental run no. 1.

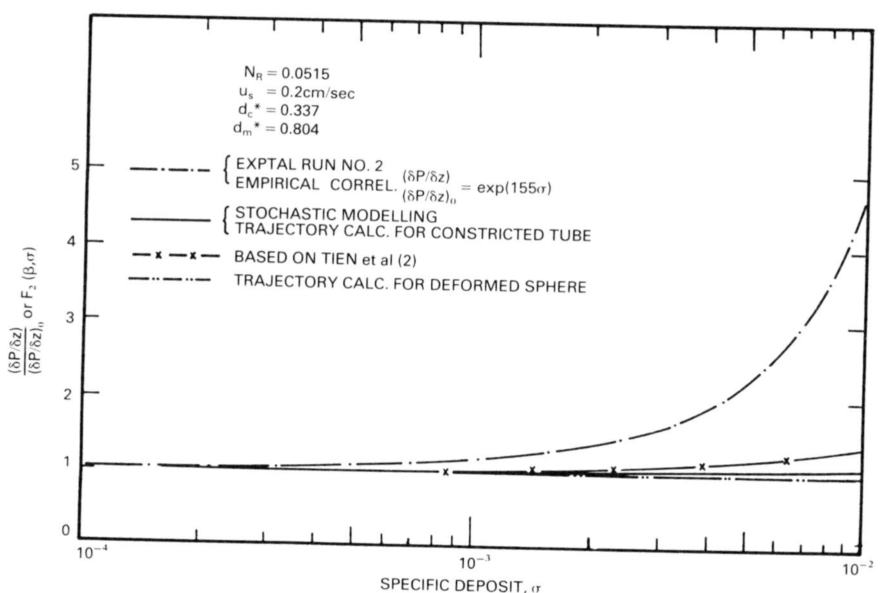

Fig. 6(b) – Comparison of predicted relationship of $F_2(\beta, \sigma)$ and σ with experimental data. Experimental run no. 2.

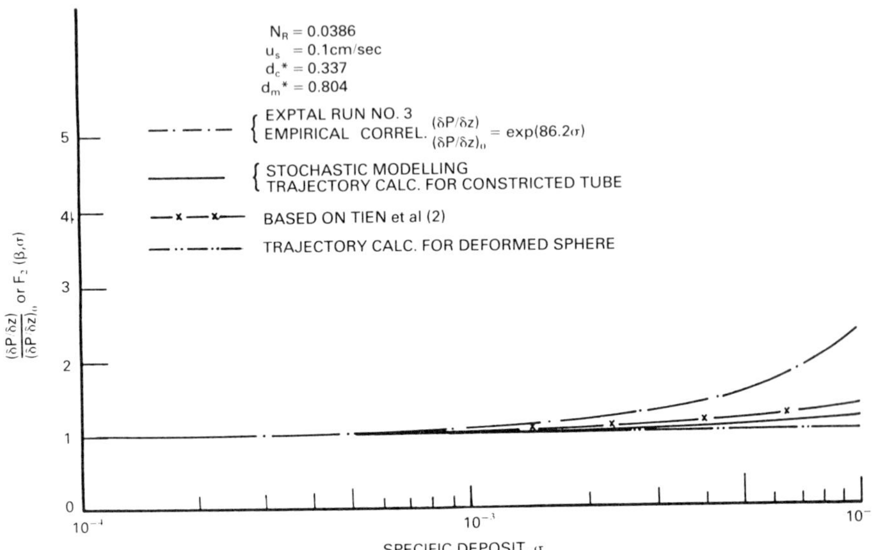

Fig. 6(c) – Comparison of predicted relationship of $F_2(\beta, \sigma)$ and σ with experimental data. Experimental run no. 3.

Three theoretically derived F_1's, corresponding to those discussed under sections 2.1, 2.2, and 2.3, were tested against the experimental results. The phenomenological model of Tien *et al.* [2] assumes a two-stage deposition process. In view of the nature of experimental data which exhibit monotonic improvement of filtrate quality, however, only F_{1_s} (or F_{2_s}) was considered. The model based on trajectory calculations (discussed under section 2.2) was used in conjunction with both the isolated-sphere model and the constricted-tube model for filter bed characterization, although the results were essentially the same no matter what collector geometry was considered. The few differences will be elaborated below.

To develop F_1 and F_2 from stochastic simulation, the adhesion probability was estimated from three different measurements [12–14]. The use of these results plus the very nature of stochastic modelling implies that the F_1 (or F_2) obtained varies with the random number set used in simulation as well as with the adhesion force data. Nevertheless, the results all display the same qualitative behaviour as shown in Figs. 5(a), 5(b), and 5(c).

The experimentally derived F_1 (or F_2) in all the cases was found to disagree with any of the theories considered. For the effect of particle deposition on the collection efficiency, both the phenomenological model and the trajectory analysis underestimate the increase in η by one order of magnitude (on the basis of $F_1 - 1$), while the stochastic simulation overestimates the increase (by a factor of 3–5). The trajectory analysis assumes that collected particles form smooth deposits. Similarly, under the conditions used in the experiments and based on the phenomenological model of Tien *et al.*, the morphology of particle deposits formed within the experimental filters, according to the values of σ, is one of a smooth coating. In either situation, the effect of deposited particles acting as additional particle collectors is minimized. The fact that both theories give essentially the same results, which are lower than experiments, is, therefore, not surprising. On the other hand, the stochastic modelling, as is formulated at present, does not take into account the effect of the changes of the flow field resulting from the presence of the deposited particles. As pointed out by Pendse and Tien [10], this omission leads to greater error (overestimation) as the particle inertia decreases. For filtration of liquid suspensions of particles with low particle density, the dominant mechanism of particle collection is interception. Consequently, the observed significant overestimation of the increase in η by stochastic modelling is not entirely unexpected.

The disagreement between the observed pressure drop increase and theories, however, is more puzzling. All the theories predict a much lower pressure drop increase than experiments actually yielded. A plausible explanation for this discrepancy is the cake formation of deposited particles. In conducting the experimental work, a run was terminated when filter cake was observed. An exact determination of the onset of cake formation, however, was impracticable with the present experimental set-up.

As an indication that the filter bed height used in the experiments did not cause error or bias in the development of the functions F_1 and F_2, the results obtained using one bed height were used to predict the behaviour of a filter bed with another bed height. An example of this is shown in Figs. 7 and 8, in which

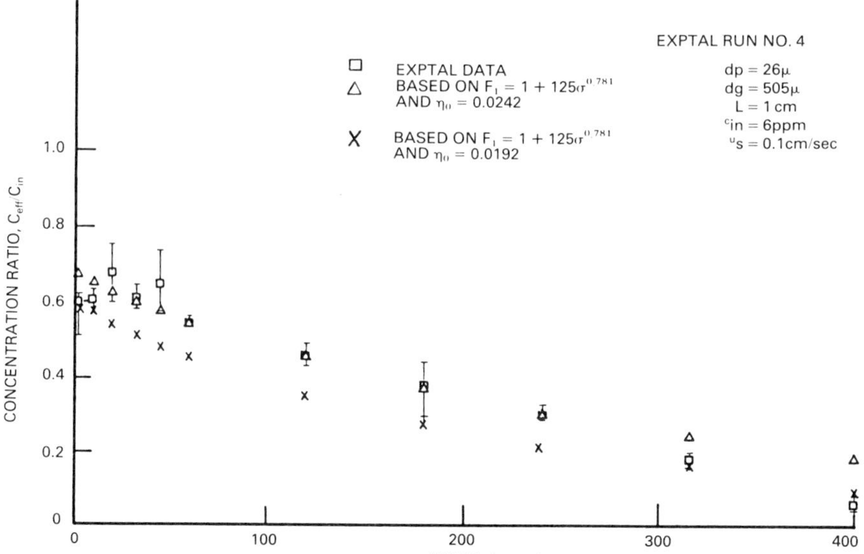

Fig. 7 – Comparison of experimental values of concentration ratio with the predicted values based on the F_1 obtained from experimental run no. 1.

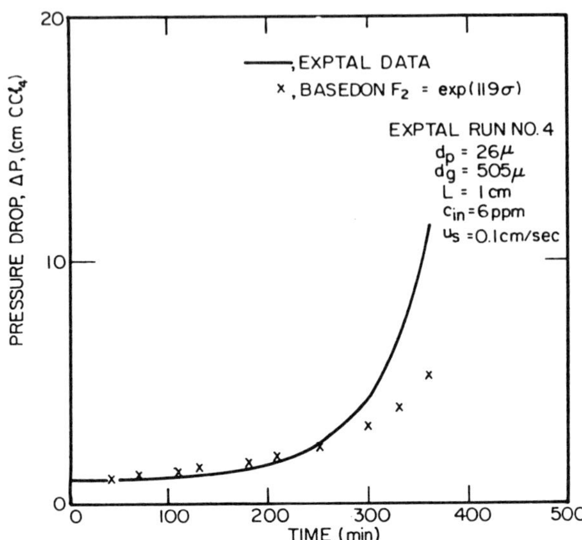

Fig. 8 – Comparison of experimental values of pressure drop with the predicted values based on the F_2 obtained from experimental run no. 1.

the predicted concentration ratio values $c_{\text{eff}}/c_{\text{in}}$ (or pressure drop) corresponding to a filter bed of 1 cm in height, based on F_1 (or F_2) developed from the data based on a shallower bed (0.5 cm height), are compared with experiments. The agreement is, in general, very satisfactory. For the case of the concentration ratio, the results indicate rather clearly the importance of both η_0 and F_1 in accurately predicting filtrate quality.

ACKNOWLEDGEMENT

This study was performed under Grant No. CPE 8110760 National Science Foundation.

NOMENCLATURE

a, b	constants in equation (12)
c_{eff}	effluent particle concentration
c_i	effluent particle concentration of the ith unit collector
c_{in}	influent particle concentration
F_1, F_2	functions defined by equation (1) and (2), respectively
J_i	particle flux
p	pressure
p_{w}^*	dimensionless wall radius of constricted tube

$p^*_{w_0}$	initial value of p^*_w
u_s	superficial velocity
v	volume of particles
t^*	dimensionless time
z	axial distance
$\alpha, \beta,$	parameter vectors of F_1 and F_2, respectively
γ	adhesion probability
δA	surface element
Δ	thickness of deposits
ϵ	bed porosity
ϵ_d	deposit porosity
η	unit collector efficiency
η_0	initial value of η
λ	filter coefficient
λ_0	initial values of λ
σ	specific deposit
σ_{tran}	transitional value of σ
ϕ	objective function
ψ^*	dimensionless stream function

REFERENCES

[1] Payatakes, A. C., Turian, R. M., and Tien, C., *Proc. 2nd World Congress on Water Resources,* V. 241 (1975).

[2] Tien, C., Turian, R. M., and Pendse, H., *AIChE J.,* **25**, 385 (1979).

[3] Rajagopalan, R. and Tien, C., *AIChE J.,* **22**, 523 (1976).

[4] Rajagopalan, R., and Tien, C., The Theory of Deep Bed Filtration, *Progess in Filtration and Separation,* Vol. 1, pp. 179–270, Elsevier (1979).

[5] Chow, J. C. F. and Soda, K., *Phys. Fluids,* **15**, 1700 (1972).

[6] Happel, J. and Brenner, H., *Low Reynolds Number Hydrodynamics,* 2nd ed., Nordhoff International Publishing, Leyden, The Netherlands (1973).

[7] Chiang, H. W., *Transient Behavior of Deep Bed Filtrations,* Ph.D. Dissertation in Progress, Syracuse University (1983).

[8] Tien, C., Wang, C. S. and Barot, D. T., *Science,* **196**, 983 (1977).

[9] Wang, C. S., Beizaie, M., and Tien, C., *AIChE J.,* **23**, 879 (1977).

[10] Pendse, H. and Tien, C., *J. Colloid. Interface Sec.,* **81**, 225 (1982).

[11] Gimbel, R. and Sontheimer, H., Recent Results on Particle Deposition in Sand Filters, in *Deposition and Filtration of Particles from Gases and Liquids,* p. 31, Society of Chemical Industry, London (1978).

[12] Sehn, P., *Untersuchengen zur Haftung von kleiner Glaskugeln auf Quarzplatten in Wassrigen Losungen* Thesis, Karlsruhe (1978).

[13] Johannisson, R., *Untersuchungen zur Veranderung der Partikelhaftung in immergierten Systemen durch Polymerzusatz,* Thesis, Karlsruhe (1979).

[14] Yoshimura, Y., *A Study of Clarification and Filtration in Packed Beds,* Ph.D. Dissertation, Kyoto (1980).

[15] Pendse, H., Tien, C., and Turian, R. M., *AIChE J.,* **27**, 364 (1980).

CHAPTER 21

Effect of Polymers on Particle Adhesion Mechanisms in Deep Bed Filtration

PETER SEHN AND ROLF GIMBEL
Department of Water Chemistry, Engler-Bunte-Institute, Karlsruhe University, West Germany

---

ABSTRACT

The positive effect of polymer addition on particle adhesion in deep bed filters is demonstrated by some experimental examples. As the increase in the degree of removal is due to a change in the surface properties of the particles and/or filter grains, the adsorption behaviour of polymers on solid surfaces is studied. A technique based on the flocculating effect of polymers is used to determine the very low residual concentrations involved. It is shown both theoretically and experimentally that the mass transfer mainly determines the polymer adsorption behaviour for the systems discussed here. The thickness of a layer adsorbed at a glass surface as measured by a permeability method depends largely on the ionic character of the polymer chain.

The effect of such sorbate layers on particle adhesion is investigated using two different methods, where the separation forces acting on small quartz particles and glass microbeads, which are in contact with a smooth quartz plate, are applied either in normal or in tangential direction. The comparison obtained from these two methods allows a further interpretation of the effects of polymers on the particle adhesion mechanisms. Furthermore, the dependence of the particle adhesion probability on the particle/substrate contact time is determined, showing a much faster increase in the presence of polymers than in the case without polymer.

Using a new model for the overall particle adhesion probability in a deep bed filter, which considers multiple particle contacts on the surface of the filter grain, the findings from the particle adhesion measurements can be transferred to the particle behaviour in the filter. This allows a better understanding of filtration results.

1. INTRODUCTION

Deep bed filtration is an important process in drinking water treatment and is of increasing importance in waste water treatment too. The deposition of turbidity within a filter bed, usually consisting of granular material, is based on two successive steps: The transport process which brings the particles into contact with the filter grain surface (or with particles previously deposited there), and the subsequent adhesion process on this surface. Most of the deterministic theories for deep bed filtration are restricted to describe the transport process whereby the adhesion probability is assumed to be 100%.

In reality the adhesion probability may be much less than 100% depending on the ratio of the adhesion forces to separation forces acting on the particle, the latter being primarily drag forces. These forces are governed by the size and structure of the particles and of the filter grains, the position where the particle comes into contact with the filter grain, the superficial velocity and the specific particle deposit in the filter bed. The adhesion forces are usually London – van der Waals forces which are effective when the surface separation distance is sufficiently small. Surface roughnesses, electrical double layer forces and hydration forces can diminish the resulting adhesion forces.

Under certain conditions, e.g. high superficial velocity and/or relatively high filter deposit, the particle adhesion behaviour may become the limiting process for the overall removal efficiency of a deep bed filter. This is, for instance, always true when a complete turbidity breakthrough occurs after a certain filtration time.

By adding small doses of a synthetic water-soluble polymer to the raw water, the removal efficiency of a deep bed filter can be considerably increased not only in the initial filtration phase, but also in the dynamic phase, where the amount of retained particles affects the particle deposition. The reason for this behaviour may be due to an enhancement of the efficiency of the particle transport mechanisms and/or may be due to an improvement in the particle adhesion conditions. If relatively high turbidity concentrations are present in the raw water an enhanced particle flocculation by the polymer may also positively affect the removal efficiency in an indirect manner. In all cases the adsorption of the polymer at the solid–liquid interface is a prerequisite for the effects of the polymers, as the pure hydrodynamic conditions in the filter bed can be considered as nearly unchanged because of the very low concentrations in the bulk phase.

To understand the effects of polymers on the particle adhesion mechanisms in a deep bed filter, first of all it is necessary to have some information about the adsorption behaviour of these substances and the resulting sorbate layers. How these influence the particle adhesion behaviour and the effects on filter efficiency are discussed in this paper.

2. SOME EXPERIMENTAL FILTRATION RESULTS

To provide a rough idea of the effects of polymers on the removal efficiency of deep bed filters, some typical results for the deposition of quartz particles in packed beds of glass beads and quartz sand are shown in Figs. 1 and 2. The filter efficiency, experimentally determined using a Royco particle counter, is expressed as the elementary removal degree γ, which describes the probability of a particle being retained in a unit bed element of the thickness of the filter grain diameter d_k. γ can be expressed in terms of the filter coefficient λ by

$$\gamma = \lambda \cdot d_k \quad \text{for} \quad \gamma \ll 1 \tag{1}$$

In all cases the polymers characterized in Table 1 are continuously added to the model suspension. The time available for the polymers to adsorb on the surface of the quartz particles before the suspension enters the filter bed amounts to about 3 min.

Table 1

Characterization of polymers
Data taken from [6] (A 972, A 968), [7] (PEI P. PEI SN) and the manufacturer

Name	A 972	A 968	222 K	444 K	PEI P	PEI SN
Delivered by	Stockhausen, Krefeld				.BASF, Ludwigshafen	
Base molecule	polyacrylamide				.polyethylenimine	
Ionic character	anionic	non-ionic	cationic		.strongly cationic	
Molecular weight in 10^6 g/mol	9.7	9.8	~0.2	~2	0.6	1.8
Coil diameter d_{pol} in nm	390	290	90*	–	90	290
determined by	viscometry				.light scattering	

* Own measurement

Figure 1 shows the dependence of the elementary removal degree in the initial filtration phase (γ_0) on the particle size with and without different polymer dosages. The solid lines represent the theoretical results from deterministic theories under favourable surface conditions, i.e. negligible electrostatic repulsion barriers, and under the assumption of an adhesion probability of 100%. The comparison of the experimental with the theoretical results shows that the discrepancies are most pronounced for relatively large particles. This can be explained by a decreasing adhesion probability with increasing particle size because of the different particle size dependencies of the adhesion and drag force acting on a deposited particle in a shear flow field [1, 4, 5].

The addition of polymers results in a significant enhancement of the filter efficiency in all cases shown in Fig. 1. For relatively small particles this is mainly due to an improvement of the particle transport mechanisms because in this case the particle adhesion probability is relatively close to 100%. Therefore the observed effects may be mainly due to a reduction of electrostatic potential barriers and some particle-catching effects of polymer filaments partly adsorbed at the surfaces of the solids. For relatively large particles the effect of the polymer is mainly due to an improvement in the particle adhesion probability. This is the case because the separation forces acting on these particles are relatively high, which results in a low adhesion probability. On the other hand, the efficiency of the particle transport mechanisms is high for the large particles in comparison with the small ones.

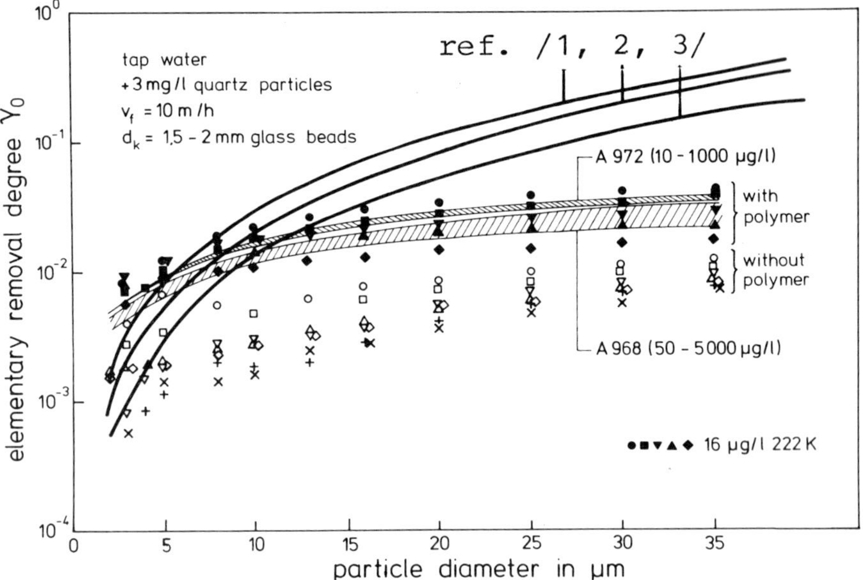

Fig. 1 – Comparison of experimental filtration results with different theories.

It is interesting to note from Fig. 1 that the polymer effects are very similar regardless whether the polymers are cationic, non-ionic or anionic. The concentration range, where the non-ionic and anionic products are effective is quite large. With the cationic polyelectrolyte an improvement in the filter efficiency has been observed over a wide concentration range too, but in this case a distinct optimum exists at about 16 μg/l for particle diameters between 2 and 40 μm. This optimum is strongly correlated with zero electrophoretic mobilities of the quartz particles. This demonstrates the important influence of charge neutralization in this case. On the other hand comparable effects on the filtration efficiency are obtained with the anionic polymer, despite the fact that the originally negative electrophoretic mobility of the particles has not been reduced by this polyelectrolyte.

Figure 2 shows a typical example of the change in elementary removal degree with increasing particle deposit, i.e. during the dynamic filtration phase. The experimental results have been obtained for a particle size range between 6 and 10 μm by continuously adding different amounts of a cationic polyelectrolyte (40 and 300 μg/l 222 K). The results for 40 μg/l, which correspond to an electrophoretic mobility equal to zero under the given experimental conditions, show the highest removal efficiencies during the initial phase of filtration. In contrast to this behaviour a higher polyelectrolyte addition (300 μg/l) results in a lower filter efficiency in the initial phase, but when relatively high deposits are present, higher efficiencies are obtained, thus resulting in a higher ultimate filter deposit. This also indicates the importance of the particle adhesion, especially with increasing deposits.

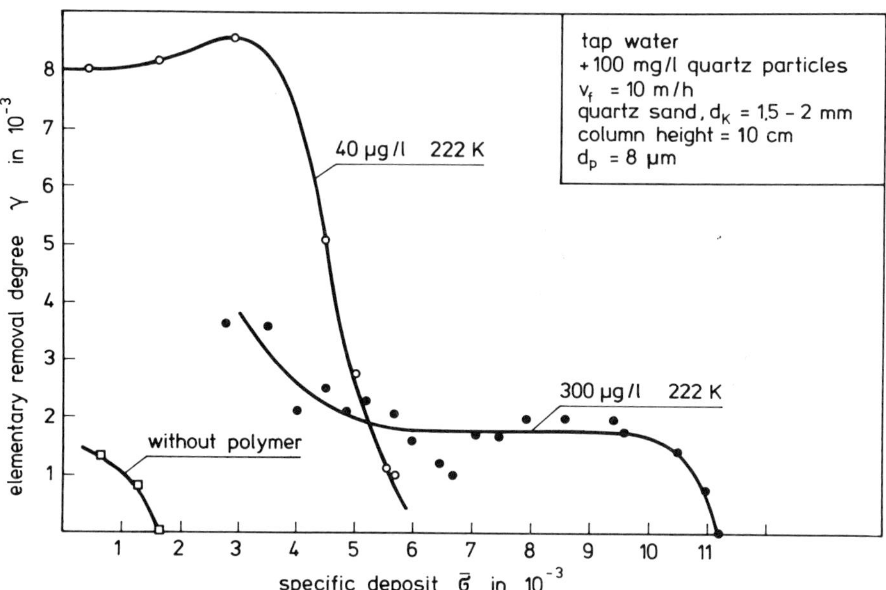

Fig. 2 – Dependence of the elementary removal degree on the specific deposit for different polyelectrolyte doses [8].

During the initial phase, charge neutralization and other effects of the cationic polymer on the transport efficiency are strongly responsible for the improvement of the overall filter efficiency. Therefore 40 μg/l 222 K lead to an optimum filter efficiency. Under such conditions the particle adhesion forces are not yet improved to the highest extent as has been determined with special experimental methods which are discussed later. A polyelectrolyte addition of 300 μg/l 222 K results in a significantly higher enhancement of the particle adhesion, but leads also to the disadvantage of again building up electrostatic potential barriers because of the change in surface charge of the quartz particles and quartz filter grains from negative to positive values, which is the reason for the lower filter efficiency in comparison to 40 μg/l. Due to the growing importance of the particle adhesion process with increasing deposit, the positive effects of a polyelectrolyte addition of 300 μg/l on the improvement of the particle adhesion forces overrides the partly negative effects on the particle transport mechanisms when relatively high deposits are pressent, thus leading to a higher ultimate specific deposit.

3. POLYMER ADSORPTION

The usual way to determine the amount of polymer adsorbed on solid surfaces is to measure the residual polymer concentration in the bulk phase. As most of the experimental work done in this field is concerned with systems that are more highly concentrated (by some orders of magnitude) than the systems relevant to deep bed filtration, a method had to be found which enables the determination of polymer concentrations in the p.p.b. range.

3.1 Determination of low polymer concentrations

The method shown in Fig. 3 is a further development of the one described by Burkert [9]. The principle is to measure the ability of the polymer-containing sample to flocculate an added standard kaolin suspension. The higher the polymer concentration, the better is the effect on flocculation (for the conditions of interest).

Ten millilitres of the standard suspension containing 1 g of purified kaolin is added to 100 ml of the sample in a cuvette situated in a Brice-Phoenix 3000 light scattering photometer. The flocculation conditions are held constant at 800 r.p.m. of a magnetic stirrer for 5 min. Because of the high solids content and the turbulent mixing of the sample this time is sufficient to achieve a practically complete adsorption of the polymer on the kaolin surface.

After 5 minutes the stirrer is stopped, the flocculated suspension starts to settle and the transmitted light at a distance of 2 cm from the bottom of the cuvette is recorded. As soon as the solids concentration in the zone of the light beam begins to decrease, a steep increase of the transmitted light results. The time t_s necessary to reach a certain value of the transmitted light can be taken as a measure for the polymer concentration using a calibration curve which has been previously determined with known polymer concentrations.

The detection limit of this method is about 2 μg/l except for PEI SN (~15 μg/l). In the case of anionic polymers a low level of bivalent cations has to be present to promote flocculation.

3.2 Adsorption isotherms

Polymers of different ionic characters and molecular weight (see Table 1) were used for the adsorption studies. The adsorbent was a fine quartz powder (Sikron

Determination of low polymer concentrations

① flocculation at defined conditions

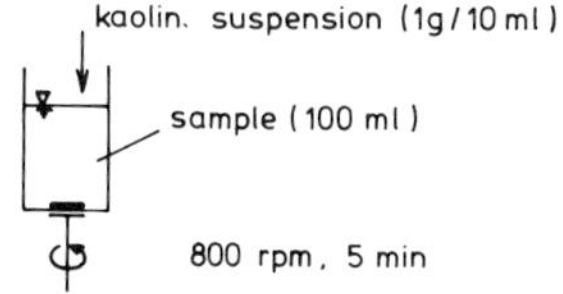

② sedimentation - transmissive light measurement

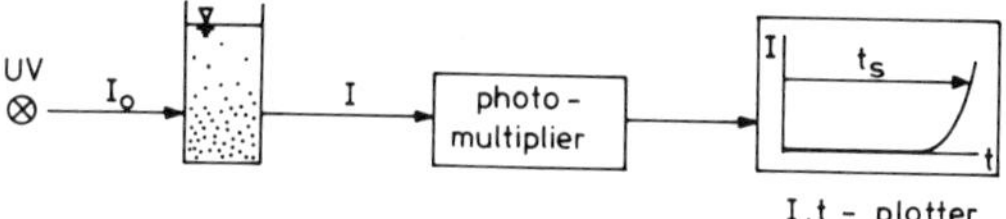

③ determination of the unknown polymer concentration by comparison with a calibration curve

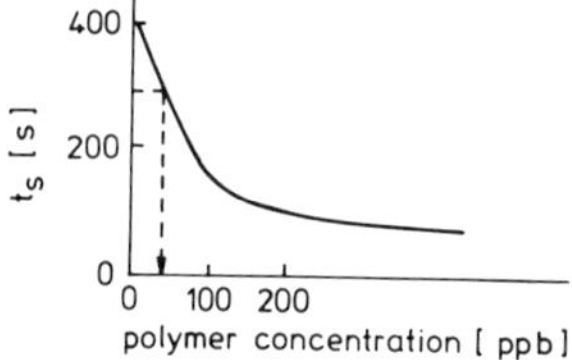

Fig. 3 – Scheme of the polymer determination method.

F 600) with a mean particle diameter of 1.5 μm (based on number distribution) and a specific surface of 1.2 m^2/g, calculated from the particle size distribution. The solvent used was membrane-filtered tap water of Karlsruhe.

Five-hundred millilitre flasks containing the quartz suspensions with different initial polymer concentrations were slowly turned (6 r.p.m.) to avoid mechanical degradation of the polymer. After 17 h adsorption time, approximate equilibrium was reached in most cases. The solids were removed by centrifugation and the residual polymer was determined.

The resulting adsorption isotherms for a solid content of 1 g/l are shown in Fig. 4.

There exists a plateau region for all the polymers studied, where the surface loadings are almost independent of the polymer concentration or only slightly increase with increasing concentration. The height of the plateau depends strongly on the ionic character of the polymer, yielding maximum loadings for the strongly positively charged chains and minimum values for the negatively charged polymer. This effect can be explained by the negative charge of the quartz surface that gives rise to a preferred adsorption of polymers of opposite charge.

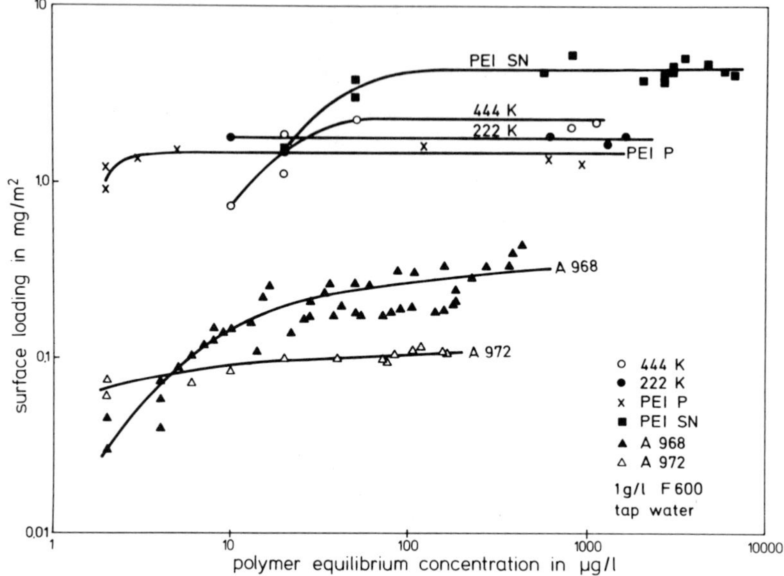

Fig. 4 – Adsorption isotherms of different polymers onto fine quartz particles (F 600) from tap water

In addition, the adsorbed amount of cationic polymers of different molecular weight is higher for the higher molecular weight. The plateau is reached at rather low equilibrium concentrations near the detection limits.

The isotherms are of the well-known 'high affinity type' and are in qualitative agreement with existing theories [10–14] based on random walk computations.

Adsorption isotherms measured at lower solid concentrations (< 1 g/l) showed increased surface loadings with decreasing solid concentration for some of the polymers. This effect, also found by others, can be explained by the theory of Scheutjens and Fleer [13, 14] as a displacement of smaller molecules from the solid surface by larger ones with higher adsorption affinities, taking into account the polydispersity of the polymer [15]. However, for applications like those discussed in this paper competitive adsorption is of minor importance because the surfaces are only partly covered by polymer. This can be concluded from the following considerations.

Parallel to determinations of residual concentrations of cationic polymers, the electrophoretic mobilities of the polymer-covered particles were also measured. The initially negative zeta potential of the quartz particles becomes zero at a corresponding surface loading of about 20% of the plateau value and becomes positive for higher loadings. As under the conditions of optimum filtration efficiency in the initial phase the particle mobilities have turned out to be zero, the surface loading of these particles is far from saturation. For flocculation with different kinds of polymers it is also known that the surfaces are only partly covered by adsorbed polymers.

Considering the filtration conditions given in Fig. 1, the optimum polymer dosage of a cationic polymer (222 K) was found to be 16 μg/l, leading to a surface loading of about 0.4 mg/m^2 (zero mobility). This means that only about 1.3% of the added polymer is adsorbed, taking the ratio of added polymer to the given particle surface (5.3 cm^2/l) into consideration. On the other hand, from the adsorption isotherm (Fig. 4) one could expect saturation loading and an equilibrium concentration of about 15 μg/l.

The reason for this discrepancy is due to the non-equilibrium of polymer adsorption under the conditions of filtration experiments. As the kinetics of polymer adsorption seem to play a major role in low concentration systems, this point is discussed in more detail in the next section.

3.3 Adsorption kinetics

The purpose of this section is to provide an idea of the most important conditions governing adsorption kinetics and to estimate the amount of adsorbed polymer with increasing adsorption time by some simplifying assumptions. Here only the rate of the initial adsorption phase is considered, i.e. the main part of the solid surface is always free of polymer. The adsorption rate is governed only by the mass transfer of polymer molecules from the bulk to the surface. The following adsorption step itself is assumed to be extremely fast and to be describable by a 'high affinity' type of isotherm, leading to a zero polymer concentration in the solution in the immediate neighbourhood of the solid surface as long as the saturation loading is not reached. Furthermore, both particles and polymer coils are considered as monosized spheres.

In the following, five different systems are considered which are encountered in the filtration and adhesion experiments.

3.3.1 Adsorption onto the surface of particles that are uniformly distributed in the polymer solution

The particles are in motion either due to Brownian movement (perikinetic) or due to a shear rate G (orthokinetic). Following Gregory [16], the kinetics of adsorption can be treated by the Smoluchowski theory [17] as a heterocoagulation of particles and polymer coils. For a first approximation, hydrodynamic and electrostatic interactions are neglected. Then the decreasing number concentration of free polymer coils with time may be calculated:

$$-\frac{dN_{\mathrm{pol}}}{dt} = k_{12} \cdot N_{\mathrm{p}} \cdot N_{\mathrm{pol}} \tag{2}$$

where the index 'pol' stands for polymer coil and 'p' for particle. k_{12} is the rate constant, given by

$$k_{12} = \frac{2kT}{3\eta} \cdot \frac{(d_p + d_{pol})^2}{d_p \cdot d_{pol}} \qquad \text{(perikinetic)} \qquad (3)$$

$$k_{12} = \frac{G}{6} \cdot (d_p + d_{pol})^3 \qquad \text{(orthokinetic)} \qquad (4)$$

for the perikinetic and orthokinetic cases. It can be seen that the adsorption rate depends linearly on the polymer and particle number concentration and on the shear gradient. The orthokinetic effect is more pronounced for larger particles or larger polymer coils, the perikinetic for smaller ones. Under the assumption of constant N_p values (low particle concentration) the fraction of polymer adsorbed in a given time t_a can be calculated by integration of equation (2).

For the conditions of the filtration experiment described before t_a is 3 min, $N_p = 10^9$ m^{-3}, $d_p = 13$ μm, $d_{pol} = 0.09$ μm (222 K, see Table 1), $G = 200$ s^{-1} and the coefficient $2\,kT/3\,\eta = 2.70 \cdot 10^{-18}$ m$^3 \cdot$ s^{-1}. The fraction of polymer adsorbed turns out to be orthokinetically controlled and is only 1.3% of the added amount. This value agrees with the one resulting from considerations discussed in section 3.2. The rather slow adsorption rate is due to the highly diluted system, in contrast to fast adsorption found by others for more concentrated systems.

3.3.2 Mass transfer onto the filter grain surface

Based on a film diffusion model the outflow concentration c of a packed bed with spherical adsorbant grains can be written as

$$\frac{c}{c_0} = \exp(-\beta_k \cdot A/\dot{V}) \qquad (5)$$

provided that dispersion effects are negligible and steady state conditions can be assumed, as is the case for the given system. A is the adsorbing surface of the packed bed, $\dot{V}$ the flow rate through it and β_k a mass transfer coefficient. Gnielinski [18] gives an empirical correlation for β_k:

$$\beta_k = \frac{D}{d_k} \quad 1 + 1.5 \cdot (1-\epsilon) \cdot (2 + \sqrt{Sh^2_{lam} + Sh^2_{turb}}) \qquad (6)$$

with

$$Sh_{lam} = 0.664 \cdot Sc^{1/3} \cdot Re^{1/2} \qquad (7)$$

$$Sh_{turb} = \frac{0.037 \cdot Re^{0.8} \cdot Sc}{1 + 2.443 \cdot Re^{-0.1}\,(Sc^{2/3} - 1)} \qquad (8)$$

and

$$Re = \frac{v_f \cdot d_k}{\epsilon \cdot \nu} \tag{9}$$

$$Sc = \frac{\nu}{D} \tag{10}$$

Equation (6) is valid for $Re \cdot Sc \geqslant 500 - 1000$ and $Sc \leqslant 12{,}000$, but for a first estimation it can also be used for higher Sc numbers as encountered with high molecular weight polymers.

The diffusion coefficient D of the polymer dissolved in water is given by the Stokes–Einstein equation

$$D = \frac{kT}{3\pi\eta d_{pol}} \tag{11}$$

The values for D are given in Table 2, together with the resulting c/c_0-values, calculated for the filtration experiments described in Fig. 1 with a packed bed of 15 cm height, 9 cm diameter and porosity $\epsilon = 0.4$.

Table 2

Diffusion coefficients and concentration ratios for the given filtration experiment

	A 972	A 968	222 K	PEI P	PEI SN
D in 10^{-12} m²/s	1.10	1.48	4.77	4.77	1.48
c/c_0	0.970	0.964	0.922	0.922	0.964

The results clearly indicate that only small amounts (a few percent) can be adsorbed while the solution flows through the filter column. For instance, to reach a mean surface loading of 0.4 mg/m² for 222 K in these filtration experiments, about 10 hours of polymer flow ($v_f = 10$ m/h, $c_0 = 16$ μg/l) are necessary, provided the adsorption rate does not decrease.

The mass transfer to the surface is governed by a number of parameters: diffusion coefficient of the polymer, surface area of the packed bed and the hydrodynamic conditions in the bed, given by the grain diameter, porosity and flow rate. As the estimates from sections 3.3.1 and 3.3.2 show, for the given conditions similar to the ones in practice, the kinetics of polymer adsorption may be controlled by mass transfer processes. To interpret the effects of polymers on deposition of turbid matter in a deep bed filter, this result has to be taken into account.

3.3.3 Mass transfer onto the surface of a setting sphere

The analogous problem in heat transfer has been treated by Schlünder [19]. Provided the bulk concentration c_0 remains constant within relatively short times, the mass transfer onto the adsorbing surface A of the sphere can be written as

$$\dot{m} = \beta_p \cdot A \cdot c_0 \tag{12}$$

where the mass transfer coefficient

$$\beta_p = \frac{Sh_p \cdot D}{d_p} \tag{13}$$

is composed of the diffusion coefficient (see Table 2), the diameter of the sphere d_p and the Sherwood number

$$Sh_p = 1.01 \left(\frac{u \cdot d_p}{D}\right)^{1/3} \tag{14}$$

with the Stokes sedimentation velocity u.

3.3.4 Mass transfer onto the surface of the wall of a rectangular channel from laminar flow

The mass transfer can be calculated by equation (12) with the adsorbing wall surface A and a mass transfer coefficient [19]

$$\beta_c = \frac{Sh_c \cdot D}{l} \tag{15}$$

where l is the length of the channel and

$$Sh_c = 1.45 \cdot \left(\frac{\dot{V}}{b \cdot D} \cdot \frac{s}{l}\right)^{1/3} \tag{16}$$

3.3.5 Mass transfer by non-steady state diffusion

The concentration field $c(t, x)$ in a distance x from an adsorbing surface obtained from Fick's second law for short times (constant bulk phase concentration) is given by

$$\frac{c}{c_0} = \operatorname{erf} \frac{x}{2 \cdot \sqrt{D \cdot t}} \tag{17}$$

The mass transport to a surface area A is derived from Fick's first law

$$\dot{m} = D \cdot A \cdot \left.\frac{\delta c}{\delta x}\right|_{x = 0} \tag{18}$$

With the differential form of (17) one gets from (18)

$$\dot{m} = \frac{A}{\sqrt{\pi}} \cdot \sqrt{\frac{D}{t}} \cdot c_0 \tag{19}$$

Integration yields the specific surface loading m/A:

$$\frac{m}{A} = 2\sqrt{\frac{D}{\pi}}\, c_0 \cdot \sqrt{t} \tag{20}$$

As can be seen, the surface loading increases with the bulk concentration c_0 and the square root of the adsorption time.

An experimental demonstration for the slow adsorption kinetics from a diluted polymer solution is given in Fig. 5. An adsorption experiment was performed using the clean cuvette glass walls of the light scattering device (see Fig. 3) as adsorbing surfaces. After different adsorption times under turbulent stirring kaolin was added to determine the residual polymer concentration. The amount of added polymer was sufficient to give a surface loading of about 25% of the saturation value when equilibrium was reached. From the shape of the isotherm an almost complete adsorption can be expected under the given conditions. In spite of the low surface loading and turbulent mixing more than two hours are necessary to yield almost complete adsorption for this polymer with a relatively high diffusion coefficient. This shows again the importance of mass transfer in diluted systems.

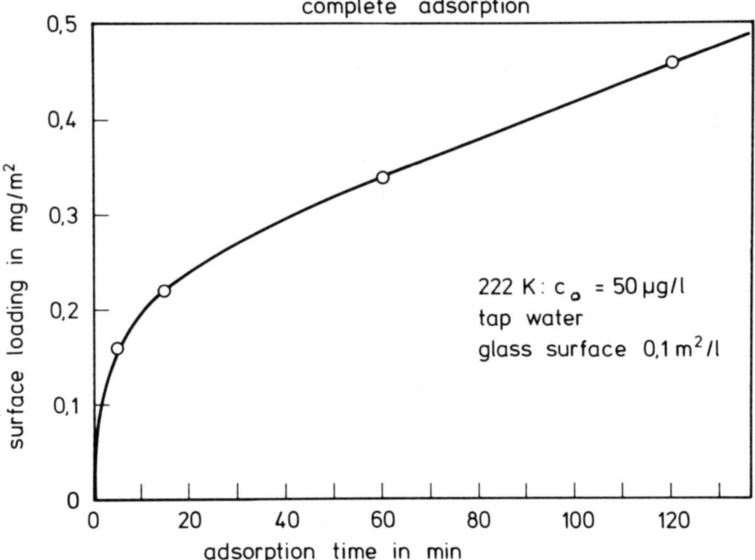

Fig. 5 – Adsorption kinetics of a cationic polymer onto cuvette glass walls under turbulent mixing condition.

However, not only are the adsorbed amounts of polymer important to understand the interactions of surfaces, the conformation and thickness of an adsorbed layer are also of consequence.

To yield high adhesion forces by polymer bridging, a high fraction of segments of a molecule should be adsorbed, resulting in a flat configuration. On the other hand, polymer loops and tails protruding into the solution are able to form bridges between surfaces which are a relatively large distance apart.

If the adsorbed amount is known, measurements of the sorbate layer thickness can give some insight into the conformation of the adsorbed macromolecules.

3.4 Thickness of adsorbed polymer layers

The principle of the method used here to determine a sorbate layer thickness is to measure the permeability decrease of a packed bed of glass microbeads by adsorbed polymers. For laminar flow the permeability p of a packed bed can be calculated using Happel's cell model [20]:

$$p = \frac{1 - \frac{3}{2}(1-\epsilon)^{1/3} + \frac{3}{2}(1-\epsilon)^{5/3} - (1-\epsilon)^2}{1 + \frac{2}{3}(1-\epsilon)^{5/3}} \cdot \frac{d_b{}^2}{18\eta(1-\epsilon)}$$

$$= \frac{\dot{V}}{\Delta p} \cdot \frac{h}{A_f} \tag{21}$$

and is measured as the flow rate $\dot{V}$ through the packed bed of porosity ϵ, height h (3.3 mm), cross section A_f (0.554 cm^2) and constant pressure drop Δp. The glass beads ($d_b^0 = 30 - 35$ μm) are sintered on their contact points to retain a reproducible packed bed for the experiments. The polymer layer that is built up when a polymer solution is flowing through the bed is assumed to be of uniform thickness $(d_b - d_b^0)/2$. Therefore the decrease in permeability can be assigned to a decrease of ϵ and an increase of the bead diameter by the adsorbed layer. A layer thickness determined in this way means a hydrodynamically active value. Its detection limit is about 25 nm for the described conditions.

These findings from these experiments indicate again that adsorption times to achieve approximately equilibrium loadings are of the order of hours. Values for the layer thickness correspond to the free coil diameters, which means they increase from cationic to anionic character and depend also on the salt concentration of the solution. As is known from the isotherms (Fig. 4) the surface loadings decrease in this direction. Therefore the sorbate layer of a cationic polymer can be assumed to be dense and thin (below 100 nm), whereas anionic polymers form thick and light ones. Measured thicknesses of the anionic and non-ionic polymer layers range from about 100 to 500 nm. The upper value corresponds to an almost constant permeability obtained after few hours of polymer flow, the lower one to the permeability obtained, when the adsorbed layer had been left some more hours without flow. This effect is probably due to a slow reconformation process of the adsorbed polymer chains, leading to a flattening of the layer. Additionally, the hydrodynamic interactions may also play a role, as an increase of the flow rate may increase the layer thickness. This rather complicated behaviour of the adsorbed molecules is the subject of current studies.

4. PARTICLE ADHESION

4.1 Experimental methods

To study the adhesion forces a model system is employed using quartz particles and glass beads, 30–35 μm in diameter, which are in contact with a smooth quartz plate in the solution under study. To come into contact with the plate, the particles settle a mean distance of 10 cm through the solution within a glass tube that is mounted vertically above the plate and removed after the particles have arrived on the plate surface. Depending whether particle adhesion is to be measured under normally or tangentially acting separation forces, two different adhesion cells are used. In the first case the cell consists of the quartz plate and a glass plate with a silicone gasket between, enclosing a volume of 0.35 ml of solution together with the deposited particles, which are usually some hundreds in number. Mounted in a holding device, this cell can be inverted and placed in a centrifuge to apply forces normal to the substrate plate. In the second case, the quartz plate is part of the wall of a small flow channel, and the force acting tangentially on the particles is applied by laminar shear flow of the solution through this channel [21, 22]. The torque due to the shear gradient can be accounted for by an additional equivalent force acting on the centre of the sphere in tangential direction. The forces involved in both cases are shown in Fig. 6. The adhering fraction of the particles is determined by counting them using a microscope, before and after the separation force has been applied.

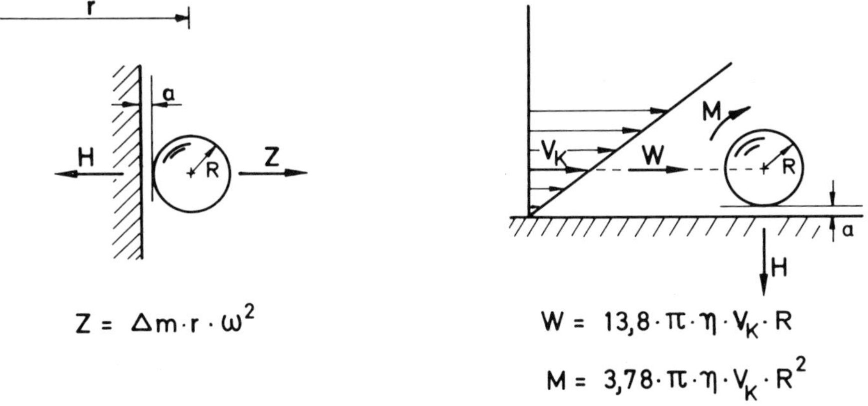

Fig. 6 – Forces and torque acting on an adhering microbead in a centrifuge and in a shear flow.

Some experimental results for a contact time of 30 minutes in tap water are shown in Fig. 7. The force that removes 50% of the adhering particles is usually called the mean adhesion force. Strictly speaking, this is only true for the case of normal separation forces, as an adhesion force can only act in a direction normal to the substrate plate, when the particle is ideally spherical and the plate is ideally smooth. Under such conditions one could therefore expect that a spherical particle is much more sensitive to a tangential separation force than to a normal one. Furthermore Fig. 7 also demonstrates that the smooth glass beads are far more easily removed by tangential forces than the irregularly shaped quartz particles, although under normal forces both kinds of particles behave quite similarly. This effect of surface roughness is discussed in more detail in refs. [5] and [23].

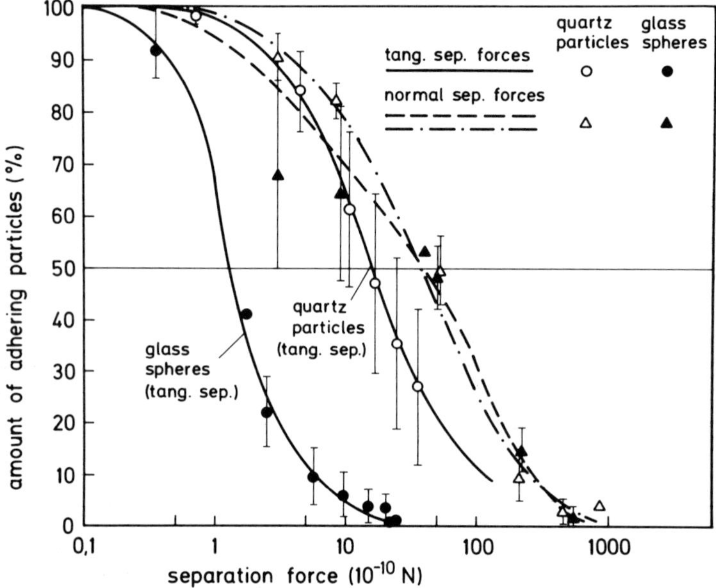

Fig. 7 – Particle adhesion probabilities on smooth quartz plates in tap water under normal and tangential separation forces ($d_p = 30–35\ \mu m$).

Nevertheless, in the case of tangential separation forces, the force necessary to remove 50% of particles is interpreted in the following as 'mean adhesion force' also.

The mean adhesion force of glass beads determined by normal and tangential separation forces is consistent with the London theory for retarded Van der Waals forces if a Hamaker constant of about 4×10^{-20} J is assumed and a geometrical configuration consisting of three contact points of surface roughnesses in the nm range is considered [22]. For the studied system (tap water) double layer forces seemed to be of minor importance. Structural or hydration forces may be important [24], but because of the problem of surface roughnesses and inability to determine an exact 'contact distance', they could not be taken into account.

4.2 Effect of polymer addition

Polymers influencing the adhesion properties of interfaces first have to be adsorbed at the interfaces. The adsorption conditions in the adhesion experiments have been chosen and held constant as follows.

Before contact, adsorption onto the particles occurs by sedimentation through a stationary polymer solution of given concentration. This case is treated in section 3.3.3. The mean sedimentation time is 2 min. After contact, during the course of the experiment additional adsorption takes place by non-steady state diffusion (see 3.3.5) onto particles and substrate plate. During the application of the shear flow additional polymer is adsorbed from the flowing polymer solution (see 3.3.4). Adsorption onto the plate before the particle contacts it is due to non-steady state diffusion for 7 min.

The surface loadings as expected from the given mass transfer conditions can be calculated using procedures in the previous sections. Under the assumption that the free surface does not significantly decrease and adsorption is only limited by mass transfer, the surface loadings of the particles q_p and the quartz plate q_q at the moment of contact are determined for different concentrations of the cationic polymer 222 K and given in Table 3.

Table 3

Surface loadings of particles (q_p) and quartz plate (q_q) at the time of contact for the conditions given in section 4.2 for 222 K

c_0 (μg/l)	q_p (mg/m^2)	q_q (mg/m^2)
2.5	0.0008	0.0001
50	0.0158	0.0025
500	0.1575	0.0253

It turns out that the surface loadings are far from the saturation value of about 2 mg/m^2. Therefore the stated assumptions are justified. Furthermore it should be noticed that the loading of the quartz plate is only about 15% of that of the particles. During a contact period of 10 minutes the plate loading is nearly doubled, whereas the increase of the particle loading is only about 15%. Up to this time the adsorption conditions are equal for both kinds of adhesion measurements.

The additional adsorption during the application of the separation forces in the flow channel as well as in the cintrifuge is low compared to the amount adsorbed before. From this considerations it can be deduced that the bulk concentrations are not markedly decreased and that the polymer layers for equal bulk concentrations are comparable for the two different methods to determine the particle adhesion.

Figure 8 shows how the mean adhesion force is changed by cationic polymers both for normally (T_N) and tangentially acting separation forces (T_T). The experiments were performed with the system glass bead/quartz plate/tap water. It can be seen that the force to remove a particle in a direction normal to the plate is always significantly higher than to remove it by rolling over the surface. In the last case a particle is regarded as removed (in a somewhat arbitrary way) when it has changed its original position at least for a distance of one particle diameter. Here the particle is still in contact with the plate, contrary to the normal force case. In the latter the contact has to be completely broken and the increase of the mean adhesion force with polymer concentration suggests that the increase in the adsorbed amount (see Table 3) is responsible for the adhesion force, probably due to an increase in the number of polymer bridges.

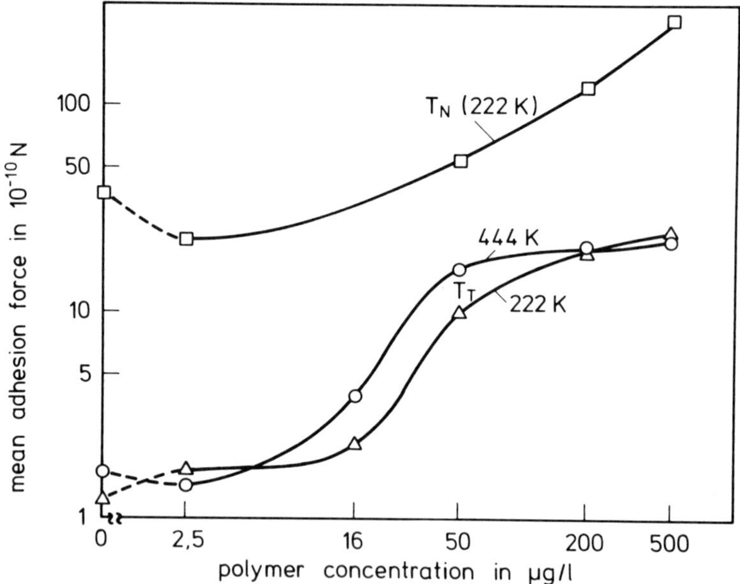

Fig. 8 – Influence of the concentration of cationic polymers on the mean adhesion force of glass beads (30–35 μm) adhering to a quartz plate in tap water for normally (T_N) and tangentially (T_T) acting separation forces in a double logarithmic scale.

When tangential forces are applied, a significant adhesion increase is established in the concentration range 10 – 50 μg/l, and is even more evident for higher concentrations and a higher-molecular weight polymer. Some interesting effects have been observed visually, especially in this concentration range.

Shortly before the glass beads were removed by the flow, some of them exhibited vibrational motions lateral to the flow direction with amplitudes of several micrometres. After they were removed, many of them adhered again after travelling only very short distances. These findings may be explained by an elasticity of the polymer bridges which perhaps consist partly of aggregates of molecules. They are able to compensate tangentially acting forces by adjusting themselves to the direction of the force. Evidence for this elasticity diminishes with increasing surface loadings.

To investigate the effect of the polymer adsorption time on the particle adhesion, both particle and quartz plate surface were separately loaded by polymer for a given adsorption time before they were brought into contact. For that purpose they were exposed to a stirred polymer solution of 10 μg/l to 100 μg/l. These concentrations are so low that from mass balance considerations the surfaces can only become partly covered. Figure 9 shows the amount of adhering quartz particles to a quartz plate in a buffer solution after a particle contact time of 30 s plotted as a function of the polymer adsorption time. The buffer solution contains 10 mmol/l NaCl, 1 mmol/l $NaHCO_3$ and 0.95 mmol/l HCl to yield a pH value similar to the one of tap water, in equilibrium with the atmosphere. The normally acting separation force is gravity (b/g = 1) or the tenfold gravity (b/g = 10).

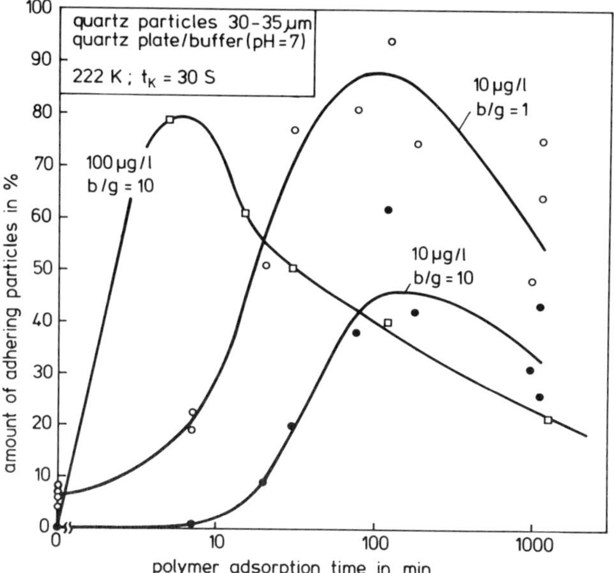

Fig. 9 – Amount of adhering particles after application of gravity (b/g=1) or centrifugation (b/g=10) as a normally acting separation force in dependence of the time of adsorption of a cationic polymer from solution (c_0 = 10 μg/l and 100 μg/l).

For short adsorption times the amount of adsorbed polymer is determined by mass transfer processes, and therefore the latter are determining the particle adhesion behaviour. After a certain adsorption time (~ 2 h for 10 μg/l, ~ 6 min for 100 μg/l) a maximum adhesion probability is reached, and for longer times adhesion decreases. For higher polymer concentrations the initial increase is faster as indicated for 100 μg/l. This was expected from the concentration dependence of mass transfer. The decrease after longer adsorption times may be due to some slow reconformation process already mentioned in section 3.4. Obviously, a non-equilibrium conformation is more favourable for good adhesion, probably because bridging is more likely when the adsorbed chain has an extended form rather than a flat one. This idea is also suggested in polymer flocculation [16].

4.3 Effect of contact time

In the experiments discussed so far, the particles were in contact with the quartz plate for relatively long times (30 min and 30 s) until a separation force was applied. However, for the deposition of a particle in a deep bed filter the adhesion condition has to be satisfied at the moment when the particle contacts the filter grain surface. That means, very short contact times have to be taken into account. Therefore the effect of contact time on adhesion probabilities of quartz particles and microbeads was investigated.

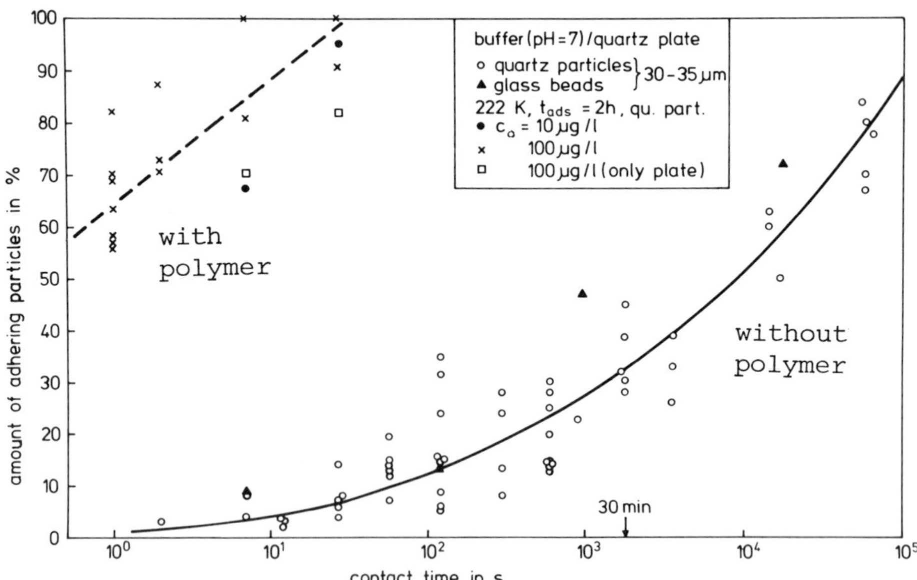

Fig. 10 – Comparison of the contact time dependences of the adhesion probabilities without polymer and with a cationic polymer for gravity acting as normal separation force.

To give contact times of a few seconds, the method for normal separation forces was modified as follows. After the particles are settled onto the glass plate, which is one of the covering plates of the particle adhesion cell, the cell is covered by the quartz plate. Then the whole cell is inverted and the particles are removed from the glass plate by a mechanical shock. They settle the distance of 2 mm from the glass to the quartz plate in about 3 seconds, contacting it almost simultaneously. After leaving them for a certain contact time, the cell is inverted again and the percentage of particles resisting the gravity force is evaluated. Figure 10 shows, how this percentage depends on contact time.

For the case without polymer very low adhesion probabilities are found for short contact times, increasing slowly until a value of about 80% is reached after one day. Possible explanations for this phenomenon are slow changes in the Sternlayers and/or in the hydration layers of the contacting surfaces that cause a decrease of the contact distance resulting in an increase of van der Waals forces, or also some deformation of roughnesses in the nm range.

For the experiments with polymer the surfaces were loaded by the cationic polymer in a separate adsorption step lasting 2 hours. Already after contact times of a few seconds, very high adhesion probabilities are reached. The results do not differ very much within the limits of scatter for different polymer concentrations and for the case when only the plate is polymer covered.

These results agree well with the observation made during the studies using the flow channel. When no polymer was present, after a contact time of 30 min, microbeads, which had just been detached from their original contact point by the shear flow, rolled along the quartz plate without any new deposition. On the contrary, with adsorbed polymer, the rolling beads were alternately detached and deposited in their motion within very short distances.

These findings demonstrate that for conditions encountered in deep bed filtration, and also in flocculation, the dynamic behaviour of the particle adhesion and the effect of polymers on it is of great importance. The previously mentioned elasticity of polymer bridges obviously plays a significant role.

For this reason a new method is currently being tested that makes it possible to study adhesion probabilities at the moment of first contact, when a settling microbead hits the surface of an inclined plate and the tangential component of the gravity force is acting as a separation force. It is also intended to investigate the deposition behaviour of particles in shear fields and the effect of polymers on it by using a high frequency microcinematographic method.

5 CONCLUSIONS – A PARTICLE ADHESION MODEL IN DEEP BED FILTRATION

These investigations demonstrate that the particle adhesion can be considerably improved by polymers adsorbed at the solid surfaces. Under the given experimental conditions, which are not far from practical ones, an adsorption equilibrium is not usually reached. Therefore a change of polyelectrolyte concentration mainly affects the particle adhesion behaviour by changing the adsorption kinetics.

The particle adhesion experiments have shown that the improved particle adhesion caused by small amounts of added polymer is especially obvious under tangential separation forces. This can be explained by a certain elasticity of polymer bridges. Furthermore, the time of contact between the solids in immersed systems has an extremely strong influence. In a system (quartz particles on quartz plates) without any polymer the adhesion force slowly increases during a time interval of hours. In contrast to this, in a system with polyelectrolyte the adhesion force reaches a relatively high value within seconds.

Such effects must have a strong influence on the particle deposition in a deep bed filter especially considering the fact that for a particle contacting the collector surface there is only a very short time available during which it must come to rest from its movement in the flowing liquid. Therefore one can expect that the adhesion probability of the particle at one particular point of contact will be very small. On the other hand, the particle may have many possibilites of coming to a stable deposition on the same collector surface because it may have many contacts with this collector.

For the initial phase of filtration this behaviour leads to the model schematically represented in Fig. 11. Non-Brownian particles ($d_p > 1$ μm) coming into contact with the collector surface have a certain adhesion probability at the first point of contact. If this first contact is not successful, i.e. if the particle does not stay at this point, it will move along the collector surface. During the particle movement each position on the collector surface will be associated with a distinct adhesion probability, resulting from the interplay of the adhesion and separation forces being present at this particular point. The particle can leave the collector surface again and will be re-entrained when it has reached a position where the adhesion probability is relatively small and the different forces acting on the particle lead to an increase of the surface separation distance between particle and collector.

The expectation of the overall adhesion probability $\overline{\gamma_{0,\,H}}$ for a particular kind of particles (constant physico-chemical properties) coming into contact with the collector surface can be determined by the following relation

$$\overline{\gamma_{0,\,H}} = \int_0^{\Theta_{k,max}} \gamma_{0,\,H}\,(\Theta_k) \cdot \psi(\Theta_k)\, d\Theta_k \tag{22}$$

To use this relation one should know the adhesion probability $\gamma_{0,H}\,(\Theta_K)$ when the particle first comes into contact with the collector surface at $\Theta = \Theta_k$ and is moving on this surface up to a critical position Θ_{max} at which it will be re-entrained. Furthermore one has to know the probability that the first point of contact will be at Θ_K, which is described by the density function $\psi(\Theta_K)$. The latter is determined by the transport mechanisms and can be calculated using appropriate theories (e.g. [1–3].

Sample calculations with available single adhesion probabilities of particles being in contact for 30 min with the substrate surface resulted in values for $\overline{\gamma_{0,H}}$ of nearly 100%. These are much higher than those which can be expected from appropriate experiments using equation (23)

$$\gamma_0 = \gamma_{0,T} \cdot \overline{\gamma_{0,H}} \tag{23}$$

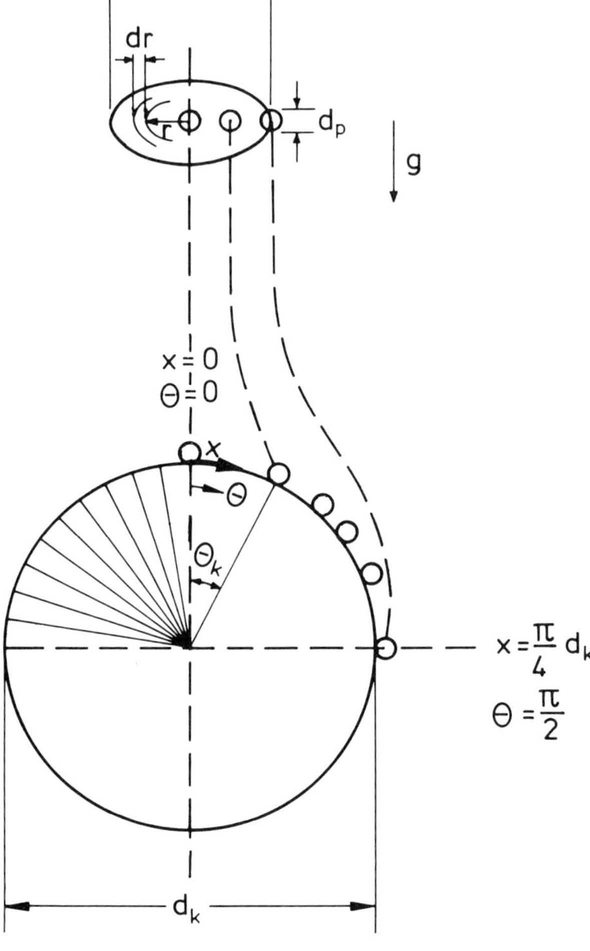

Fig. 11 – Model of a spherical collector.

where $\gamma_{0,\,T}$ is the transport efficiency which can be determined theoretically and γ_0 is the measured elementary removal degree (compare for example the results given in Fig. 1, from which one should expect a $\overline{\gamma_{0,\,H}}$ value of about 10% for the 30–35 μm particles). More details about the method to determine $\overline{\gamma_{0,\,H}}$ are given in ref. [23]. This discrepancy also demonstrates the importance of considering the dynamic adhesion behaviour when a particle comes into contact with the collector surface in a deep bed filter.

ACKNOWLEDGEMENT

The authors gratefully acknowledge the financial support for these investigations by the Deutsche Forschungsgemeinschaft.

LIST OF SYMBOLS

a	contact distance
A	adsorbing surface
A_f	cross section of packed bed
b	width of flow channel
b/g	ratio of normally acting separation force to gravity force
c	polymer concentration
c_0	initial or inlet polymer concentration
d	diameter
D	diffusion coefficient
h	height of packed bed
H	net adhesion force
I	intensity of transmissive light
k	Boltzmann's constant
k_{12}	rate constant
l	length of flow channel
$\dot{m}$	polymer flux
M	torque from shear flow
N	number concentration
p	permeability
q	surface loading
r	radius in centrifuge
R	particle radius
s	height of flow channel
t	time
t_K	contact time
T	absolute temperature
T_N	normally acting separation force
T_T	tangentially acting separation force

u	Stokes-sedimentation velocity
v_f	superficial velocity
v_K	flow velocity corresponding to particle centre
$\dot{V}$	flow rate
W	drag force
x	distance from adsorbing surface
Z	centrifugal force
Re	Reynolds number
Sc	Schmidt number
Sh	Sherwood number
β	mass transfer coefficient
γ	elementary removal degree
γ_0	elementary removal degree in the initial filtration phase
$\overline{\gamma_{0,H}}$	mean particle adhesion probability
$\gamma_{0,T}$	particle transport efficiency
Δm	mass of particle minus displaced water
Δp	pressure drop
ϵ	porosity of packed bed
η	dynamic fluid viscosity
Θ	angle coordinate
λ	filter coefficient
ν	kinematic fluid viscosity
ψ	probability density for contact angle
ω	angular velocity

Indices

a	adsorption
b	glass bead
k	filter collector assumed as sphere
lam	laminar
p	particle
pol	polymer coil
turb	turbulent

REFERENCES

[1] Gimbel, R., *Untersuchungen zur Partikelabscheidung in Schnellfiltern.* Dissertation, Universitat Karlsruhe (1978).

[2] Spielman, L. A., and Fitzpatrick, J. A., Theory for Particle Collection under London and Gravity Forces, *J. Colloid and Interface Sci.*, **42**, 607–623, (1973).

[3] Rajagopalan, R., and Tien, C., Trajectory Analysis of Deep-Bed Filtration with the Sphere in Cell Porous Media Model, *AIChE J.,* **22**, 523–533, (1976).

[4] Gimbel, R., and Sontheimer, H., *Recent Results on Particle Deposition in Sand Filters,* Symp. Paper on Deposition and Filtration of Particles from Gases and Liquids, Loughborough, 6–8 Sept. 1978, London SCI, (1978).

[5] Gimbel, R., *Influence of the Filter Grain Surface Structure on the Transport and Adhesion Mechanisms in Deep Bed Filters,* Symp. Paper on Water Filtration, 21–23 April 1982, Antwerp, Belgium, (1982).

[6] Kaczmar, B. U., On the Effects of $CaCl_2$ upon Partially Hydrolyzed Polyacrylamides in Solution. *Report SFB 80/E/174,* Universitat Karlsruhe, (1980).

[7] Horn, D., Polyethylenimine – Physicochemical Properties and Applications, in *Polymeric Amines and Ammonium Salts,* (Goethals, E., ed), Pergamon Press, Oxford, (1980).

[8] Grigoraki, E., *Einfluß eines kationischen Polymeren auf das dynamische Verhalten von Tiefenfiltern,* Studienarbeit, Universitat Karlsruhe, (1983).

[9] Burkert, H., Die Bestimmung von Spuren Polyacrylamid im Wasser, *gwf-Wasser/Abwasser,* **111**, 282–286, (1970).

[10] Silberberg, A., Adsorption of Flexible Macromolecules, *J. Chem. Phys.,* **48**, 2835–2851, (1968).

[11] Hoeve, C. A. J., Theory of Polymer Adsorption at Interfaces, *J. Polym. Sci. C,* **34**, 1–10, (1971).

[12] Roe, R. J., Multilayer Theory of Adsorption from a Polymer Solution, *J. Chem. Phys.,* **60**, 4192–4207, (1974).

[13] Scheutjens, J. M. H. M., and Fleer, G. J., Statistical Theory of the Adsorption of Interacting Chain Molecules. 1. Partition Function, Segment Density Distribution, and Absorption Isotherms, *J. Phys. Chem.,* **83**, 1619–1635, (1979).

[14] Scheutjens, J. M. H. M., and Fleer, G. J., Statistical Theory of the Adsorption of Interacting Chain Molecules. 2. Train, Loop, and Tail Size Distribution, *J. Phys. Chem.,* **84**, 178–190, (1980).

[15] Cohen-Stuart, M. A., Scheutjens, J. M. H. M., and Fleer, G. J., Polydispersity Effects and the Interpretation of Polymer Adsorption Isotherms, *J. Polym. Sci.: Polym. Phys. Ed.,* **18**, 559–573, (1980).

[16] Gregory, J., Polymer Flocculation in Flowing Dispersions, in *The Effects of Polymers on Dispersion Properties,* (Tadros, T. F., Ed.), Academic Press, London (1982).

[17] Smoluchowski, M., Versuch einer Mathematischen Theorie der Koagulationskinetik kolloidaler Losungen, *Zeitschr. Phys. Chem.,* **92**, 129, (1917).

[18] Gnielinski, V., Gleichungen zur Berechnung des Warme- und Stoffaustausches in durchstromten ruhenden Kugelschuttungen bei mittleren und großen Pecletzahlen, *Verf. Techn.,* **12**, 363–366, (1978).

[19] Schlunder, E.-U., *Einfuhrung in die Warme- und Stoffubertragung.* Vieweg, Braunschweig, (1972).

[20] Happel, J., Viscous Flow in Multiparticle Systems; Slow Motion of Fluids Relative to Beds of Spherical Particles, *AIChE J.*, **4**, 197–201, (1958).

[21] Kneilmann, R., *Untersuchungen zur Haftung kleiner Teilchen auf ebenen Unterlagen in waßrigen Losungen,* Diplomarbeit, Universitat Karlsruhe, (1977).

[22] Sehn, P., *Untersuchungen zur Haftung von kleinen Glaskugeln auf Quarzplatten in waßrigen Losungen,* Diplomarbeit, Universitat Karlsruhe, (1978).

[23] Zitzmann, W., Sehn, P., and Gimbel, R., Bedeutung der Partikelhaftung bei der Tiefenfiltration, *Veroffentlichungen des Bereichs und Lehrstuhls fur Wasserchemie,* vol. 20, Karlsruhe, (1982).

[24] Israelachvili, J. N., Forces between Surfaces in Liquids, *Adv. Colloid and Interface Sci.,* **16**, 31–47, (1982).

CHAPTER 22

Filtration of Dilute Suspensions Using Non-Woven Cloths

M. S. ABDEL-GHANI AND G. A. DAVIES
Department of Chemical Engineering, UMIST, Manchester M60 1QD, UK

1. INTRODUCTION

Non-woven material is widely used as a medium for filtration. It is a relatively low cost material and, because of its construction, can be used to filter small particles (~5 μm). This paper is concerned with characterizing some of the physical properties of non-woven materials particularly relating to filtration of particles from dilute suspensions.

In filtration the medium acts as a selective membrane through which the filtrate passes and the solid material is held back. Considering this basic process the medium may be assessed relative to three important criteria, viz.:

(a) minimum resistance to filtrate flow – this is related to the pore size and free volume of the medium.
(b) ability to bridge solids across the pores of the medium in the surface in contact with the feed slurry – this is again related to the pore size distribution of the medium and the particle size distribution of the suspension. This factor is most important in cake filtration or filtration of high concentration slurries. It directly affects the initial filtration and is measured by the quantity of solids bleed.
(c) rate of entrapment of solids within the intertices of the medium – this is a measure of the blinding of the medium and will directly affect the filtration rate.

There are of course many other factors relating to media properties to consider when selecting for a particular application – mechanical strength, resistance to chemical attack, mechanical wear, ease of removal of solids from the media after filtration, cost, etc. In this paper we will focus attention on the three factors stated and consider how non-woven materials can be assessed relative to them. This is not to say that, if a given material satisfies these three factors then this is a necessary and sufficient condition for final selection. On the contrary, other factors such as wear, chemical resistance and cost may preclude a given choice. These three factors, (a) – (c), are singled out in this paper since they are controlled by the structure of the medium and the basic structure is fundamental to the filtration process.

2. METHODS OF ASSESSING NON-WOVEN MEDIA

In comparing various media it is normal for the following properties to be quoted:

(i) weight per unit area,
(ii) weight of a given length of filament (from which the mean filament diameter can be determined),
(iii) thickness of the material,
(iv) free (or void) volume,
(v) mean pore size,
(vi) air permeability,
(vii) bubble point pressure.

In considering the criteria set out in Section 1, properties (v), (vi) and (vii) must be examined. We will deal with them in reverse order.

The bubble point pressure is determined by immersing a sample of medium in a liquid (usually water), air is then introduced underneath the medium and the air pressure required to form a bubble at the upper surface of the medium in the water is determined. The excess pressure in a bubble is, from Laplace's equation:

$$\Delta P = 2\sigma H \tag{1}$$

where σ is the surface or interfacial tension and H is the principal radius of curvature of the bubble. If the pressure is determined at the point of bubble release and if the pore in the medium at which the bubble is formed is assumed to be circular of radius r then:

$$\Delta P = \frac{2\sigma}{r} \tag{2}$$

Since ΔP is inversely proportional to r the minimum bubble point pressure will give an indication of the *maxium* pore radius in the surface of the media. In non-woven media the pores are formed between intersections of filaments and, as we will discuss later, these filaments are randomly arranged so that there will be a distribution of pore sizes. The bubble point pressure at best only gives a measure of the maximum pore radius in the surface.

In carrying out bubble point pressure experiments, if the measurements were extended to include air permeation rates, J, for a range of excess pressures then an estimate of the distribution of pore radii could be made. If the distribution of pore radii in the surface is $N(r)$ then air flow per unit area of surface may be calculated, assuming the pores are independent and extend between the inlet and outlet surfaces of the media, by applying Poiseuille's equation for flow:

$$J = \frac{\pi \Delta P}{8\mu l} \int_{\frac{2\sigma}{\Delta P}}^{\infty} r^4 N(r) \mathrm{d}r \tag{3}$$

Differentiating equation (3) with respect to ΔP and rearranging gives:

$$N(r) = \left[\frac{\mathrm{d}J}{\mathrm{d}(ln\Delta P)} - J\right] \Big/ \left[\frac{\sigma\pi}{4\mu l}\left(\frac{2\sigma}{\Delta P}\right)^4\right] \tag{4}$$

where l is the thickness of the media sample and μ is the viscosity of air.

In applying this to a non-woven medium, assumptions are made which are not really valid. The pores are not circular in section, they are not independent and they do not form circular tubes stretching between the two surfaces of the medium. Before commenting further, let us consider air permeability measurements (vi) above. The viscous flow of a fluid through a porous medium is described by Darcy's law. The flow per unit area, J' is related to the pressure drop across the medium:

$$J' = \frac{K}{\mu} [\Delta P] \tag{5}$$

where μ is the viscosity of the fluid and K is a constant, the permeability, K, can be calculated from measurements of air flow at different differential pressures. This is usually quoted in $\mathrm{cm}^3/\mathrm{sec}^2$ or l/min^2.

Again, if the assumptions on pore geometry as outlined above are invoked, an estimate of the mean pore radius can be made from the measurements.

The information usually available on non-woven media which relates to the filtration performance will, at best, be an estimate of the maximum pore radius at the surface, a mean pore radius and permeability to air. It would be possible to obtain (subject to considerable assumptions) information on pore distribution.

In applying this to the criteria set out in section 1 what information can be obtained?

Bridging and blinding of filter media is a result of an interaction between a distribution of particles and a distribution of pores. In order to investigate both phenomena this interaction must be taken into account, and a stochastic model can be used to simulate the process [1]. A simplified approach is to use the concept of 'cut-off' size. In this approach, particles are compared in relation to a pore size. If the particle is greater than the pore diameter it will be retained on the surface; if it is less than the pore diameter it will penetrate or even pass through the pore and thus bleed through the filter. If data on air permeability are analysed to obtain information on pore size or if, using equation (4), bubble pressure experiments are used to obtain pore distribution data, then the analyses assume cylindrical pores which would imply that any particle entering the pore would be swept through. This is not the case, since the structure of the medium is random and particle penetration will vary. Notwithstanding these comments an estimate on particle blinding can be made by applying results from work on screening reported by Rose and English [2]. They concluded that a pore would be blinded by a particle becoming jammed in the inlet if the particle/pore diameter ratio, d_s/d_p, was within the range $1.0 < d_s/d_p < 1.1$.

The conditions are illustrated in Fig. 1. A mean pore size is shown superimposed on the cumulative particle size distribution of the feed. The three regions of (a) particles filtered out and deposited on the surface, (b) permanently trapped in the surface pores, (c) entrained through the media, are shown. If the mean pore diameter (rather than a pore distribution) is used, then the information on regions (a) and (b) cannot be extended, other than considering particle shape, even if more information on the structure of the media were available. The conclusions on region (c) are doubtful but reflect the primitive model used to represent the media. No information on particle penetration into the media can be made.

3. A MODEL TO SIMULATE TYPES OF NON-WOVEN MEDIA

Non-woven felts and cloths are made up of a random array of fibres. The spaces contained within these arrays are random in pattern, shape and dimensions.

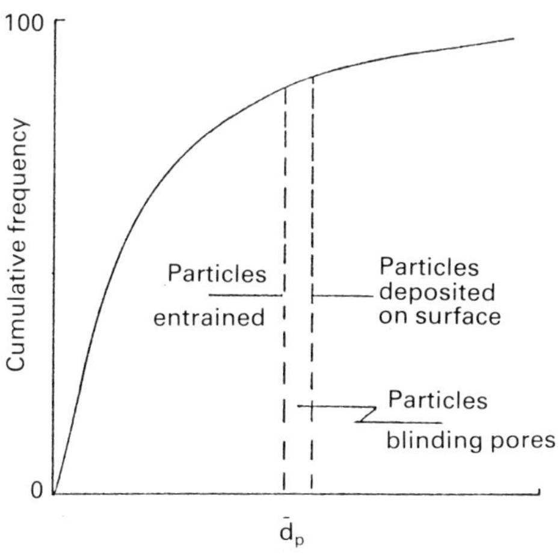

Fig. 1 – Particle size distribution showing the mean pore diameter of a filter medium and its relation to filtration.

Many non-woven materials are formed by random arrays of filaments being laid one above another so that the penetration of fibres between layers is small. (This results in low shear strength and, to improve this, fibre penetration is attained, thereby physically connecting adjacent layers by needling.) When entanglement between layers can be neglected a model can be developed to simulate fibre arrangements. The model rests on the proposal that the media is made up of layers of fibres, each layer being made up of a random network of fibres. The layers are arranged at random one above another so that no predetermined pattern is produced. If random networks can be produced to simulate fibres to a given surface area and the layers incorporated to a given free volume then this structure could be used to investigate pore properties, filtration and particle capture.

Two assumptions are made, the first is that the fibres approximate to cylindrical rods and the second that the ratio of the length to diameter of the fibres is large. The cylindrical rods are straight in the x–y plane (plane of the layer) but can deform in the z plane so that one filament can bend around another. With the first assumption, a plane view of any layer of the medium would then appear as a set of intersecting lines of various thickness. The second assumption is not essential to the analysis, it is incorporated to simplify the computer calculations and does not infringe on most practical cases since $l/d \sim 2 \times 10^2 - 3 \times 10^3$. The first assumption is essential and may not be appropriate to some fibres which curl. Inspection of many fibres however shows that they are almost straight filaments. These ideas were first put forward by Piekaar and Clarenburg [3] as a means of analysing dust filters. The analyses were, however, restricted to uniformaly thick fibres and only extended to analysis of network dimensions to relate to pore characteristics. Ghani [4] has investigated the description, formation and analysis of random networks in much more detail and applied the results to interpret coalescence of droplets from liquids and gases and particle penetration in fibrous media. In the present paper we will outline the properties of these random networks and examine results to simulate some typical non-woven filter media and apply the data to filtration of dilute suspensions.

The plan view of a layer of the material would be, according to the model, as in Fig. 2(a). Assuming that the surface area of the material is constant between layers and that the free volume is uniform within the media the number of fibres in a unit area of a layer, n, is determined

$$S = \sum_{i}^{n} \pi d_i \, l_i \tag{6}$$

where d_i and l_i are the diameters, S is constant, d_i and l_i vary so that the network is analysed to determine the number of fibres required. The network is built up fibre by fibre, the information on each fibre is stored and the total surface area S calculated after each fibre addition. The layer simulation is stopped when the

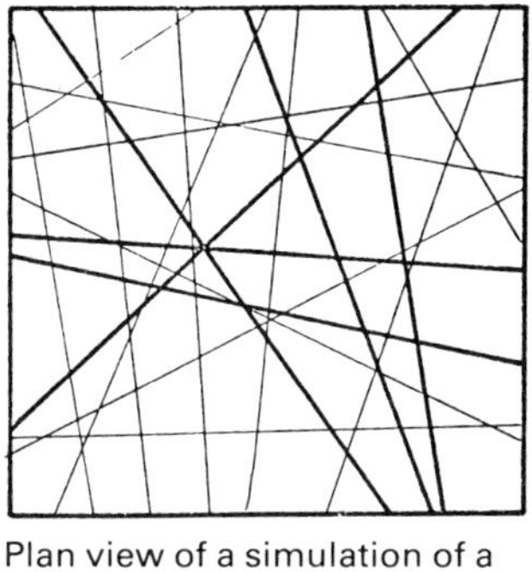

(a) Plan view of a simulation of a single layer of media

(b) Cross section of a single layer

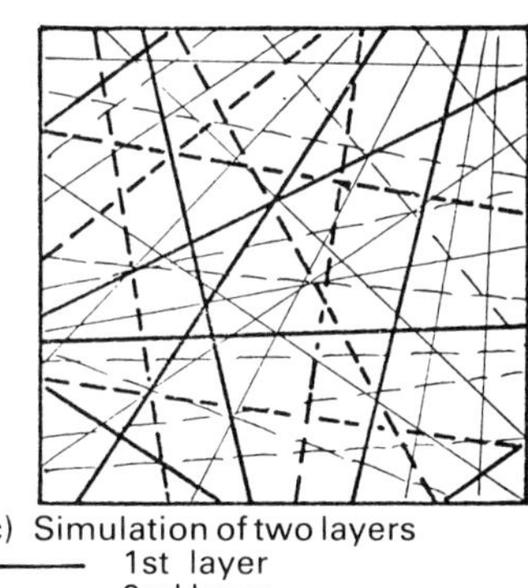

(c) Simulation of two layers
——— 1st layer
– – – 2nd layer

Fig. 2.

required value of S is reached. The area enclosed by the fibres represents a cross section of the pores at that layer. Because the fibres are assumed to be cylindrical rods these areas are external polygons whose order varies from 3, i.e. triangles, up to 9 or 10. Each layer is produced so that the surface area of fibre per unit layer area is constant. The layers will all be different (although the overall properties will be the same). Therefore as one layer is placed above another the pore structure, as observed from above (similar to measurements by light transmission), will change. A sketch is shown in Fig. 2(c).

In non-woven media the filament size is not constant but varies with method of manufacture. This was taken into account in the simulation. For each material tested the filament size distribution was measured and used in subsequent computation. Details of the methods used for generation and analysis of the networks are given elsewhere and we will summarize the properties. Computer programmes were written to generate sets of random filaments to conform to fixed values of free area. Each was stored in the order generated, so that when the layers were combined to simulate a given depth of material the random structure was preserved.

Programmes were written to analyse each network. The filament intersections and position were determined. Referring to Fig. 3, the distance between successive intersections on each fibre was computed, l_{ij}, this represents the length of a side of each pore. Complete information on the simulated pore structure was obtained by determining the order, n, and area of each polygon, a_I. From this, order and area distributions can be obtained and the mean pore area,

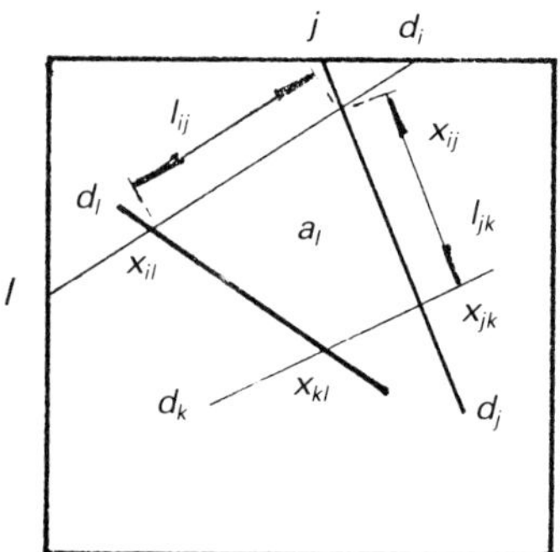

Fig. 3 – Detail of a pore I in the media.

mean hydraulic radius, mean pore perimeter and the mean distance between fibre intersections computed. The minimum thickness of a layer of the material can be determined by computing the maximum thickness of a junction made up of filaments i and j where the thickness at the intersection point is d_i and d_j. The data on all intersections in the network is stored and the maximum value of $(d_i + d_j)$ for all the intersections computed.

Graphical outputs to display the simulated networks were produced on microfiche. An example is shown in Fig. 4. This shows a simulation of the plan view of a fibre mat (perpendicular to the plane of the fibre layers). The diagrams are to scale, the variation in the thickness of the lines represents the variation in fibre diameter. The diagrams represent the plan as successive layers are added. Moving from (a) to (l) represents 1 to 12 layers.

3.1 Predictions on pore structure

Statistical analysis of the structures formed by random lines in a plane has been the subject of study. Amongst the theorems derived is the conclusion that the fractional number of triangles produced is 0.3551 [5]. The present work provided data on the polygon distribution. A typical set of results for a network of 100 lines to simulate a layer of material made up from glass fibre, mean fibre diameter

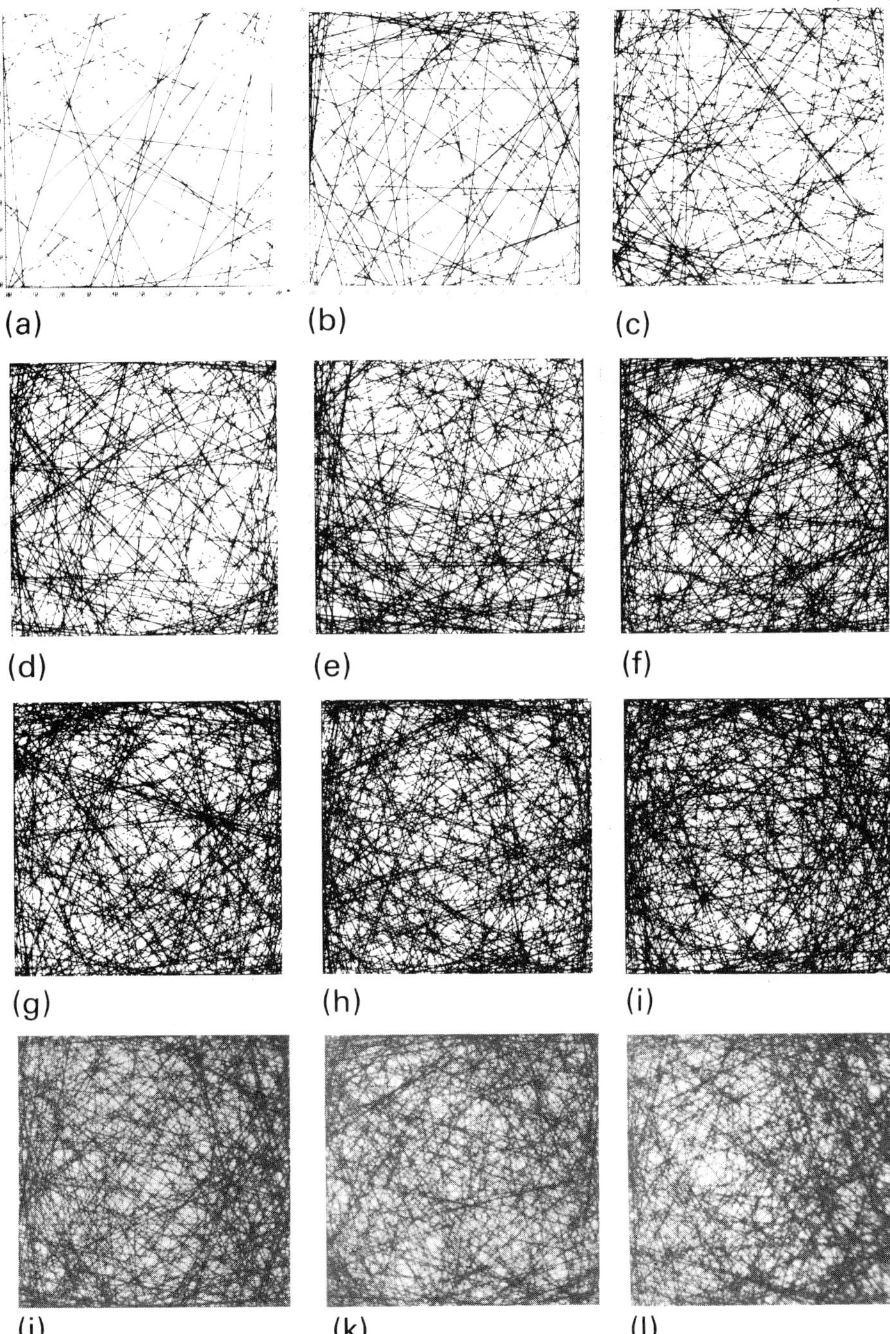

Fig. 4 – Simulation of a plan view of a fibre matt, the number of layers increasing from *a* to *l*.

31.5 μm and free area 0.9805 is shown in Fig. 5. The simulation was for an area of material 225 cm^2 and included 2657 polygons – pores. In Fig. 5 the polygon distribution from n=3 to 8 is shown. In this particular network no polygons with $n > 8$ were present. An interesting observation is that the pore order distribution is approximately log-normal. Comparing the calculated value for the fractional number of triangles in the distribution with that predicted from statistical theory [5] the mean for 50 networks analysed, covering a range of systems for mean fibre diameter from 3–31 μm and free areas from 0.88 to 0.98, was 0.3661 with a standard deviation of 0.0254. Although the number of triangular pores is high their contribution to the free area is much lower at 8–9%. The distribution of pore area with polygon order is also shown in Fig. 5.

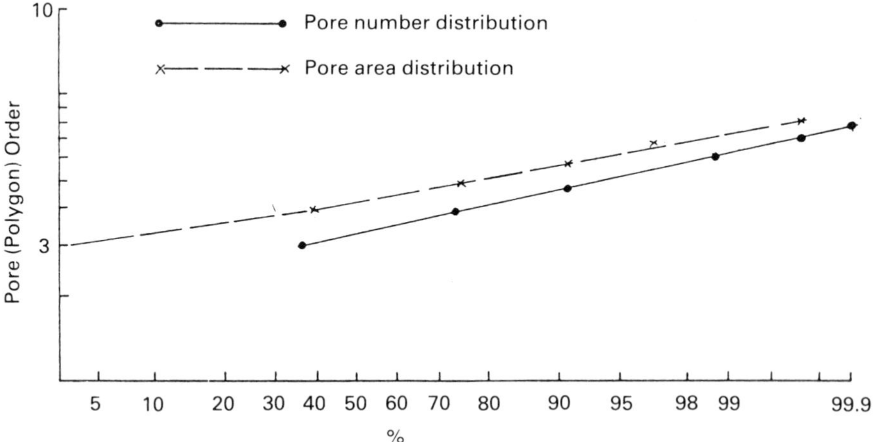

Fig. 5 – Pore shape and pore area distributions produced by a random fibre network.

For the present work, information on the relationship between the mean pore area and network or fibre properties such as filament diameter, free area, etc. is most important. If we consider an area of material A with a free area ξ then, if the mean pore area is $\bar{a}_p$ and the number of pores in area A is n_p,:

$$\frac{n_p}{A}\bar{a}_p = \xi \tag{7}$$

The area occupied by the filaments, A_f, is:

$$A_f = n_f \bar{d}_f \bar{l}_f - \beta \tag{8}$$

where n_f is the number of fibres, $\bar{d}_f$ and $\bar{l}_f$ are the mean fibre diameter and length respectively. β is the common area between any two intersecting filaments at the point of intersection.

$$\beta \approx n_{\mathrm{f}} \frac{n_{\mathrm{c}}}{2} \bar{d}_{\mathrm{f}}^{2}$$

where n_{c} is the average number of intersections per filament in area A. As a first approximation and in 'open' media $n_{\mathrm{c}} \bar{d}_{\mathrm{f}} \ll 2\bar{l}_{\mathrm{f}}$ then:

$$A - A_{\mathrm{f}} = A\xi \qquad (9)$$

or

$$(1-\xi) = \frac{n_{\mathrm{f}}}{A} \bar{d}_{\mathrm{f}} \bar{l}_{\mathrm{f}} \qquad (10)$$

The ratio of the mean length of a filament to the side of the square in a given area is constant, the constant depends on the method of network generation, i.e. randomness. For the methods used in this work Ghani [4] has shown that

$$\bar{l}_{\mathrm{f}} = 0.942 \sqrt{A}$$

Therefore rearranging equation (10)

$$\frac{n_{\mathrm{f}}}{A} = \frac{(1-\xi)}{0.942\, \bar{d}_{\mathrm{f}} \sqrt{A}} \qquad (11)$$

In addition it has been shown that the number of pores is related to the number of fibres [4]

$$n_{\mathrm{f}} = 2.298\, n_{\mathrm{p}}^{0.4888} \qquad (12)$$

Rounding the exponent in equation (12) to 0.5:

$$n_{\mathrm{f}} = k_1 n_{\mathrm{p}}^{0.5} \qquad (13)$$

Then substituting equations (7) and (13) into (11) and rearranging

$$\bar{a}_{\mathrm{p}} = 4.488 \frac{\xi}{(1-\xi)^2} \bar{d}_{\mathrm{f}}^{2} \qquad (14)$$

This equation can be used to compute the mean pore area for a given fibre material provided the mean fibre diameter and free area are known. The computed results for $\bar{a}_{\mathrm{p}}$ are shown plotted against $(d_{\mathrm{f}} \sqrt{\xi}/(1-\xi))$ in Fig. 6. A linear regression analysis of the results plotted on a log–log graph produced the equation:

$$\bar{a}_{\mathrm{p}} = 3.945 \left[\frac{\sqrt{\xi}}{(1-\xi)} \bar{d}_{\mathrm{f}}\right]^{1.948} \qquad (15)$$

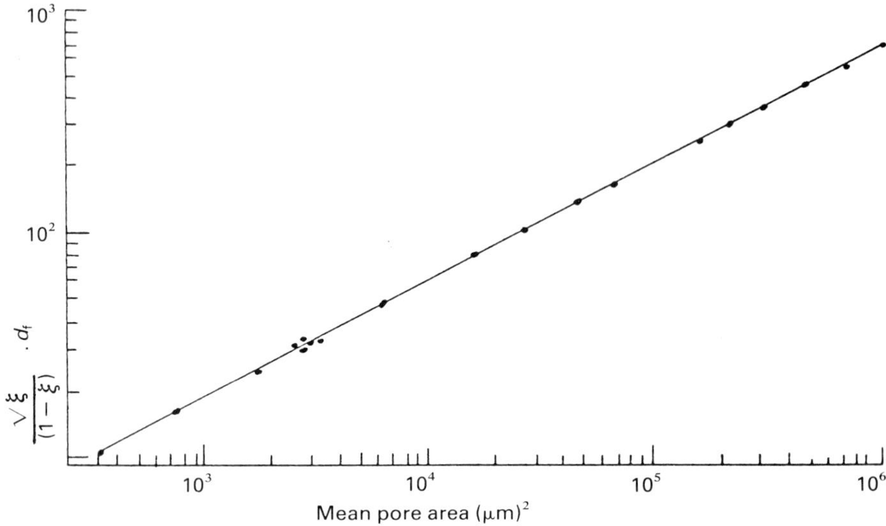

Fig. 6 – Correlation for mean pore area.

The average distance between fibre intersections $\bar{x}$, where

$$\bar{x} = \frac{\sum_{1}^{i} \sum_{1}^{j} l_{ij}}{\sum_{1}^{i} \sum_{1}^{j} ij} \tag{16}$$

is shown as a function of pore area $\bar{a}_p$ in Fig. 7. A straight line with slope 0.5 can be drawn through the data. A linear regression analysis results in

$$\bar{x} = 0.845\ \bar{a}_p^{0.5} \tag{17}$$

The mean pore perimeter $\bar{p}$ was shown to be related to $\bar{x}$ by:

$$\bar{p} = 4.166\,\bar{x} \tag{18}$$

From equations (17) and (18) the mean hydraulic radius of a pore can be determined since

$$r_H = \frac{\bar{a}_p}{\bar{p}} \tag{19}$$

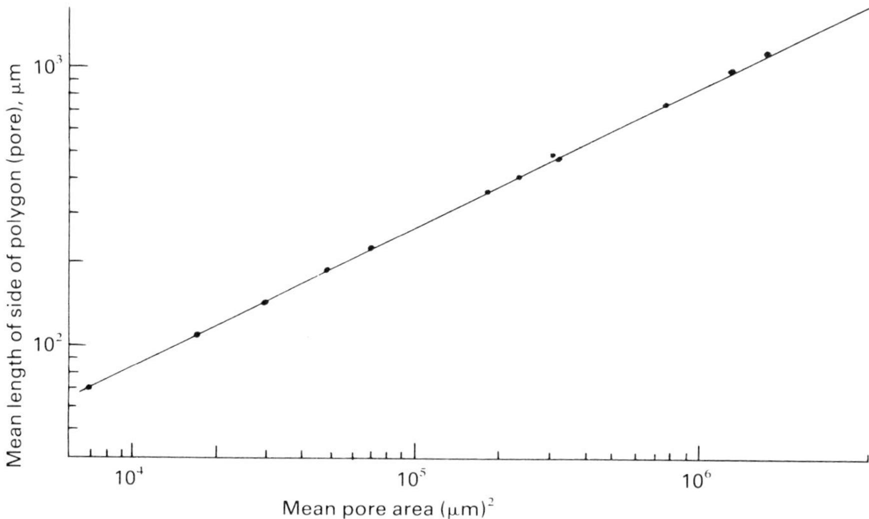

Fig. 7 – Relationship between mean pore area and mean length of a side of a pore.

Thus using the equations presented above details of the pore properties can be calculated providing primary data on the fibre material is known. This is a considerable advance on the methods described in section 2 which rely on a very much simpler description of the pore structure of the media. The results of the present work can now be applied to an aspect of filtration.

4. APPLICATION TO FILTRATION–BLINDING OF FILTER MEDIA

The data on pore properties which can be derived from the equations presented in section 3 can be used to assess particle blinding and particle penetration in non-woven media particularly under two conditions. These are:

(a) During the first seconds when a slurry is exposed to the cloth and when the first layer of particles is deposited on the surface. In more concentrated slurries it is in this period when bleeding and the major part of blinding takes place.

(b) Filtration of dilute suspensions where the number of particles is relatively small and can be compared to the number of pores present in the surface of the cloth. The build up of a thick cake does not then take place.

Considering the conditions of these two cases the procedure to assess blinding discussed earlier in section 2 can be extended, using the information on pore structure, to include particle penetration.

Consider a non-woven filter medium for which the mean fibre diameter and free area are known. Then using equation (14) the mean pore area for pores in the surface layer can be calculated. The equivalent mean circular diameter of these pores and hydraulic radius can also be calculated. Then, if the particle size distribution of the solids in the suspension to be filtered is known, invoking the criterion of a 'cut-off' analysis, the fraction of particles which will bleed through the first layer can be calculated.

$$n_b = \int_{d_{smin}}^{d_p} f(d_s)\, dd_s \tag{20}$$

d_{smin} is the diameter of the smallest particle in the feed and $f(d)$ is the particle distribution function for the feed dispersion. This is shown graphically in Fig. 8.

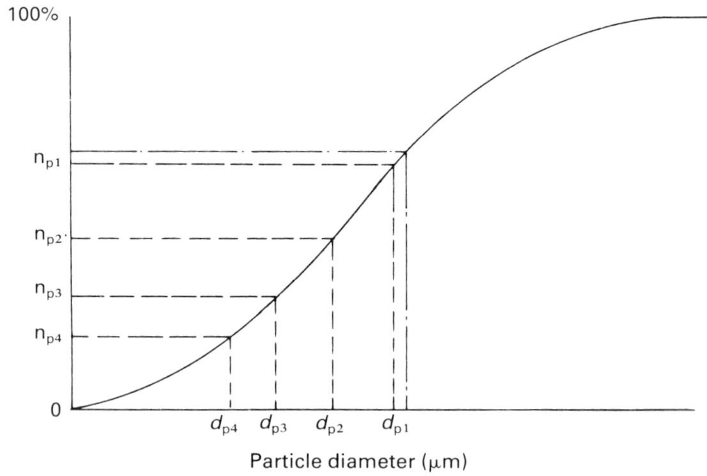

Fig. 8 – Particle size distribution showing the fractional number of particles penetrating successive layers of non-woven media.

If one then considers the first two layers of media the apparent pore distribution when observed from above is similar to the illustrations in Fig. 4. If the medium has n_f fibres per unit area in each layer, the pore area in each layer is related to n_f. From equations (7) and (13):

$$\bar{a}_p = \frac{A \xi k_1^2}{n_f^2} \tag{21}$$

In this filtration application the diameter of the particles in the feed is of the order of magnitude of the diameter of the pores. It is essentially a screening process and contrasts with the depth filtration applications when particles are much smaller than the pores. In the processes considered here diffusion of particles within the media due to Brownian motion can be ignored. The inertia forces of the particles ensure that they follow near linear trajectories. Consider now particles which bleed through the first layer. The apparent pore size distribution of the top two layers is produced by the sum of the filaments in these layers. (This may be illustrated by examining the plan of area of media of different thicknesses. Examples from the simulation programme described here are shown in Fig. 4). Particles which bleed through the top two layers may be estimated using equation (16) but now the upper limit of the integral is reduced to the apparent mean pore diameter for two layers, d_{p^2}. To calculate this the total number of filaments is now $2n_f$. Therefore using equation (11) the effective free area for the two layers can be calculated and then this used to calculate $\bar{a}_p$ in equation (14). A value for d_{p^2} can then be obtained.

Proceeding in this way the apparent mean pore area for layer m of the media can be obtained, this will involve mn_f filaments and an effective free area ξ_m. The procedure is illustrated in Fig. 9. Actual calculations are summarized in Table 1. The results are for filtration of a dilute slurry using a non-woven felt made up of filaments with a mean filament diameter of 12.0 μm and free area 0.9605. It corresponds to a glass fibre mat. The particle size distribution (mean spherical diameter) is log normal with a mean value of 1mm.

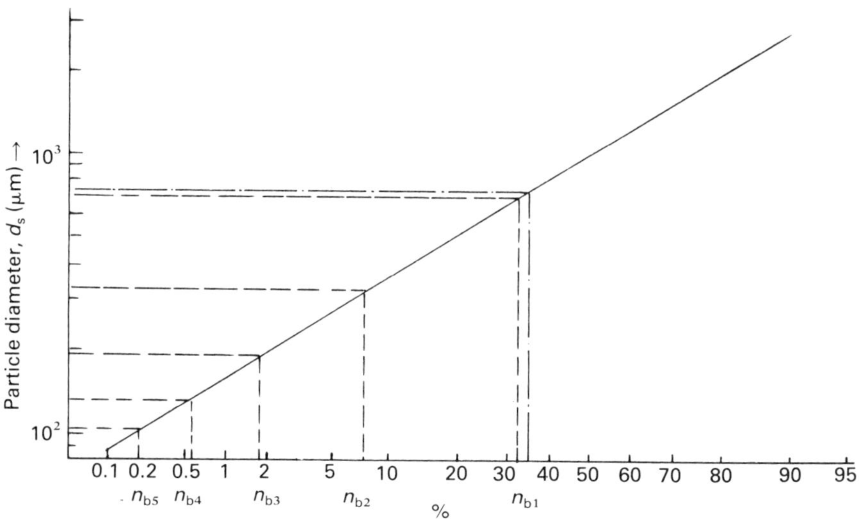

Fig. 9 – Particle size distribution showing the fractional number of particles penetrating successive layers of a glass-fibre filter.

For a given set of filtration conditions, that is for a given feed solids concentration and filtration rate per unit area of cloth, the fractional blockage of medium can be estimated. Thus if the rate of arrival of particles to the first layer of the medium is N (number per unit area, time) then using data as in Table 1 the rate of blinding of the media can be estimated. The blinding rate measured as the rate of loss of filtration area is:

$$-\frac{dA}{dt} = \frac{N.n_{b1}}{n_0}\frac{A}{n_p} \tag{22}$$

Also the rate of blinding within the media can be estimated by substituting the appropriate value (n_{bi}/n_0) (Table 1) into equation (22).

The model proposed to simulate fibre assemblies in some types of non-woven media can provide information on the geometrical properties of pores, particularly mean pore area, perimeter and hydraulic radius. This can be used to estimate particle penetration into non-woven media when the mean particle diameter is at least of the order of the mean pore diameter of a sample layer of media.

Table 1

Mean filament diameter 12 μm free area 0.9605

Layer number i	Total number of filaments per cm²	Effective free area	Mean pore area (μm²)	Mean pore diameter (μm)	Fractional number of particles deposited n_{bi}/n_0
1	34.94	0.9605	3.98×10^5	7.11×10^2	0.69
2	69.88	0.9210	0.954×10^5	3.48×10^2	0.24
3	104.82	0.8815	0.406×10^5	2.27×10^2	0.054
4	139.76	0.8420	0.218×10^5	1.66×10^2	0.012
5	174.7	0.8025	0.133×10^5	1.30×10^2	0.004

REFERENCES

[1] Wilkinson, E., Ghani, M. G., and Davies, G. A. (to be published).
[2] English, J. E., *Filtration and Separation*, March/April, 195 (1974).
[3] Piekaar, H. W., and Clarenburg, L. A., *Chem. Eng. Sci.*, **22**, 1399 (1967).
[4] Ghani, M. G., Ph.D. Thesis, University of Manchester (1983).
[5] Miles, R. E., *Proc. Nat. Acad. Sci.*, **52**, 901 (1964).

NOMENCLATURE

L = length, M = Mass, T = time

a_p	pore area	L^2
$\bar{a}_p$	mean pore area	L^2

A	area	L^2
A_f	plan area of fibre	L^2
d_f	filament diameter	L
$\bar{d}_f$	mean filament diameter	L
d_i	diameter of filament i	L
d_p	pore diameter	L
d_p	mean pore diameter	L
d_s	particle diameter	L
d_{smin}	minimum particle diameter	L
$f(d)$	particle distribution	
H	sum of principal radii of curvature	L^{-1}
i, j	elements i and j of a set	–
J	volume flux of air	$L\,T^{-1}$
J'	volume flux	$L\,T^{-1}$
k_1	constant equation (13)	–
K	permeability	$L^3\,T^{-2}$
l_f	length of a filament or fibre	L
$\bar{l}_f$	mean length of a filament or fibre	L
l_i	length of component i in a set	L
l_{ij}	length between intersection i and j on a fibre	L
$\bar{l}$	mean length of a fibre in a unit area of media	L
n_{bi}	number of particles blocking pores in layer i of media	–
n_c	number of intersections of filaments in media	–
n_f	number of filaments or fibres	–
n_p	number of pores	–
$N(r)$	pore radii distribution	–
p	pore perimeter	L
$\bar{p}$	mean pore perimeter	
ΔP	excess pressure	$M\,L^{-2}$
r_H	hydraulic radius of a pore	L
$\bar{r}_h$	mean hydraulic radius	L
S	surface area	L^2
x	distance between successive intersections	L
$\bar{x}$	mean intersection distance	L
β	correction factor to surface area	–
μ	fluid viscosity	$ML^{-1}\,T^{-1}$
ξ	fractional free area	–
σ	surface tension	MT^{-2}

Index

G

H

I

J

K

L